Computing Essentials

2001–2002

COMPLETE EDITION

Computing Essentials
2001–2002

COMPLETE EDITION

Timothy J. O'Leary
Arizona State University

Linda I. O'Leary

McGraw-Hill
Irwin

Boston   Burr Ridge, IL   Dubuque, IA   Madison, WI   New York   San Francisco   St. Louis
Bangkok   Bogotá   Caracas   Kuala Lumpur   Lisbon   London   Madrid   Mexico City
Milan   Montreal   New Delhi   Santiago   Seoul   Singapore   Sydney   Taipei   Toronto

McGraw-Hill Higher Education

A Division of The **McGraw-Hill** Companies

COMPUTING ESSENTIALS, 2001–2002, COMPLETE EDITION
Published by McGraw-Hill/Irwin, an imprint of The McGraw-Hill Companies, Inc. 1221 Avenue of the Americas, New York, NY, 10020. Copyright © 2001, by The McGraw-Hill Companies, Inc. All rights reserved. No part of this publication may be reproduced or distributed in any form or by any means, or stored in a database or retrieval system, without the prior written consent of The McGraw-Hill Companies, Inc., including, but not limited to, in any network or other electronic storage or transmission, or broadcast for distance learning. Some ancillaries, including electronic and print components, may not be available to customers outside the United States.

This book is printed on acid-free paper.

1 2 3 4 5 6 7 8 9 0 QPD/QPD 0 9 8 7 6 5 4 3 2 1

ISBN 0072479973
ISSN 1067-8549

Publisher: *David Brake*
Sponsoring editor: *Jodi McPherson*
Developmental editor: *Alexandra Arnold/Melissa Forte*
Marketing manager: *Nicole Young*
Senior project manager: *Mary Conzachi*
Production supervisor: *Melonie Salvati*
Cover design: *Artemio Ortiz, Jr.*
Interior design: *Artemio Ortiz, Jr.*
Photo research coordinator: *Ira Roberts*
Supplement coordinator: *Marc Mattson*
New media: *David Barrick*
Compositor: *Cecelia G. Morales*
Typeface: *10/12 ITC New Baskerville*
Printer: *Quebecor World Dubuque Inc.*

www.mhhe.com

Information Technology at McGraw-Hill/Irwin

InformationTechnology

At McGraw-Hill Higher Education, we publish instructional materials targeted at the higher education market. In an effort to expand the tools of higher learning, we publish texts, lab manuals, study guides, testing materials, software, and multimedia products.

At McGraw-Hill/Irwin (a division of McGraw-Hill Higher Education), we realize that technology has created and will continue to create new mediums for professors and students to use in managing resources and communicating information with one another. We strive to provide the most flexible and complete teaching and learning tools available as well as offer solutions to the changing world of teaching and learning.

McGraw-Hill/Irwin is dedicated to providing the tools for today's instructors and students to successfully navigate the world of Information Technology.

- **Seminar series**—McGraw-Hill/Irwin's Technology Connection seminar series offered across the country every year demonstrates the latest technology products and encourages collaboration among teaching professionals.

- **McGraw-Hill/Osborne**—This division of The McGraw-Hill Companies is known for its best-selling Internet titles *Harley Hahn's Internet & Web Yellow Pages* and the *Internet Complete Reference*. Osborne offers an additional resource for certification and has strategic publishing relationships with corporations such as Corel Corporation and America Online. For more information visit Osborne at **www.osborne.com.**

- **Digital solutions**—McGraw-Hill/Irwin is committed to publishing digital solutions. Taking your course online doesn't have to be a solitary venture, nor does it have to be a difficult one. We offer several solutions that will allow you to enjoy all the benefits of having course material online. For more information visit **www.mhhe.com/ solutions/index.mhtml.**

- **Packaging options**—For more about our discount options, contact your local McGraw-Hill/Irwin sales representative at 1-800-338-3987 or visit our Web site at **www.mhhe.com/it.**

BRIEF CONTENTS

CONTENTS

5 Input and Output 98

8 The Internet and the Web 178

9   Multimedia, Web Authoring, and More   210

10   Privacy, Security, Ergonomics, and the Environment   234

11   Databases   262

14 Programming and Languages 334

15 Your Future and Information Technology 362

Introduction

The twentieth century not only brought us the dawn of the Information Age, but continued to bring us rapid changes in information technology. There is no indication that this rapid rate of change will be slowing—it may even be increasing. As we begin the twenty-first century, computer literacy will undoubtedly become prerequisite in whatever career a student chooses. The goal of *Computing Essentials, 2001–2002* is to provide students with the basis for understanding the concepts necessary for success in the Information Age. *Computing Essentials* also endeavors to instill in students an appreciation for the effect of information technology on people and our environment, and a basis for building the necessary skill set to succeed in this new, twenty-first century.

About the Authors

Tim and Linda O'Leary live in the American Southwest and spend much of their time engaging instructors and students in conversation about learning. In fact, they have been talking about learning for over 25 years. Something in those early conversations convinced them to write a book, to bring their interest in the learning process to the printed page. Today, they are as concerned as ever about learning, about technology, and about the challenges of presenting material in new ways, both in terms of content and the method of delivery.

A powerful and creative team, Tim combines his years of classroom teaching experience with Linda's background as a consultant and corporate trainer. Tim has taught courses at Stark Technical College in Canton, Ohio, Rocherster Institute of Technology in Upper New York state, and is currently a professor at Arizona State University in Tempe, Arizona. Tim and Linda have talked to and taught students from 8 to 80, all of them with a desire to learn something about computers and the applications that make their lives easier, more interesting, and more productive.

Each new edition of an O'Leary text, supplement, or learning aid has benefited from these students and their instructors who daily stand in front of them (or over their shoulders). *Computing Essentials, 2001–2002* is no exception.

About Our Book

Times are changing, technology is changing, and this text is changing too. Do you think the students of today are different from yesterday? Mine are and I'll wager that yours are as well. On the positive side, I am amazed how much effort students put toward things that interest them and things they are convinced are relevant to them. Their effort directed at learning application programs and exploring the Web seems at times limitless. On the other hand, it is difficult to engage them in other equally important topics such as personal privacy and technological advances.

I've changed the way I teach and this book reflects that. I no longer *lecture* my students about how important certain concepts like microprocessors, input devices, and utility programs are. Rather, I begin by *engaging* their interest by presenting practical tips related to the key concepts, by *demonstrating* interesting  applications that are relevant to their lives, and by *focusing* on outputs rather than processes. Then, I *discuss* the concepts and processes.

Motivation and relevance are the keys. This text has several features specifically designed to engage students and to demonstrate the relevance of technology in their lives. These elements are combined with a thorough coverage of the concepts and sound pedagogical devices.

Features of this Edition

- The Introductory version of *Computing Essentials, 2001–2002* contains two additional chapters: *Privacy, Security, Ergonomics, and the Environment* and *Your Future and Information Technology.*

- One of the most significant features of this edition is the three-level approach to learning in both the text and the supplements. Based upon a proven learning model, this three-level format is engineered to help students take ownership of the text material.

  Level 1 questions and exercises test recall of the basic information and terminology in the chapter.
  Level 2 questions and exercises review students' understanding of concepts and ability to integrate ideas presented in different parts of the chapter.
  Level 3 questions and exercises test students' critical thinking skills and ability to apply the concepts they have mastered to solve problems.

- Within the text material, we have incorporated pedagogical features that follow the three-level approach and enhance both instruction and learning.

  In-text Concept Checks serve to reinforce text material by calling upon students to recall facts and concepts relating to what they have just read.

  Making IT Work is a feature that visually demonstrates how technology is applied in our everyday lives. Topics covered include Web-based Applications; Active Desktop; TV Tuner Cards; Voice Recognition Systems; Music from the Internet; Home Networks; Instant Messaging; Personal Web Sites; Virus Protection; and Finding Jobs Online. These "gallery" style boxes combine text and art to take students step-by-step through technological processes that are both interesting and useful. This edition now includes a total of 10 making IT Work for You topics.

- To facilitate student learning, the **Chapter Review** follows the three-level approach described above.

  Level 1 includes matching, true/false, multiple choice, and completion exercises.
  Level 2 includes short answer and concept mapping exercises.
  Level 3 includes critical thinking questions and problems.

- The **Testbank** follows the same three-level approach that is introduced in the text to provide a valuable testing and reinforcement tool. Each question is assigned a category: **Level 1** definition, **Level 2** concept, and **Level 3** application. Also included are a text page reference and a rationale for each question and answer.

- For each chapter, the student **Online Learning Center** offers both a review of the text material, in the form of an **e-learning session,** and additional exercises organized around the following themes: Group/team projects, Internet/Web-related content, mini case studies of actual companies, and profiles of careers that are influenced by information technology. The content and activities for the exercises further establish O'Leary's three-level learning approach.

Resources for Instructors

We understand that, in today's teaching environment, offering a textbook alone is not sufficient to meet the needs of the many different instructors who use our books. To teach effectively, instructors must have a full complement of supplemental resources to assist them in every facet of teaching from preparing for class; to conducting and lecture; to assessing students' comprehension. *Computing Essentials, 2001–2002* offers a complete, fully integrated supplements package, as described below.

Instructor's Resource Kit

The Instructor's Resource Kit contains an updated CD-ROM containing the Instructor's Manual in both MS Word and .pdf formats, PowerPoint slides, and Brownstone's Diploma test generation software with accompanying test item files for each chapter. The distinctive features of each component of the Instructor's Resource Kit are described below.

- **Instructor's Manual:** The Instructor's Manual contains a schedule showing how much time is required to cover the material in the chapter; a list of the chapter competencies; tips for covering difficult material; and answers to the Concept Checks. Also included are references to corresponding topics on the Interactive Companion CD-ROM, answers to all the exercises in the Chapter Review section and answers to the On the Web Exercises. The manual also includes a helpful introduction that explains the features, benefits, and suggested uses of the IM and an index of concepts and corresponding competencies.

- **PowerPoint Presentation:** The PowerPoint presentation is designed to provide instructors with a comprehensive resource for use during lecture. It includes a review of key terms and definitions, figures from the text, along with several new illustrations, anticipated student questions with answers, and additional resources which can be accessed in Internet-enabled classrooms. Also included with the presentation are comprehensive speaker's notes.

- **Testbank:** The *Computing Essentials 2001–2002* edition testbank contains over 3,000 questions categorized by level of learning (definition,

concept, and application). This is the same learning scheme that is introduced in the text to provide a valuable testing and reinforcement tool. The test questions are identified by text page number to assist you in planning your exams, and rationales for each answer are also included. Additional test questions, which can be used as pretests and posttests in class, can be found on the Online Learning Center, accessible through our supersite (**www.mhhe.com/it**).

Business Week Edition of *Computing Essentials*

An exciting new supplement with this edition of *Computing Essentials* is our *Business Week Edition*. With the purchase of a *Business Week Edition* of a McGraw-Hill/Irwin textbook, students will receive a 15-week subscription to *Business Week* for only $8.25 more than the price of the book alone. Professors who adopt the *Business Week Edition* will enjoy a complimentary subscription for a full year to *Business Week* magazine and complimentary access to the *Business Week* Resource Center Web site as well as *Business Week* Online through the duration of their subscription.

Students will also enjoy free access to the *Business Week* Resource Center Web site (**www.resourcecenter. businessweek.com**) for the duration of their magazine subscription. The *Business Week* Resource Center Web site contains a wealth of supplemental materials, including the *Business Week* Online Archives. Students will have instant access to any business topic from the past nine years of *Business Week*—from 1991 to 1999. From the Resource Center, students may also access *Business Week* Online (**www.business week.com**) for current issues, online-only features, and career tips. Access to these sites provides a marvelous opportunity to increase students' Internet literacy as instructors explore new ways to integrate the Web into a wide array of student exercises and research projects.

Interactive Companion CD-ROM

This free student CD-ROM, designed for use in class, in the lab, or at home by students and professors alike includes a collection of interactive tutorial labs on some of the most popular topics in information technology. By combining video, interactive exercises, animation, additional content, and actual "lab" tutorials, we expand the reach and scope of the textbook.

Digital Solutions to Help You Manage Your Course

PageOut—PageOut is our Course Web Site Development Center that offers a syllabus page, URL, McGraw-Hill Online Learning Center content, online exercises and quizzes, gradebook, discussion board, and an area for student Web pages. For more information, visit the PageOut Web site (**www.pageout.net**).

Online Learning Centers—The Online Learning Center that accompanies *Computing Essentials* is accessible through our Information Technology Supersite (**www.mhhe.com/it**). This site provides additional learning and instructional tools developed using the same three-level approach found in the text and supplements. This offers a consistent method for students to enhance their comprehension of the concepts presented in the text.

Online Courses Available—OLCs are your perfect solutions for Internet-based content. Simply put, these Centers are "digital cartridges" that contain a book's pedagogy and supplements. As students read the book, they can go online and take self-grading quizzes or work through interactive exercises. These also provide students appropriate access to lecture materials and other key supplements.

Online Learning Centers can be delivered through any of these platforms:

McGraw-Hill Learning Architecture (TopClass)

Blackboard.com

ECollege.com (formally Real Education)

WebCT (a product of Universal Learning Technology)

O'Leary Series Applications Lab Manuals

Available separately, or packaged with *Computing Essentials* is the O'Leary Series computer applications lab manuals for Microsoft Office. The O'Leary Series offers a step-by-step approach to developing computer applications skills and is available in both brief and introductory levels. The introductory level manuals are MOUS Certified and prepare students for the Microsoft Office User Certification Exam.

Skills Assessment

McGraw-Hill/Irwin offers two innovative systems to meet your skills assessment needs. These two products are available for use with any of our applications manual series.

ATLAS (Active Testing and Learning Assessment Software)—ATLAS is one option to consider for an application skills assessment tool from McGraw-Hill. ATLAS allows students to perform tasks while working live within the Microsoft applications environment. ATLAS provides flexibility for you in your course by offering:

- Pre-testing options
- Post-testing options
- Course placement testing
- Diagnostic capabilities to reinforce skills
- Proficiency testing to measure skills

ATLAS is Web-enabled, customizable, and is available for Microsoft Office 2000.

SimNet (Simulated Network Assessment Product)—SimNet is another option for a skills assessment tool that permits you to test students' software skills in a simulated environment. SimNet is available for Microsoft Office 97 (deliverable via a network) and Microsoft Office 2000 (deliverable via a network and the Web). SimNet provides flexibility for you in your course by offering:

- Pre-testing options
- Post-testing options
- Course placement testing
- Diagnostic capabilities to reinforce skills
- Proficiency testing to measure skills

PowerWeb for Concepts

PowerWeb is an exciting new online product available for *Computing Essentials 2001–2002*. A nominally priced token grants students access through our Web site to a wealth of resources—all corresponding to the text. Features include an interactive glossary; current events with quizzing, assessment, and measurement options; Web survey; links to related text content; and WWW searching capability via Northern Lights, an academic search engine. Visit PowerWeb at **www.dushkin.com/powerweb.**

Student's Guide to the O'Leary Learning System

Recently, at the end of the semester, some of my students stopped by my office to say they enjoyed the class and that they "learned something that they could *actually* use." High praise indeed for a professor! Actually, I had mixed feelings. Of course, it felt good to learn that my students enjoyed the course. However, it hurt a bit that they were *surprised* that they learned something useful.

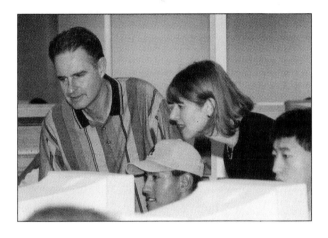

Here's my promise to you: *In the following pages you will find things that you can actually use now as well as provide a foundation to understand future technological advances.*

As you read the text, notice the "Tips" scattered throughout the book. These tips offer suggestions on a variety of topics from the basics of cleaning a monitor to how to make your computer run faster and smoother. Also, notice the "Making IT Work for You" sections that demonstrate some specific computer applications you might find interesting. For example, one demonstrates how to capture and use television video clips for electronic presentations and another shows how to capture, save, and play music from the Internet.

Many learning aids are built into the text to ensure your success with the material and to make the process of learning rewarding. On the pages that follow, we call your attention to the key features in the text. We also show you supplemental materials, such as the student Online Learning Center, that you should take advantage of to ensure your success in this course.

CHAPTER

8

The Internet and the Web

COMPETENCIES

After you have read this chapter, you should be able to:

1. Describe Internet providers, connections, and protocols.

2. Discuss e-mail, mailing lists, newsgroups, chat groups, and instant messaging.

3. Describe electronic commerce including Web storefronts, auctions, and electronic payment.

4. Describe Internet services: Telnet, FTP, Gopher, and the Web.

5. Discuss browsers, Web pages, and Web portals.

6. Compare the two types of search tools: indexes and search engines.

7. Discuss the two types of Web utilities: plug-ins and helper applications.

8. Describe intranets, extranets, and firewalls.

W ant to communicate with a friend across town, in another state, or even in another country? Perhaps you would like to send a drawing, a photo, or just a letter. Looking for travel or entertainment information? Perhaps you're researching a term paper or

exploring different career paths. Where do you start? For these and other information-related activities, try the Internet and the Web. They are the 21st-century information resources designed for all of us to use.

The Internet is like a highway that connects you to millions of other people and organizations. Unlike typical highways that move people and things from one location to another, the Internet moves your *ideas* and *information*. Rather than moving through geographic space, you move through **cyberspace**—the space of electronic movement of ideas and information. The Web provides an easy-to-use, exciting, multimedia interface to

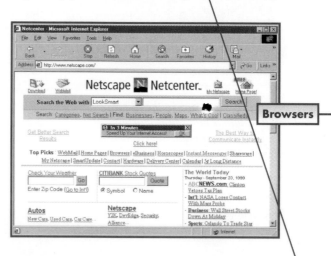

178

Visual Chapter Openers

Each chapter begins with a unique presentation of text and graphics that introduces you to the key concepts in the chapter. A list of competencies will help you structure your reading of the chapter and keep you focused on the key points.

Browsers

Key Terms

Throughout the text, the most important terms are presented in bold type and are defined within the text. You will also find a list of key terms at the end of each chapter and in the glossary at the end of the book.

Term	Description
FAQ	Frequently asked question
Flaming	Insulting, putting down, or attacking
Lurking	Reading news but not joining in to contribute
Saint	Someone who aids new users by answering questions
Thread	A sequence of ongoing messages on the same subject
Wizard	Someone who has comprehensive knowledge about a subject

FIGURE 8-12
Selected discussion-group terms

Terms

Before you submit a contribution to a discussion group, it is recommended that you observe or read the communications from others. This is called **lurking.**

By lurking, you can learn about the culture of a discussion group. For example, you can observe the level and style of the discussions. You may decide that a particular discussion group is not what you were looking for— in which case, unsubscribe. If the discussions are appropriate and you wish to participate, try to fit into the prevailing culture. Remember that your contributions will likely be read by hundreds of people.

For a list of some other commonly used discussion group terms, see Figure 8-12.

CONCEPT CHECK

✔ Give examples of the types of discussion groups you can find on the Internet.

✔ What is the most popular chat service on the Internet?

Electronic Commerce

Web storefronts offer goods and services. Web auctions are like traditional auctions. Electronic payment options include check, credit card, and electronic cash.

Electronic commerce, also known as **e-commerce,** is the buying and selling of goods over the Internet. Have you ever bought anything over the Internet? If you have not, there is a very good chance that you will within the next year or two. Shopping on the Internet is growing rapidly and there seems to be no end in sight. (See Figure 8-13.)

Web Storefronts

Web storefronts are virtual stores where shoppers can go to inspect merchandise and make purchases. (See Figure 8-14.) A new type of program

TIPS What if you don't know or have forgotten someone's e-mail address? You can go to e-mail address directories, also known as e-mail "white pages." These directories can be used much like you would use the telephone white pages. Here are three e-mail address directories you might try:

www.bigfoot.com

www.people.yahoo.com

www.infospace.com

Tips
Tips offer practical and timely advice on a variety of topics related to computers and information technology. These tips answer your questions about how to clean your mouse; how to protect your privacy on the Web; and what to do to avoid catching a virus (a computer virus, that is!).

ON THE WEB EXPLORATIONS
Electronic commerce is one of the fastest growing Web applications. To learn more about it, visit our Web site at
http://www.mhhe.com/it/oleary/ explore.mhtml

On the Web Explorations
Throughout each chapter, you will notice On the Web Explorations notes that encourage you to visit well-established and informative sites to learn more about topics presented.

Concept Checks
These questions, located at the end of major sections in each chapter, encourage you to stop and assess your understanding of the concepts you have read in the preceding section. The questions are designed to be answered quickly and easily. If the questions seem difficult, you should probably review that material before continuing.

188 CHAPTER 8 The Internet and the Web

Instant Messaging

Like chat groups, **instant messaging** allows one or more people to communicate via direct, "live" communication. Instant messaging, however, provides greater control and flexibility than chat groups. (See Making IT Work for You: Instant Messaging, shown below.)

To use instant messaging, you specify a list of friends, or "buddies," and register with an instant messaging server. Whenever you connect to the Internet, you use special software to tell your messaging server that you are online too. It notifies you if any of your buddies are online. At the same time, it notifies your buddies that you are online. You can then send messages back and forth to one another instantly.

MAKING IT WORK FOR YOU

Instant Messaging

Do you enjoy chatting with your friends? Are you working on a project and need to collaborate with others in your group? Perhaps instant messaging is just what you're looking for. It's easy and free with an Internet connection and the right software.

How It Works

Users register with an instant messaging server and identify friends and colleagues (buddies). Whenever a user is online, the instant messaging server notifies the user of all buddies who are also online and provides support for direct "live" communication.

Getting Started

The first step is to connect to one of the many Web sites that support instant messaging. Once at the site, register, download, and install instant messaging software, and create your buddy list.

For example, you can set up AOL Instant Messenger as shown below.

1. Enter aim.aol.com in the Location box of your browser.

2. Select the link to register as a new user.

3. Complete the registration form.

4. After installing the instant messaging software, create your "buddy list."

Making IT Work

Making IT Work is a new feature that visually demonstrates how technology is used in our everyday lives. Topics covered include TV Tuner Cards; Voice Recognition Systems; Music from the Internet; Home Networks; Instant Messaging; and Finding Jobs Online. These "gallery" style boxes combine text and art to take you step-by-step through technological processes that are both interesting and useful.

FIGURE 8-32
Connecting people and organizations

A Look to the Future
These boxes involve interesting discussions about future trends in information technology and their effect on our lives.

A LOOK TO THE FUTURE

Internet2 will be a private high-performance Internet.

Have you ever been unable to connect to the Internet? Have you ever had a long wait before a Web page or a graphic appeared on your screen? Almost all of us have experienced busy servers and slow access. Unfortunately, Internet service is expected to get worse. For organizations that depend on the Internet to reach customers and conduct other business activities, this trend is very concerning.

To address this concern, a separate, private Internet called Internet2 is being developed. It will be a high-speed network capable of dazzling feats that far exceed today's Internet capabilities. Expected to be fully operational by the end of 2002, Internet2 will have limited access to those willing to pay more to get more. Access to today's Internet will remain public and available for a nominal fee.

The primary beneficiaries of Internet2 will be federal agencies and major corporations. Each will pay an annual fee of $500,000 for access to this network that combines high performance with tightly controlled security. One of the first to take advantage of Internet2 will be online publishers of books, photographs, and original artwork. Advanced virtual reality interfaces, called **nanomanipulators,** are expected to be available. Researchers from different parts of the world will be able to share devices such as atomic microscopes and to jointly study, experience, and move within realistic virtual subatomic environments.

Will moving power users to Internet2 increase the performance of the public Internet? We will have to wait and see.

202 CHAPTER 8 The Internet and the Web

VISUAL SUMMARY The Internet and the Web

INTERNET APPLICATIONS

The most common Internet applications are **communicating, shopping, researching,** and **entertainment.**

ACCESS

Once connected to the Internet, your computer seemingly becomes an extension of a giant computer that branches all over the world.

Providers
The most common access is through a **provider** or **host computer.** Three widely used providers are:
- **Colleges and universities**—often offer free Internet access through their LANs.
- **Internet service providers (ISPs)**—offer access for a fee.
- **Online service providers**—offer access and a variety of other services for a fee.

Connections
To access the Internet, you need to connect to a provider. Three types of connections are **direct (dedicated), SLIP** and **PPP,** and **terminal.**

TCP/IP
TCP/IP is the standard **protocol** of the Internet.

E-MAIL

Sending and receiving **e-mail** is the most common Internet activity.

Basic Elements
E-mail messages have three basic elements:
- **Subject**—one line description.
- **Addresses**—sender, receiver, and anyone else receiving copies.
- **Attachments**—files.

Addresses
The Internet uses the **domain name system (DNS)** for e-mail. The first part is the **user name,** followed by the **domain name,** and then the **domain code.**

DISCUSSION GROUPS

Discussion groups support electronic communication between individuals. Four types exist:
- **Mailing lists** use e-mail **subscription** and **list addresses.**
- **Newsgroups** are organized by major topic areas and use the **UseNet** network.
- **Chat groups** allow direct "live" communication.
- **Instant messaging** is for communicating and collaborating.

Visual Summaries
Visual Summaries incorporate text and graphics to provide you with a dynamic review of all the material presented in the chapter. The Visual Summary also serves as a useful text reference.

Chapter Review

The Chapter Review now follows a three-level format and includes exercises that reinforce a review of terms, a review of concepts, and application of concepts. This is achieved through various types of questions and exercises including matching; true/false; multiple choice; completion; short answer; concept matching; and critical thinking.

Level 1 questions and exercises help you test your recall of the basic information and terminology in the chapter.

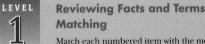

Chapter Review

2001 2002

LEVEL 1

Reviewing Facts and Terms
Matching

Match each numbered item with the most closely related lettered item. Write your answers in the spaces provided.

1. The space of electronic movement of ideas and information. ____
2. The standard protocol for the Internet. ____
3. Security system designed to protect an organization's network against external threats. ____
4. Special software used by your computer to access the Web. ____
5. Identifies a unique person or computer at the listed domain. ____
6. Programs that are automatically uploaded and operate as part

a. Applets
b. Browser
c. Cyberspace
d. DNS
e. E-commerce
f. Firewall
g. HTML
h. Hyperlinks
i. Intranet
j. Java
k. Mailing list
l. Packets

208 CHAPTER 8 **The Internet and the Web**

Level 2 questions and exercises help you review your understanding of concepts and your ability to integrate ideas presented in different parts of the chapter.

Reviewing Concepts

2001 2002

LEVEL 2

Open-Ended

On a separate sheet of paper, respond to each question and statement.

1. Discuss four frequently used Internet services.
2. Describe how addresses are assigned under the domain name system.
3. What are the two types of Internet search tools? How do they differ?
4. Describe some typical Web utilities and how they can help you.
5. Discuss the similarities and differences between intranets and extranets.

Concept Mapping

On a separate sheet of paper, draw a concept map or a flowchart showing how the following terms are related. Show all relationships. Include any additional terms you can think of.

applets	FTP	plug-ins
browsers	host computer	TCP/IP
chat groups	HTML	Telnet
cyberspace	hyperlinks	uploading/downloading
e-mail	intranets	Web site
end user	ISP	World Wide Web
firewall	mailing lists	

Critical Thinking Questions and Projects

LEVEL 3

Read each exercise and answer the related questions on a separate sheet of paper.

1. *Going on an Internet scavenger hunt.* Use the Internet to find information about the following topics. (Record the URLs where you found the information, and write a short description of your findings.)
   a. Hotels in London
   b. Cast members for one of your favorite television programs
   c. MTV's news for this week
   d. Employment opportunities in a career of your choice
   e. Painting known as the *Mona Lisa*
   f. The weather conditions for your city (or the nearest large city)

Level 3 questions and exercises test your critical thinking skills and ability to apply the concepts you have mastered to solve problems.

4. *Instant Messaging:* Instant messaging is a popular way to have "live" communication over the Internet. (See Making IT Work for You: Instant Messaging on pages 188–189.)

   a. Have you used instant messaging? If so, describe how and why you used it and whether you found it useful.

   b. Discuss the differences between instant messaging and chat groups. What are the relative advantages and disadvantages?

   c. Can you foresee any situations in which you would not want to be on another person's buddy list? If so, describe the situation.

   d. Describe three situations in which you might use instant messaging. Be specific.

   e. Do you think that the popularity of instant messaging will continue in the future? Why or why not?

On the Web Exercises

1. Off-Line Browsing

Off-line browsing offers a great promise of increased productivity. What could be better than to have your computer searching the Web for you while you sleep? To learn more about off-line browsing, visit our site at http://www.mhhe.com/it/oleary/exercise.mhtml to link a site that specializes in off-line browsing. Once connected, explore and learn more about off-line browsing. Print out the most informative Web page you find and write a brief paragraph summarizing the pros and cons of off-line browsing.

2. Adventure Tours

Travel to exotic places around the world without ever leaving home. To do this, visit our site at http://www.mhhe.com/it/oleary/exercise.mhtml to link to a site specializing in exotic virtual travel. Once connected to the site, explore and learn more about adventure tours. Print out the Web page you find the most informative, and write a brief paragraph describing a virtual tour and discuss what you think the future is for this type of activity.

3. E-Cash

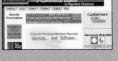

Several companies, including Cyber-Cash and Digi-Cash, are working on different variations of e-cash. Visit our site at http://www.mhhe.com/it/oleary/exercise.mhtml to link to a site that specializes in e-cash. Once connected, explore and learn more about e-cash and its alternatives. Print out the most informative Web page you find and write a brief paragraph describing how e-cash works.

4. Instant Messaging

Instant messaging is widely used to directly communicate with friends, to conduct meetings, and to share files on the Internet. (See Making IT Work for You: Instant Messaging on pages 188–189.) Most servers that support instant messaging require participants to use the same instant messaging software. To learn more about different instant messaging software, visit the Yahoo site at http://www.yahoo.com and look at the subject area "Computers and Internet: Software: Internet: Instant Messaging." Explore and print out the Web page you found the most informative. Write a paragraph comparing the different instant messaging programs that are available.

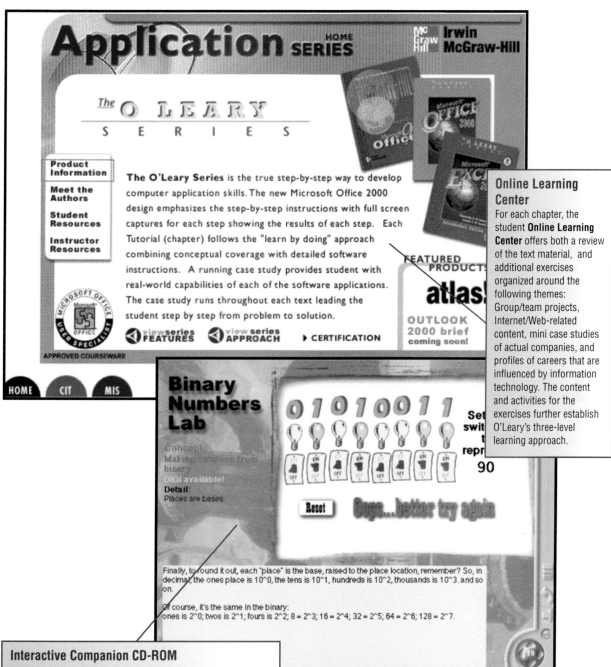

The O'Leary Series is the true step-by-step way to develop computer application skills. The new Microsoft Office 2000 design emphasizes the step-by-step instructions with full screen captures for each step showing the results of each step. Each Tutorial (chapter) follows the "learn by doing" approach combining conceptual coverage with detailed software instructions. A running case study provides student with real-world capabilities of each of the software applications. The case study runs throughout each text leading the student step by step from problem to solution.

Online Learning Center
For each chapter, the student **Online Learning Center** offers both a review of the text material, and additional exercises organized around the following themes: Group/team projects, Internet/Web-related content, mini case studies of actual companies, and profiles of careers that are influenced by information technology. The content and activities for the exercises further establish O'Leary's three-level learning approach.

Binary Numbers Lab

Finally, to round it out, each "place" is the base, raised to the place location, remember? So, in decimal, the ones place is 10^0, the tens is 10^1, hundreds is 10^2, thousands is 10^3, and so on.

Of course, it's the same in the binary:
ones is 2^0; twos is 2^1; fours is 2^2; $8 = 2^3$; $16 = 2^4$; $32 = 2^5$; $64 = 2^6$; $128 = 2^7$.

Interactive Companion CD-ROM
Use this free CD-ROM to explore some of the most popular topics in information technology. Video, interactive exercises, animation and actual "labs" expand the reach and scope of the textbook.

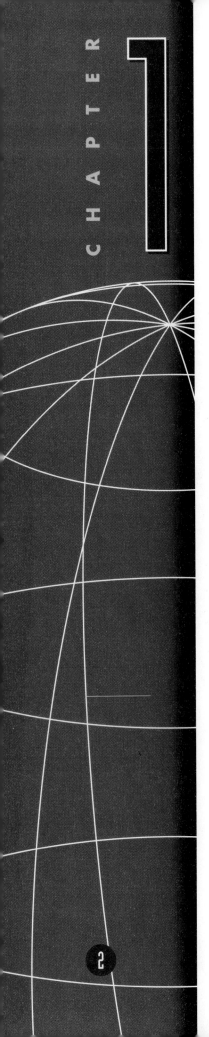

CHAPTER 1

Your Future and Computer Competency

COMPETENCIES

After you have read this chapter, you should be able to:

1. **Explain the five parts of an information system: people, procedures, software, hardware, and data.**

2. **Distinguish application software from system software.**

3. **Distinguish four kinds of computers— microcomputer, minicomputer, mainframe, and supercomputer—and describe hardware devices for input, processing, storage, output, and communications.**

4. **Describe document, worksheet, database, and presentation files.**

5. **Explain computer connectivity, the Internet, and the Web.**

*C*omputer competency: This notion may not be familiar to you, but it's easy to understand. The purpose of this book is to help you become *competent* in computer-related skills. Specifically, we want to help you walk into a job and immediately be valuable to an employer. In this chapter, we first present an overview of what makes up an information system: people, procedures, software, hardware, and data. Competent end users need to understand these basic parts and how connectivity through the Internet and the Web expands the role of information technology (IT) in our lives. In subsequent chapters, we will describe these parts of information systems in detail.

Fifteen years ago, most people had little to do with computers, at least directly. Of course, they filled out computerized forms, took computerized tests, and paid computerized bills. But the real work with computers was handled by specialists—programmers, data-entry clerks, and computer operators.

Then microcomputers came along and changed everything. Now it is easy for nearly everybody to use a computer. People who use microcomputers are called end users. Today:

- Microcomputers are common tools in all areas of life. Writers write, artists draw, engineers and scientists calculate—all on microcomputers. Students and businesspeople do all this, and more.

- New forms of learning have developed. People who are homebound, who work odd hours, or who travel frequently may take courses on the Web. A college course need not fit within the usual time of a quarter or a semester.

- New ways to communicate, to find people with similar interests, and to buy goods are available.

All kinds of people are using electronic mail, electronic commerce, and the Internet to meet and to share ideas and products.

What about you? How can microcomputers enhance *your* life?

Many interesting and practical uses of information technology have recently surfaced to make our personal lives richer and more entertaining. These applications range from recording video clips to creating personalized Web sites. (See Making IT Work for You below.) What about you? How can information technology and microcomputers enhance your life?

Competent end users need to know the five parts of an information system made up of people, procedures, software, hardware, and data. Additionally, they need to understand connectivity through the Internet and the Web and to recognize the role of information technology in their professional and personal lives.

MAKING IT WORK FOR YOU

TV Tuner Cards and Video Clips
Want to watch your favorite television program while you work? Perhaps you would like to include a video clip from television in a class presentation. It's easy using a TV tuner card.

Voice Recognition Systems and Dictating a Paper
Tired of using your keyboard to type term papers and to control programs? Voice recognition maybe just what you're looking for.

CD-R Drives and Music from the Internet
Did you know that you could use the Internet to locate music, download it to your computer, and create your own compact discs? All it takes is the right software, hardware, and a connection to the Internet.

Instant Messaging
Do you enjoy chatting with your friends? Are you working on a project and need to collaborate with others in your group? Perhaps instant messaging is just what you're looking for.

Personal Web Site
Do you have anything to share with the world? Would you like a personal Web site, but don't want to deal with learning HTML and paying for server time? Many services are available to get you started for FREE!

Information Systems

An information system has five parts: people, procedures, software, hardware, and data.

When you think of a microcomputer, perhaps you think of just the equipment itself. That is, you think of the monitor or the keyboard. There is more to it than that. The way to think about a microcomputer is as part of an information system. An **information system** has five parts: *people, procedures, software, hardware,* and *data.* (See Figure 1-1.)

- **People:** It is easy to overlook people as one of the five parts of a microcomputer system. Yet that is what microcomputers are all about—making **people,** end users like yourself, more productive.

- **Procedures: Procedures** are rules or guidelines for people to follow when using software, hardware, and data. Typically, these procedures are documented in manuals written by computer specialists. Software and hardware manufacturers provide manuals with their products.

- **Software: Software** is another name for a program or programs. A **program** consists of the step-by-step instructions that tell the computer how to do its work. The purpose of software is to convert *data* (unprocessed facts) into *information* (processed facts).

- **Hardware:** The **hardware** consists of the equipment: keyboard, mouse, monitor, system unit, and other devices. Hardware is controlled by software. It actually processes the data to create information.

- **Data: Data** consists of the raw, unprocessed facts, including text, numbers, images, and sounds. Examples of raw facts are hours you worked and your pay rate. After data is processed through the computer, it is usually called **information.** An example of such information is the total wages owed you for a week's work.

FIGURE 1-1
The five parts of an information system

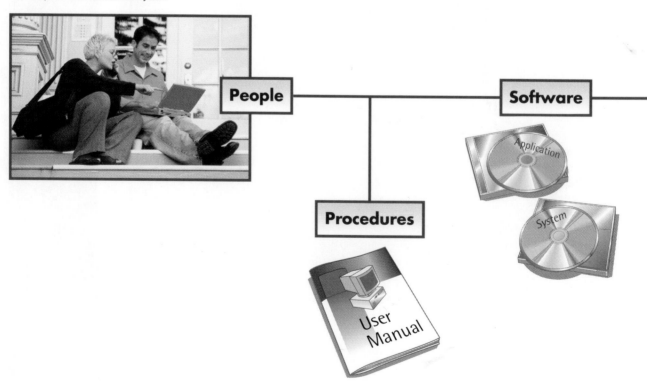

People

Software

Procedures

In large computer systems, there are specialists who deal with writing procedures, developing software, and capturing data. In microcomputer systems, however, end users often perform these operations. To be a competent end user, you must understand the essentials of **information technology (IT),** including software, hardware, and data.

CONCEPT CHECK

✔ What are the five parts of an information system?

✔ To be a competent end user, you must understand the essentials of information technology, including _____, _____, and _____.

People |
ProcEDures |
SoftwarE |
HARDWARE |
Data

SOFTWARE |
HARDWARE |
Data

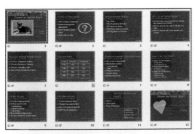

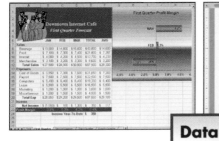

Data

Hardware

Connectivity

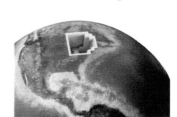

People

People are the most important part of an information system. Examples include people in entertainment, medicine, education, and business.

Although easy to overlook, people are surely the most important part of any information system. Our lives are touched every day by computers and information systems. Many times the contact is direct and obvious such as when we create documents using a word processing program or when we connect to the Internet.

Other times, the contact is not as obvious. Nonetheless, computers and information systems touch our lives hundreds of times every day. Consider just the following four examples. (See Figure 1-2.)

People just like you are making information technology work for them every day. Throughout this book you will find several features designed to make technology work for you. Three specific features are Making IT Work for You topics, Tips, and On the Web Explorations. (See pages 8 and 9.)

Entertainment

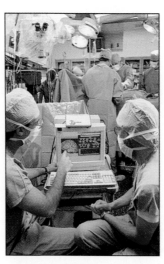

Medicine

Business

Education

FIGURE 1-2
Computers in entertainment, medicine, business, and education

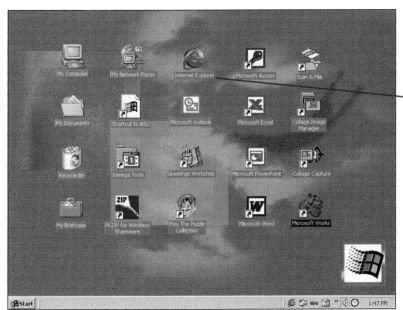

FIGURE 1-3
Windows 2000 operating system

Browser icon

Application Software is End User Software.

System Software enables the application Software to interact w/ the computer hardware. Background Software

Software

Software is of two kinds: system software and application software.

Software, as we mentioned, is another name for programs. Programs are the instructions that tell the computer how to process data into the form you want. In most cases, the words *software* and *programs* are interchangeable.

There are two major kinds of software—*system software* and *application software*. You can think of application software as the kind you use. Think of system software as the kind the computer uses.

System Software

The user interacts with application software. **System software** enables the application software to interact with the computer hardware. System software is "background" software that helps the computer manage its own internal resources.

The most important system software program is the **operating system,** which interacts with the application software and the computer. The operating system handles such details as running ("executing") programs, storing data and programs, and processing data. Windows 2000 is one of the best-known operating systems for today's microcomputer users. (See Figure 1-3.)

Application Software

Application software might be described as "end-user" software. Application software performs useful work on general-purpose tasks such as word processing and data analysis.

There are certain general-purpose programs or basic applications. These programs are widely used in nearly all career areas. They are the kind of programs you *have* to know to be considered computer competent. One of these basic applications is a browser to navigate, explore,

TIPS Have you used the Internet? If so, then you probably already know how to use a browser. For those of you who do not, here are a few tips to get you started.

1. *Start browser.* Typically, all you need to do is double-click the browser's icon on the desktop.

2. *Enter URL.* In the browser's location box, type the URL (uniform resource locator, or address) of the Internet or Web location (site) that you want to visit.

3. *Press ENTER.* On your keyboard, press the ENTER key to connect to the site.

4. *Read and explore.* Once connected to the site, read the information displayed on your monitor. Click on underlined text to explore other locations.

5. *Close browser.* Once you are done exploring, click on your browser's CLOSE button.

Information Technology Topics

Information technology touches our lives everyday in a personal way. Many interesting and practical uses of IT have recently surfaced to make our lives richer and more entertaining. In the following chapters, you will find several of these applications presented in detail.

Web-based Applications

Would you like access to free general-purpose applications from anywhere in the world? What about accessing your data files from any location? You can have it all with Web-based applications. See page 30.

Active Desktop

Want to add some interest to your desktop? Would you like to see the most recent sports scores, news, or stock market updates? You can customize your desktop to provide that information and much more. See page 62.

TV Tuner Cards and Video Clips

Want to watch your favorite television program while you work? Perhaps you would like to include a video clip from television in a class presentation. It's easy using a video TV card. See page 84.

Voice Recognition Systems and Dictating a Paper

Tired of using your keyboard to type term papers and to control programs? Voice recognition may be just what you're looking for. See page 108.

CD-R Drives and Music from the Internet

Did you know that you could use the Internet to locate music, download it to your computer, and create your own compact discs? All it takes is the right software, hardware, and a connection to the Internet. See page 136.

Home Networking

Computer networks are not just for corporations and schools anymore. If you have more than one computer, you can use a home network to share files and printers, to allow multiple users access to the Internet at the same time, and to play interactive computer games. See page 166.

Instant Messaging

Do you enjoy chatting with your friends? Are you working on a project and need to collaborate with others in your group? Perhaps instant messaging is just what you're looking for. See page 188.

Personal Web Site

Do you have anything to share with the world? Would you like a personal Web site but don't want to deal with learning HTML and paying for server time? Many services are available to get you started for FREE! See page 216.

Virus Protection

Ever been attacked by a computer virus? If not, chances are that you will in the near future. Fortunately, special software is available to protect you against computer viruses. See page 244.

Online Job Opportunities

Did you know that you could use the Internet to find a job? You can browse through job openings, post your resume, and even use special programs that will search for the job that's just right for you.

Tips and Web Explorations

Two other features in this book that make technology work for you are a variety of Tips and On the Web Explorations.

TIPS We all can benefit from a few tips or suggestions. Throughout this book you'll find numerous tips ranging from the basics of cleaning a monitor to how to efficiently locate information on the Web. Just a few of these tips are:

- *Inserting audio clips.* Want to add some interest and a personal touch to your correspondence? You can by including an audio clip of your voice in a text document. See page 33.

- *Improving slow computer operations.* Does your computer seem to be getting slower and slower? Consider a few suggestions that might add a little zip to your current system. See page 82.

- *Playing music on your computer.* Do you like to listen to music while working on your computer? If you have a CD-ROM drive you can use it to play your favorite CDs while you work. See page 134.

Other tips include:

- Recovering from a power failure, Cleaning your computer, Increasing hard disk space, Creating a multimedia presentation, Locating information on the Web, Ensuring your privacy on the Web, and Protecting children from the Web's dark side.

ON THE WEB
EXPLORATIONS

Numerous outstanding and informative sites are presented in the On the Web Explorations scattered throughout this book. Just a few of the many On the Web Exploration sites are:

Hotmail   offers free e-mail service
America Online   provides an array of online services
Cosmos Software   creates virtual reality applications on the Web

For links to each Web Exploration site, go to:

http://www.mhhe.com/it/oleary/explore.mhtml

and find information on the Internet. (See Figure 1-4.) The two most widely used browsers are Microsoft's Internet Explorer and Netscape's Navigator. For a summary of the basic applications, see Figure 1-5.

There are many other applications that are more specialized and widely used within certain career areas. They are the kind of programs you *should* know to be truly computer competent in the future. One of the most exciting applications is multimedia, which allows users to integrate video, music, voice, and graphics to create interactive presentations. For a summary of these specialized applications, see Figure 1-6.

CONCEPT CHECK

✔ Name the two major kinds of software.

✔ The most important system software program is the _____ _____.

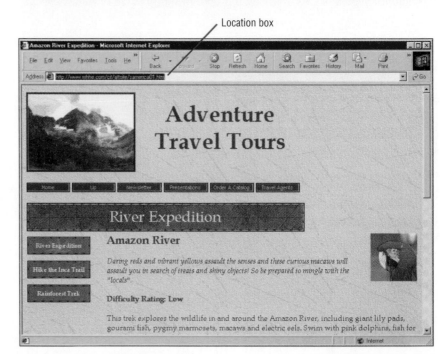

FIGURE 1-4
Browser (Internet Explorer)

Type	Description
Word processor	Prepare written documents
Spreadsheet	Analyze and summarize numerical data
Database management system	Organize and manage data and information
Presentation graphics	Communicate a message or persuade other people
Browser	Navigate, explore, and find information on the Internet
Information managers	Maintain electronic calendars, address books, and to-do lists

FIGURE 1-5
Basic applications

Type	Description
Multimedia	Integrate video, music, voice, and graphics to create interactive presentations
Web publishers	Create interactive multimedia Web pages
Graphics programs	Create professional publications: draw, edit, and modify images
Virtual reality	Create realistic three-dimensional virtual or simulated environments
Artificial intelligence	Simulate human thought processes and actions
Project managers	Plan projects, schedule people, and control resources

FIGURE 1-6
Specialized applications

Hardware

Four types of computers are supercomputer, mainframe computer, mini-computer, and microcomputer. Microcomputer hardware consists of the system unit, input/output, secondary storage, and communications devices.

Computers are electronic devices that can follow instructions to accept input, process that input, and produce information. This book focuses principally on microcomputers. However, it is almost certain that you will come in contact, at least indirectly, with other types of computers.

Types of Computers

There are four types of computers: supercomputers, mainframe computers, minicomputers, and microcomputers.

- **Supercomputers:** The most powerful type of computer is the **super-computer.** These machines are special, high-capacity computers used by very large organizations. For example, NASA uses supercomputers to track and control space explorations.

- **Mainframe computers:** These large computers occupy specially wired, air-conditioned rooms. Although not nearly as powerful as supercomputers, **mainframe computers** are capable of great processing speeds and data storage. (See Figure 1-7.) For example, insurance companies use mainframes to process information about millions of policyholders.

FIGURE 1-7
Mainframe computer
(IBM ES/9000)

- **Minicomputers:** Also known as **midrange computers, minicomputers** are desk-sized machines. Medium-sized companies or departments of large companies typically use them for specific purposes. For example, production departments use minicomputers to monitor certain manufacturing processes and assembly-line operations.

- **Microcomputers:** Although the least powerful, **microcomputers** are the most widely used and fastest-growing type of computer. Apple recently introduced their iMac computers. (See Figure 1-8.) Categories of microcomputer include *desktop, notebook,* and *personal digital assistants.* **Desktop computers** are small enough to fit on top of or alongside a desk yet are too big to carry around. **Notebook computers** are portable, weigh between 4 and 10 pounds, and fit into most briefcases. (See Figure 1-9.) **Personal digital assistants (PDAs)** are also known as **palmtop computers** or **handheld computers.** They combine pen input, writing recognition, personal organizational tools, and communications capabilities in a very small package. (See Figure 1-10.)

Microcomputer Hardware

Hardware for a microcomputer system consists of a variety of different devices. See Figure 1-11 for a typical system. This physical equipment falls into four basic categories: system unit, input/output, secondary storage, and communication devices. Because we discuss hardware in detail later in this book, we will present just a quick overview here.

- **System unit:** The **system unit** is electronic circuitry housed within the computer cabinet. (See Figure 1-12.) Two important components of the system unit are the *microprocessor* and *memory.* The **microprocessor** controls and manipulates data to produce information. **Memory,** also known as **primary storage** or **random access memory (RAM),** holds data and program instructions for processing the data. It also holds the processed information before it is output. Memory is sometimes referred to as *temporary storage* because its contents will typically be lost if the electrical power to the computer is disrupted.

- **Input/output devices: Input devices** translate data and programs that humans can understand into a form that the computer can process.

ON THE WEB EXPLORATIONS

Intel is a leading manufacturer of microprocessors. To learn more about this company, visit our Web site at

http://www.mhhe.com/it/oleary/ explore.mhtml

FIGURE 1-8
Colorful desktop computers from Apple (iMac)

FIGURE 1-9
Notebook computers

FIGURE 1-10
Personal digital assistant

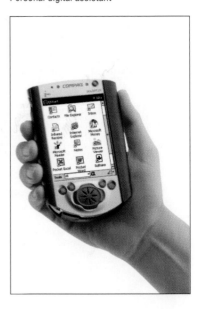

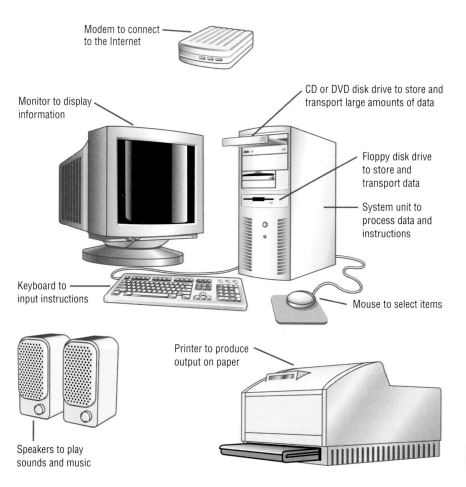

Modem to connect to the Internet

Monitor to display information

CD or DVD disk drive to store and transport large amounts of data

Floppy disk drive to store and transport data

System unit to process data and instructions

Keyboard to input instructions

Mouse to select items

Printer to produce output on paper

Speakers to play sounds and music

FIGURE 1-11
Microcomputer system

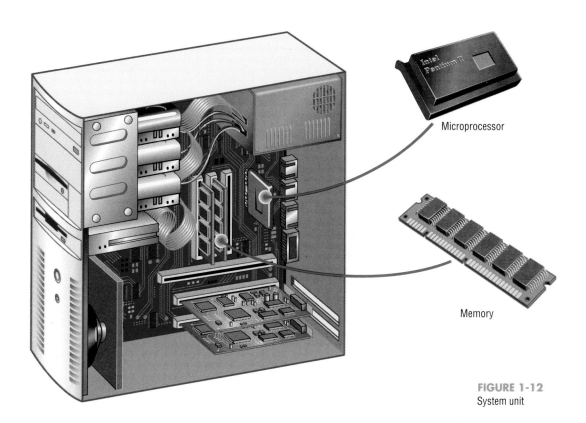

Microprocessor

Memory

FIGURE 1-12
System unit

FIGURE 1-14
Monitor

FIGURE 1-15
A 3½" floppy disk

FIGURE 1-16
An optical disk

FIGURE 1-13
Keyboard and mouse

The most common input devices are the **keyboard** and the **mouse.** (See Figure 1-13.) **Output devices** translate the processed information from the computer into a form that humans can understand. The most common output devices are **monitors** or **video display screens** (see Figure 1-14) and **printers.**

- **Secondary storage devices:** Unlike memory, **secondary storage devices** hold data and programs even after electrical power to the computer system has been turned off. The most important kinds of secondary media are *floppy, hard,* and *optical disks.* **Floppy disks** are widely used to store and transport data from one computer to another. (See Figure 1-15.) They are called floppy because data is stored on a very thin flexible, or floppy, plastic disk. **Hard disks** are typically used to store programs and very large data files. Using a rigid metallic platter, hard disks have a much greater capacity and are able to access information much faster than floppy disks. **Optical disks** use laser technology and have the greatest capacity. (See Figure 1-16.) The two basic types of optical disks are **compact discs (CDs)** and **digital versatile (or video) discs (DVDs).**

- **Communications devices: Communications hardware** sends and receives data and programs from one computer or secondary storage device to another. Many microcomputers use a **modem** to convert electronic signals from the computer into electronic signals that can travel over a telephone line and onto the Internet.

CONCEPT CHECK

✔ List the four types of computers.

✔ Name the four categories of microcomputer hardware.

Data

Data is stored in document, worksheet, database, and presentation files.

Data is used to describe facts about something. When stored electronically in files, data can be used directly as input for the information system.

Four common types of files (see Figure 1-17) are:

- **Document files,** created by word processors to save documents such as memos, term papers, and letters.

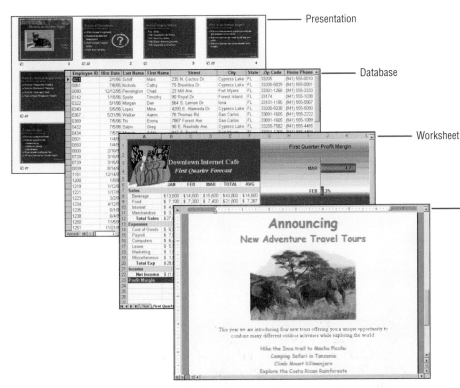

— Presentation
— Database
— Worksheet
— Document

FIGURE 1-17
Four types of files: document, worksheet, database, and presentation

- **Worksheet files,** created by electronic spreadsheets to analyze things like budgets and to predict sales.
- **Database files,** typically created by database management programs to contain highly structured and organized data. For example, an employee database file might contain all the workers' names, social security numbers, job titles, and other related pieces of information.
- **Presentation files,** created by presentation graphics programs to save presentation materials. For example, a file might contain audience handouts, speaker notes, and electronic slides.

FIGURE 1-18
Computers connected together can share information

CONCEPT CHECK

✔ What is data?

✔ Name four file types in which data can be stored.

Connectivity and the Internet

Connectivity is the microcomputer's ability to communicate with other computers and information sources. The Internet is the largest network in the world.

Connectivity is the capability of your microcomputer to share information with other computers. (See Figure 1-18.) Data and information can be sent over telephone lines or cable and through the air. Thus, your microcomputer can be *connected* to other computers. It can connect you to the Internet and to many computerized data banks and other sources of information that lie well beyond your desk.

Connectivity is a very significant development, for it expands the uses of the microcomputer severalfold. Central to the concept of connectivity is the **computer network.** A network is a communications system connecting two or more computers. Networks connect people as close as the next office and as far away as halfway around the world.

The largest network in the world is the **Internet.** It is like a giant highway that connects you to millions of other people and organizations located throughout the world. (See Figure 1-19.) The Internet is a huge computer network available to nearly everyone with a microcomputer and a means to connect to it. The **Web,** also known as the **World Wide Web** or **WWW,** is an Internet service that provides a multimedia interface to numerous resources available on the Internet.

FIGURE 1-19
The Internet connects millions of people worldwide

CONCEPT CHECK

✔ Define connectivity.

✔ What is the Internet?

A LOOK TO THE FUTURE

Computer competency is understanding the rules and the power of microcomputers. Competency lets you take advantage of increasingly productive software, hardware, and the connectivity revolution that are expanding the microcomputer's capabilities.

The purpose of this book is to help you be computer competent not only in the present but also in the future. Having competency requires your having the knowledge and understanding of the rules and the power of the microcomputer. This will enable you to benefit from three important information technology developments: more powerful software, more powerful hardware, and connectivity to outside information resources. It will also help you remain computer competent and continue to learn in the future.

Powerful Software
The software now available can do an extraordinary number of tasks and help you in an endless number of ways. More and more employers are expecting the people they hire to be able to use it. Basic application software and system software are discussed in Chapters 2 and 3. More advanced application software is presented in Chapter 9.

Powerful Hardware
Microcomputers are now much more powerful than they used to be. Indeed, the newer models have the speed and power of room-size computers of only a few years ago.

continues

However, despite the rapid change of specific equipment, their essential features remain unchanged. Thus, the competent end user should focus on these features. Chapters 4 through 6 explain what you need to know about hardware: the central processing unit, input/output devices, and secondary storage. A Buyer's Guide and an Upgrader's Guide are presented at the end of this book for those considering the purchase or upgrade of a microcomputer system.

Connectivity, the Internet, and the Web

No longer are microcomputers and competent end users bound by the surface of the desk. Now they can reach past the desk and link with other computers to share data, programs, and information. The Internet and the Web are considered by most to be the two most important technologies for the 21st century. Accordingly, we devote Chapters 7 and 8 to discussing connectivity, communications, the Internet, and the Web.

Security and Privacy

What about people? Is there a downside to all these technological advances? Experts agree that we as a society must be careful about the potential of technology to negatively impact our personal privacy and security. Additionally, we need to be aware of potential physical and mental health risks associated with using technology. Finally, we need to be aware of negative effects on our environment caused by the manufacture of computer-related products. Thus, Chapter 10 explores each of these critical issues in detail.

Changing Times

Are the times changing any faster now than they ever have? Most people think so. Those who were alive when radios, cars, and airplanes were being introduced certainly lived through some dramatic changes. Has technology made our own times even more dynamic? Whatever the answer, it is clear we live in a fast-paced age. The Evolution of the Computer Age section presented at the end of this book tracks the major developments since computers were first introduced.

Most businesses have become aware that they must adapt to changing technology or be left behind. Many organizations are now making formal plans to keep track of technology and implement it in their competitive strategies. Nearly every corporation in the world has a presence on the Internet. Delivery services such as Federal Express and UPS provide customers with the ability to personally track the delivery of their packages. Retail stores such as JCPenney and Wal-Mart provide catalog support and sales. Banks such as Wells Fargo and Citibank support home banking and electronic commerce. You can even purchase tickets to music, theater, and sporting events on the Internet.

Clearly, such changes do away with some jobs—those of many bank tellers and cashiers, for example. However, they create opportunities for other people. New technology requires people who are truly capable of working with it. These are not the people who think every piece of equipment is so simple they can just turn it on and use it. Nor are they those who think each new machine is a potential disaster. In other words, new technology needs people who are not afraid to learn it and are able to manage it. The real issue, then, is not how to make technology better. Rather, it is how to integrate the technology with people.

After reading this book, you will be in a very favorable position compared with many other people in industry today. You will learn not only the basics of hardware, software, connectivity, the Internet, and the Web. You will also learn the most *current* technology. You will therefore be able to use these tools to your advantage—to be a winner.

VISUAL SUMMARY

Your Future and Computer Competency

INFORMATION SYSTEMS

The way to think about a microcomputer is to realize that it is one part of an **information system.**

Five parts of an information system:

1. **People** are an often-overlooked part. The purpose of information systems is to make people more productive.
2. **Procedures** are rules or guidelines to follow when using software, hardware, and data.
3. **Software** (**programs**) provides step-by-step instructions to control the computer.
4. **Hardware** consists of the physical equipment.
5. **Data** consists of unprocessed facts including text, numbers, images, and sound. **Information** is processed data.

PEOPLE

People are the most important part of an information system. People are touched hundreds of times daily by computers.

Some examples:

Entertainment

Medicine

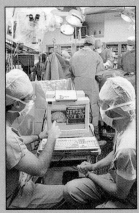

Education

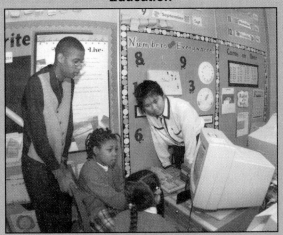

Business

To prepare for your future as a competent end user, you need to understand the basic parts of an information system: people, procedures, software, hardware, and data. Also you need to understand connectivity through the Internet and the Web and to recognize the role of technology in our professional and personal lives.

SOFTWARE

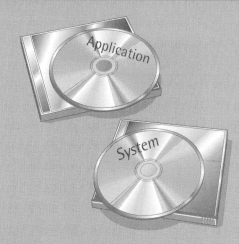

Software or **programs** consist of system and application software.

System Software

System software—*background* software that manages internal resources. An example is an operating system such as Windows 2000.

Application Software

Application software—software that performs useful work on general-purpose problems. Basic applications include:

- **Word processors** to prepare written documents.
- **Spreadsheet** to analyze and summarize numerical data.
- **Database management system** to organize and manage data.
- **Presentation graphics** to communicate or persuade.

Many other more specialized applications are used within certain career paths.

HARDWARE

Hardware is the physical equipment in an information system.

Types of Computers

There are four types of computers:

- **Supercomputers**—the most powerful.
- **Mainframe**—used by large companies.
- **Minicomputers**—also known as **midrange computers.**
- **Microcomputers**—the fastest growing. Categories include **desktop, notebook,** and **personal digital assistant.**

Microcomputer Hardware

The four categories of devices are:

- The **system unit** contains the electronic circuitry, including the CPU and memory.
- **Input/output devices** are translation units that convert human instruction into machine-readable processes.
- **Secondary storage devices** store data and programs. Typical media include **floppy, hard,** and **optical disks.**
- **Communication devices** send and receive data and programs from one computer to another. A **modem** is widely used to connect to the Internet.

DATA

Data describes something and is typically stored electronically in a file.

Common types of files are:

Document

Worksheet

Database

Presentation

CONNECTIVITY AND THE INTERNET

Connectivity

Connectivity is a concept describing the ability of end users to tap into resources well beyond their desktops. Computer **networks** are connected computers that share data and resources.

Internet

The **Internet** is the world's largest computer network. The **World Wide Web (WWW)** is an Internet service that provides a multimedia interface to resources available on the Internet.

Key Terms

application software (7)	input device (12)	personal digital assistant
communications hardware (14)	Internet (16)	(PDA) (12)
compact disc (CD) (14)	keyboard (14)	presentation file (15)
computer network (16)	mainframe computer (11)	primary storage (12)
connectivity (15)	memory (12)	printer (14)
data (5)	microcomputer (12)	procedures (4)
database file (15)	microprocessor (12)	program (4)
desktop computer (12)	midrange computer (12)	random access memory
digital versatile (or video) disc	minicomputer (12)	(RAM) (12)
(DVD) (14)	modem (14)	secondary storage device (14)
document file (14)	monitor (14)	software (4)
floppy disk (14)	mouse (14)	supercomputer (11)
handheld computer (12)	notebook computer (12)	system software (7)
hard disk (14)	operating system (7)	system unit (12)
hardware (4)	optical disk (14)	video display screen (14)
information (5)	output device (14)	Web (16)
information system (4)	palmtop computer (12)	worksheet file (15)
information technology (IT) (5)	people (4)	World Wide Web (WWW) (16)

Chapter Review

LEVEL 1

Reviewing Facts and Terms
Matching

Match each numbered item with the most closely related lettered item. Write your answers in the spaces provided.

1. The largest network in the world. _L_

2. Portable data storage made from very thin, flexible plastic. _H_

3. Computers small enough to fit on top of or alongside a desk yet too large to carry around. _G_

4. Used to describe facts about something. _F_

5. Holds data and program instructions for processing data. _l_

6. Most important system software program, interacting between the application software and the computer. _R_

7. Midrange, desk-sized computers. _Q_

8. The capability of your microcomputer to share information with other computers. _E_

9. "End-user" software used for general-purpose tasks. _A_

10. Translates the processed information from the computer into a form that humans can understand. _X_

11. It has five parts: people, procedures, software, hardware, and data. _J_

12. Large, powerful computers capable of great processing speeds and data storage. _M_

a. Application software
b. CD
c. Communications hardware
d. Computer network
e. Connectivity
f. Data
g. Desktop computer
h. Floppy disk
i. Hard disk
j. Information system
k. Input device
l. Internet
m. Mainframe
n. Memory
o. Microcomputer
p. Microprocessor
q. Minicomputer
r. Operating system
s. Optical disk
t. Output device

13. A communications system connecting two or more computers. _D_

14. Secondary storage device that uses laser technology. _S_

15. Controls and manipulates data to produce information. _P_

16. Compact disc. _B_

17. Translates data and programs that humans can understand into a form that the computer can process. _K_

18. Sends and receives data and programs from one computer or secondary storage device to another. _C_

19. Secondary storage device typically used to store programs and very large data files. _N_

20. Most widely used and fastest-growing type of computer. _P_

True/False

In the spaces provided, write T or F to indicate whether the statement is true or false.

1. Microcomputers are common tools in all areas of life. _T_

2. Hardware consists of a monitor, a keyboard, and software. _F_

3. Windows 2000 is an application program. _F_

4. Memory is also known as primary storage. _T_

5. A modem is used to send electronic signals over telephone lines. _T_

Multiple Choice

Circle the letter of the correct answer.

1. Computers are electronic devices that accept instructions, process input, and produce:
   a. information
   b. prewritten programs
   c. data
   d. end users
   e. system software

2. High-capacity computers used primarily by very large organizations or for research purposes are:
   a. microcomputers
   b. minicomputers
   c. mainframes
   d. supercomputers
   e. personalcomputers

3. The microprocessor is located in the:
   a. hard disk
   b. system unit
   c. memory
   d. monitor
   e. keyboard

4. Also called *temporary storage*, its contents will be lost if electrical power is disrupted or cut off:

   a. secondary storage

   b. basic tools

   c. memory

   d. operating system

   e. hard disk

5. Files containing highly structured and organized data are:

   a. documents

   b. worksheets

   c. databases

   d. graphics

   e. communications

Completion

Complete each statement in the spaces provided.

1. Microcomputer _____, the physical equipment, falls into four categories: the system unit, input/output, secondary storage, and communication devices.

2. Also known as midrange computers, _____ are frequently used by departments within larger organizations.

3. _____ are guidelines or rules to follow when using software, hardware, and data.

4. _____ storage devices are used to store data and programs even after electrical power has been turned off.

5. _____ are application software programs used to navigate, explore, and find information on the Internet.

Reviewing Concepts

Open-Ended

On a separate sheet of paper, respond to each question and statement.

1. Describe the five parts of an information system.

2. How would you distinguish between system software and application software?

3. Compare the four types of computers.

4. What is the difference between memory and secondary storage?

5. What are connectivity, the Internet, and the Web?

Concept Mapping

On a separate sheet of paper, draw a concept map or flowchart showing how the following terms are related. Show all relationships. Include any additional terms you can think of.

application software	desktop computer	optical disk
computer	document file	output device
computer network	floppy disk	hardware
connectivity	hard disk	information system
data	notebook	input device
database file	operating system	mainframe

memory	minicomputer	palmtop computer
microcomputer	monitor	people
microprocessor	mouse	software

Critical Thinking Questions and Projects

LEVEL 3

Read each exercise and answer the related questions on a separate sheet of paper.

1. *Your reasons for learning computing:* How are you already using computer technology? What's happened in the computer-related world in the last six months that you might have read about or seen on television? How are companies using computers to stay on the cutting edge? These are some questions you might discuss with classmates to see why computers are an exciting part of life.

    You might also consider the reasons why you want to gain computer competency. Imagine your dream career. How do you think microcomputers, from what you already know, can help you do the work you want to do? What kind of after-hours interests do you have? Assuming you could afford it, how could a microcomputer bring new skills or value to those interests?

2. *The Internet and the Web:* The Internet and Web are the most exciting connectivity developments today. If you have used the Internet or the Web, discuss your experiences by describing what you used it for, what you liked about it, and what you did not like. If you have not used the Internet or the Web, do you think you will in the near future? What would you use it for?

3. *Privacy and security:* Computer technology offers unlimited opportunities and challenges. We can do so many things faster and better. We can instantly connect to and communicate with people around the world. But have you ever thought about the other side of the coin? Will all of the changes be positive? Will everyone benefit? Discuss these issues and any others that come to mind.

4. *Making IT Work for You:* Review the list of Making IT Work for You topics on page 8. Have you used any of these? If so, identify the topic(s), describe how, when, and whether you found it useful. Assign a rank to each of the ten topics based on your interest. (Use 1 to indicate the most interesting.) For each of the first three topics in your ranking, describe why it is interesting to you and how you might use it. Be as specific as possible.

On the Web Exercises

1. Virtual Libraries

It might surprise you to learn that you can visit libraries on the Web where you can browse through the stacks, research selected topics, and check out books. Several virtual libraries have *e-texts*, which are entire books on computer, that anyone can download and use. Visit our Web site at http://www.mhhe.com/it/oleary/exercise.mhtml to link to one of these libraries. Find a text on a topic that interests you, print out its first page, and write a paragraph on the benefits and shortfalls of using virtual libraries as a resource for your school papers.

2. Ticket Master

Some of the hottest sites on the Web offer the latest music news, present live concerts, and provide updates on your favorite band. Visit our Web site at http://www.mhhe.com/it/oleary/exercise.mhtml to link to one of the most popular music sites. Once connected to that site, check out your favorite band and print out its tour dates or information about its latest album. Write a paragraph describing how you located the information and discuss how the band could better use the Internet to promote its music.

3. Virtual Shopping Malls

Like any community, the Internet community has shopping malls. Visit our Web site at http://www.mhhe.com/it/oleary/exercise.mhtml to link to one of the most popular virtual malls. Browse through the mall, find a product that you are familiar with, and print out the information provided on that item. Write a paragraph describing how the prices, services, and selection of the virtual mall compare with those of a traditional mall in your community.

4. Making IT Work for You

Although large organizations have been using information technology for years, individuals like you have only recently begun to use the technology in their every day lives. The list of Making IT Work for You topics on page 8 presents several everyday uses of technology. Further information about each topic is presented in this book and on our Web site. Select one topic that is new to you. Locate and review the Making IT Work for You coverage in this book. Visit our Web site at http://www.mhhe.com/it/oleary/exercise.mhtml and link to the topic you have selected. Print out the information presented. Write a paragraph or two describing how you might use this technology in your everyday life.

CHAPTER 2

Application Software

COMPETENCIES

After you have read this chapter, you should be able to:

1. **Explain the features common to most types of software.**

2. **Describe word processors.**

3. **Describe spreadsheets.**

4. **Describe database management systems.**

5. **Describe presentation graphics.**

6. **Describe software suites and integrated software.**

Not long ago, trained specialists were required to perform many of the operations you can now perform with a microcomputer.

Computer scientists surfed the Internet and created Web pages. Secretaries used typewriters to create professional-looking business correspondence. Market analysts used calculators to project sales. Graphic artists drew by hand. Data processing clerks created and stored files of records on large computers. Now you can do all these tasks—and many others—with a microcomputer

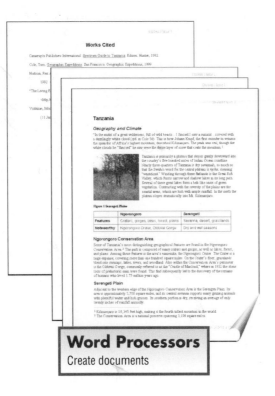

Word Processors
Create documents

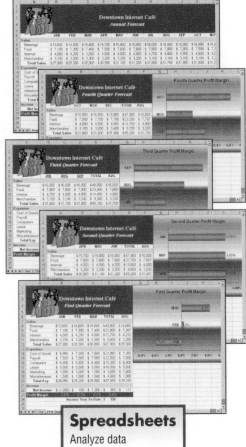

Spreadsheets
Analyze data

and the appropriate application software.

Think of the microcomputer as an *electronic* tool. You may not consider yourself very good at typing, calculating, organizing, presenting, or managing information. A microcomputer, however, can help you to do all these things—and much more. All it takes is the right kinds of software—system software and application software.

While most end users today own and run their own application software, an emerging trend is to use Web-based applications. These are programs you can access from the Internet and run on your microcomputer.

Competent end users need to understand the capabilities of basic application software, which includes word processors, spreadsheets, database management systems, and presentation programs.

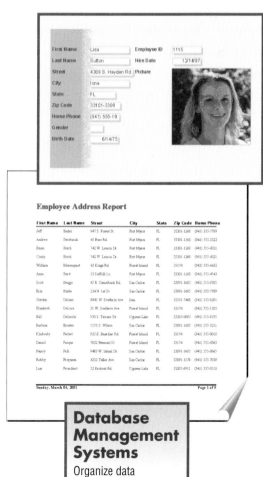

Database Management Systems
Organize data

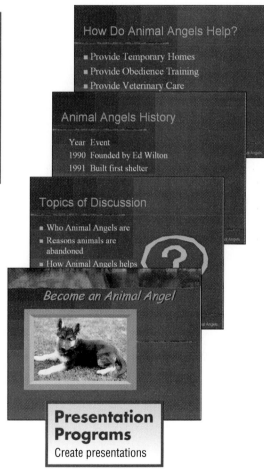

Presentation Programs
Create presentations

General-Purpose Applications

Some features are common to all kinds of applications.

Word processors, spreadsheets, database management systems, and presentation graphics are **general-purpose applications.** That is, they are designed to be used by many people to do the most common kinds of tasks. This is why they are also known as **basic applications.** These and many other types of applications and operating systems have common features. (See Figure 2-1.) The following are the most important.

FIGURE 2-1
Features

Windows	Rectangular areas that contain information and/or applications
Menus	Lists of commands
Toolbars	Series of buttons and menus representing various commands
Help	Source for reference information and assistance

Windows

A **window** is simply a rectangular area that can contain a document, program, or message. (Do not confuse the term *window* with the various versions of Microsoft's Windows operating systems, which are programs.) Many operating systems and application programs use windows to display information and request input. More than one window can be opened and displayed on the computer screen at one time. For example, one window might contain a browser, another a word processing program, and another a graphic image. Windows can generally be resized, moved, and closed.

Menus

Almost all software packages have **menus** to present commands. Typically, the menus are displayed in a menu bar at the top of the screen. When one of the menu items is selected, a pull-down menu appears. This is a list of commands associated with the selected menu.

Toolbars

Toolbars typically are below the menu bar. They contain buttons and menus that provide quick access to commonly used commands. The **standard toolbar** and the **formatting toolbar** are common to most applications.

Help

For most application packages, one of the commands on the menu bar is **Help.** When selected, the help options appear. These options typically include a table of contents, an index, and a find or search feature to locate information.

To learn more about the common features of most software, study Figure 2-2.

Typically, application programs are owned by users and stored on their hard disk drive. An emerging trend, however, is to free users from owning and storing applications by using Web-based applications. (See Making IT Work for You: Web-based Applications on pages 30–31.)

TIPS Have you ever been working on a document when the power goes off? Your original document and all your recent changes are gone. Fortunately, most applications automatically save your work every few minutes to a temporary recovery or backup file.

Here are a few tips that might help you to quickly get back underway again.

1. *Restart your computer.*
2. *Restart the application.*
3. *Open the recovery file.* If the recovery file is not automatically opened, select the option to open it.
4. *Verify recovery file contents.* The recovery file should contain the document current up to the last automatic save. Compare the contents of the recovery file to the contents of the original file (if you have one).
5. *Save the recovery file.* If the recovery file contains the information you want, save it using the original file name or some other appropriate name.

CONCEPT CHECK

✔ What are general-purpose applications?

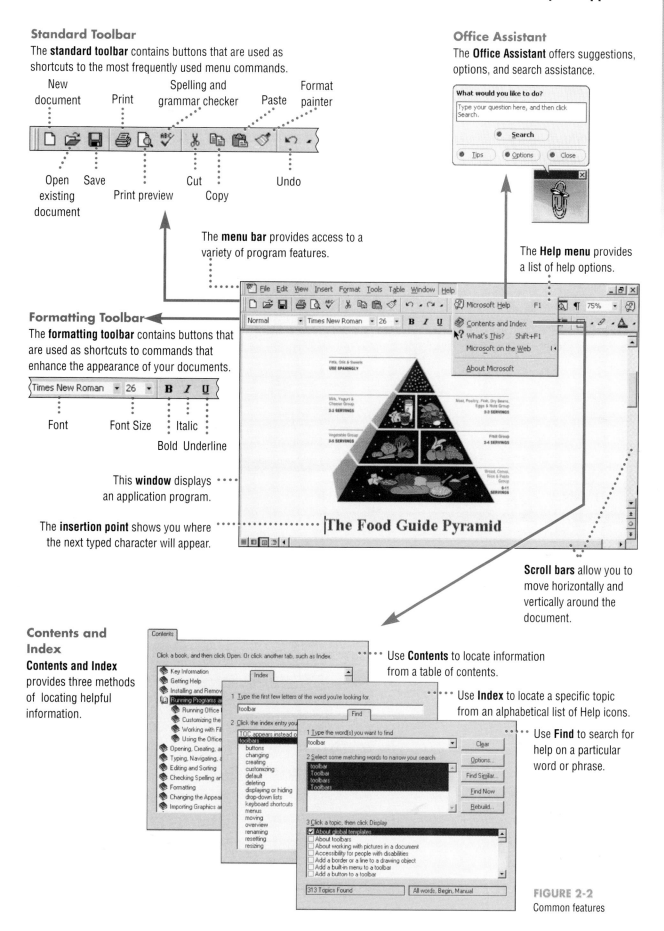

Standard Toolbar

The **standard toolbar** contains buttons that are used as shortcuts to the most frequently used menu commands.

New document — Print — Spelling and grammar checker — Paste — Format painter

Open existing document — Save — Print preview — Cut — Copy — Undo

The **menu bar** provides access to a variety of program features.

Formatting Toolbar

The **formatting toolbar** contains buttons that are used as shortcuts to commands that enhance the appearance of your documents.

Font — Font Size — Italic — Bold — Underline

This **window** displays an application program.

The **insertion point** shows you where the next typed character will appear.

Office Assistant

The **Office Assistant** offers suggestions, options, and search assistance.

The **Help menu** provides a list of help options.

Scroll bars allow you to move horizontally and vertically around the document.

Contents and Index

Contents and Index provides three methods of locating helpful information.

Use **Contents** to locate information from a table of contents.

Use **Index** to locate a specific topic from an alphabetical list of Help icons.

Use **Find** to search for help on a particular word or phrase.

FIGURE 2-2
Common features

Web-based Applications

Would you like access to free general-purpose applications from anywhere in the world? What about accessing your data files from any location? You can have it all with Web-based applications.

How It Works

A server on the Web, known as an Application Service Provider (ASP), provides access to programs such as word processors, spreadsheets, and more. After registering with an ASP, you can use the Web to access these applications and store data files at the ASP rather than on your hard disk. You can run programs and access data using the Web from any location in the world.

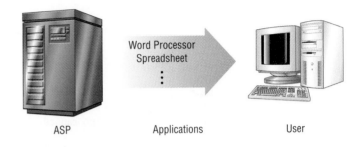

ASP   Applications   User

Registering

Several ASP sites exist on the Web and some of them offer free services. One of the best-known sites is WebOS. Their only requirement is that you register for their service.

1. Connect to www.WebOS.com.

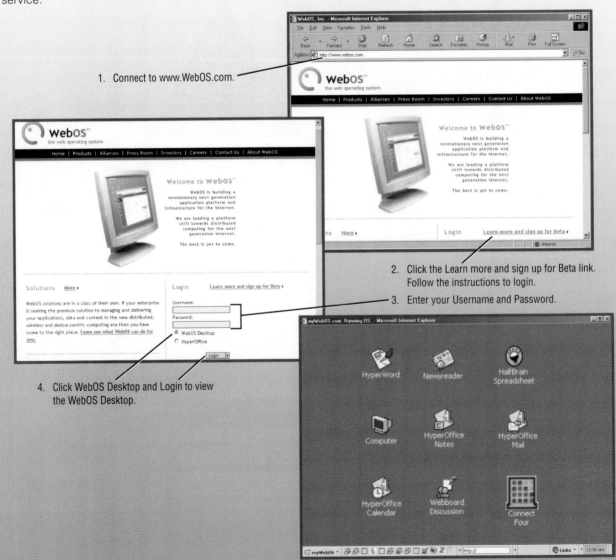

2. Click the Learn more and sign up for Beta link. Follow the instructions to login.

3. Enter your Username and Password.

4. Click WebOS Desktop and Login to view the WebOS Desktop.

Accessing Applications

Each time you connect to the WebOS site and log in, your Web-based Desktop will appear. It will display numerous icons that can be used to access Web-based applications. These include a word processor, spreadsheet, personal information manager, and a variety of games.

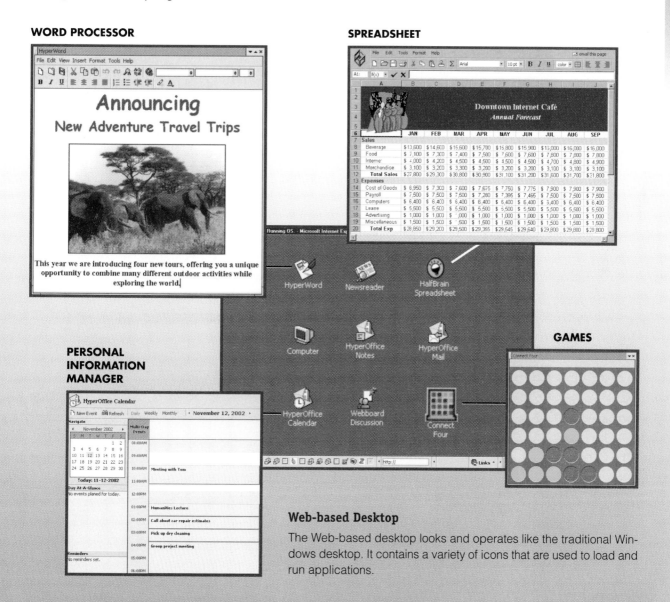

WORD PROCESSOR

SPREADSHEET

PERSONAL INFORMATION MANAGER

GAMES

Web-based Desktop

The Web-based desktop looks and operates like the traditional Windows desktop. It contains a variety of icons that are used to load and run applications.

Some suggest that Web-based applications may replace traditional application software some time in the future. They point out that Web-based applications promise many advantages to users beyond universal access to software and to data. One advantage is that users may no longer need to upgrade software on their hard disk when a new version becomes available. Of course, there are some potential disadvantages or challenges of Web-based applications. One challenge relates to privacy and security of personal data stored at an ASP.

The Web is continually changing and some of the specifics presented in this Making IT Work for You may have changed. See our Web site at http://www.mhhe.com/it/oleary/IT.mhtml for possible changes and to learn more about this application of technology.

Word Processors

Word processing software is used to create text-based documents.

Word processing software creates text-based **documents** such as reports, letters, and memos. Word processors are one of the most flexible and widely used software tools. Students and researchers use word processors to create reports. Organizations of all types create newsletters to communicate with their members. Businesses create form letters to reach new and current customers. All types of people and organizations use word processors to create personalized Web pages.

The three most widely used word processing programs are Microsoft Word, Corel WordPerfect, and Lotus Word Pro.

Features

One basic word processing feature is **word wrap.** A word processor automatically moves the insertion point to the next line once the current line is full. As you type, the words "wrap around" to the next line. To begin a new paragraph or leave a blank line, you press the **Enter key.**

Spelling can be checked by running a **spelling checker** program. Incorrectly spelled words are identified and alternative spellings suggested. In a similar manner, a **grammar checker** will identify poor wording, excessively long sentences, and incorrect grammar.

You can quickly locate any character, word, or phrase in your document using the **search** or **find** commands. In addition, you can replace the located

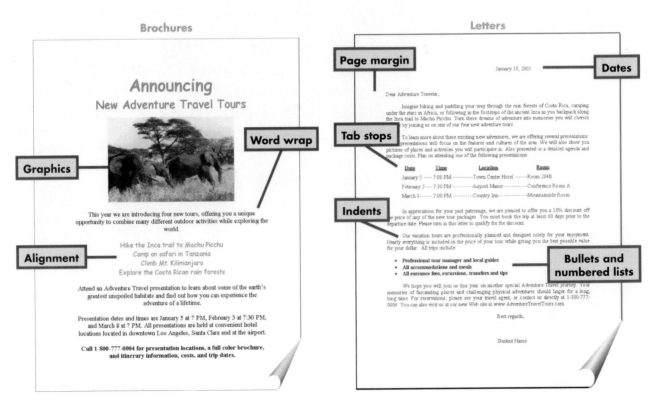

FIGURE 2-3   **FIGURE 2-4**

text with other text you specify using the **replace** command. For example, you could quickly locate each occurrence of the word *Chicago* and replace it with the word *Denver*.

Hypertext links, or **hyperlinks,** can be created to cross-reference information within the current document and between documents. Clicking on a hypertext link jumps your screen display to the specified location in the referenced document. This is most useful for viewing documents online. For example, if you cross-reference another chapter, you can add a hypertext link so the reader can jump to that chapter. Hypertext links can also reference sources on the Web.

Case

Assume that you have accepted a job as the advertising coordinator for a company called Adventure Travel. Your job is to create and coordinate the company's promotional materials, including brochures, form letters, travel reports, and the company Web page. To learn more about word processor features and how they could assist you as an advertising coordinator, see Figures 2-3 through 2-6.

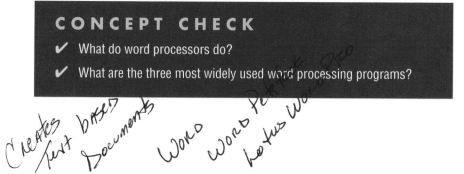

CONCEPT CHECK

✔ What do word processors do?

✔ What are the three most widely used word processing programs?

Creates Text based Documents

Word Word Perfect Lotus WordPro

TIPS Want to add some interest and a personal touch to your correspondence? You can by including an audio clip of your voice in the text document. If you are using Word 97 or 2000:

1. *Position cursor.* Place the cursor where you want the voice attachment to appear.

2. *Start recording.* From the menu bar choose Insert/Object/Wave/ Sound OK/Record.

3. *Start talking.* Dictate a short message into the microphone.

4. *Stop recording.* Choose Stop/File/Exit&Return.

A speaker icon appears in your document. To hear the audio clip, double click the icon.

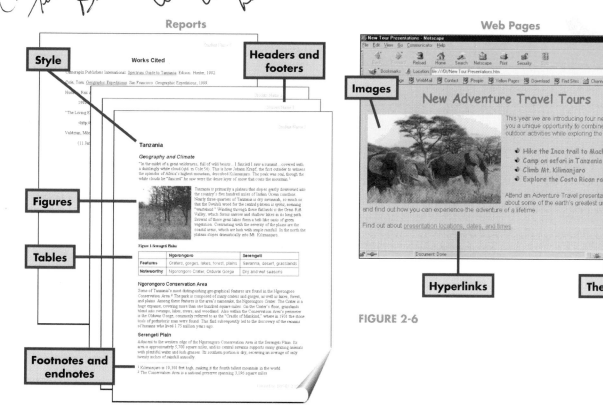

FIGURE 2-5

FIGURE 2-6

Spreadsheets

A spreadsheet is an electronic worksheet used to organize, manipulate, and graph data.

Spreadsheet programs organize, manipulate, and graph numeric data. Once used almost exclusively by accountants, spreadsheets are now widely used by people in nearly every profession. Teachers record students' grades. Students record their grades and calculate grade point averages. Financial analysts evaluate stock market trends. Personal trainers record the progress of their clients. Marketing professionals evaluate sales trends.

The three most widely used spreadsheet programs are Microsoft Excel, Corel Quattro Pro, and Lotus 1-2-3.

Features

Labels are often used to identify information in a worksheet. Usually a label is a word or symbol, such as a pound sign (#). A number in a cell is called a **value.** Labels and values can be displayed or formatted in different ways. For example, a label can be centered in the cell or positioned to the left or right. A value can be displayed to show decimal places, dollars, or percent (%). The number of decimal positions (if any) can be altered, and the width of columns can be changed.

Formulas are instructions for calculations. They calculate results using the number or numbers in referenced cells. For example, the formula C8 + C9 means add the value in C8 to the value in C9. **Functions** are prewritten

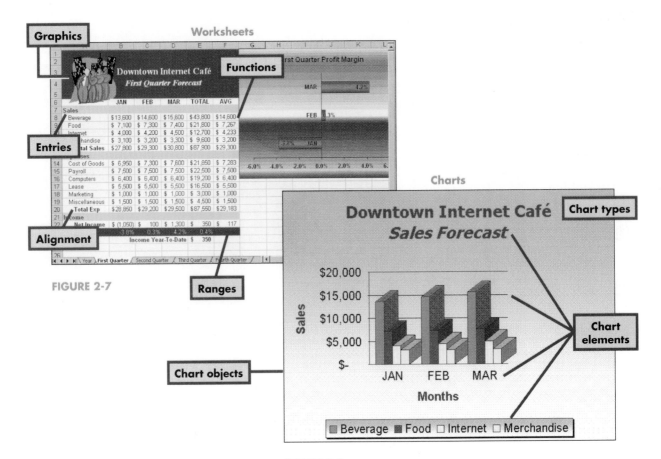

FIGURE 2-7

FIGURE 2-8

formulas that perform calculations automatically. For example, the function @SUM(C8:C11) adds all the values in the range of cells from C8 to C11.

If you change one or more numbers in your spreadsheet, all related formulas will recalculate automatically. Thus, you can substitute one value for another in a cell and observe the effect on other related cells in the spreadsheet. This is called **what-if analysis.**

To help visualize the data in your spreadsheets, you can create **analytical graphs** or **charts.** For example, you could display the numerical data in a worksheet as a pie chart, bar chart, or a line graph. The graphs automatically update when the data in the underlying worksheet changes.

Case

Assume that you have accepted a job as manager of the Downtown Internet Café. This café provides a variety of flavored coffees as well as Internet access. One of your responsibilities is to create a financial plan for the next year. To learn more about spreadsheet features and how they could assist you as a manager of an Internet café, see Figures 2-7 through 2-10.

CONCEPT CHECK

✔ What are spreadsheet programs?

✔ What are the three most widely used spreadsheet programs?

FIGURE 2-9

FIGURE 2-10

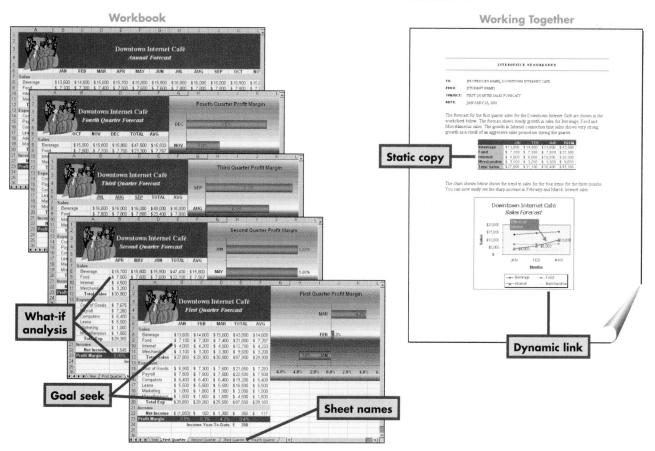

Database Management Systems

A database manager organizes a related collection of data so that information can be retrieved easily.

A *database* is a collection of related data. A **database management system (DBMS)** or **database manager** is a program that sets up, or structures, a database. It also provides tools to enter, edit, and retrieve data from the database. Database managers are used by all kinds of people, from teachers to police officers.

The **relational database** is the most widely used database structure. Data is organized into related **tables.** Each table is made up of rows called **records** and columns called **fields.** Each record contains fields of data about some specific item. For example, in an address table, a record would contain fields of data such as a person's last name, first name, and street address.

Three of the most widely used database management systems are Microsoft Access, Corel Paradox, and Lotus Approach.

Features

A basic feature of all database programs is the capability to quickly locate or find records in the database. For example, you could instruct the database manager to find and display all records in which a last name of Smith appears.

Rearranging or **sorting** is a common feature of database managers. For example, you might want to print out an entire alphabetical list of employees by last name. Built-in math formulas are used for data analysis. Using a

ON THE WEB EXPLORATIONS

Corel is one of the leaders in software development. To learn more about the company, visit our Web site at

http://www.mhhe.com/it/oleary/ explore.mhtml

FIGURE 2-11

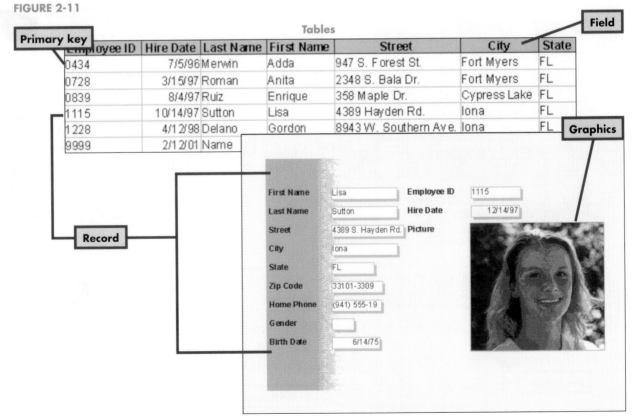

Tables

Employee ID	Hire Date	Last Name	First Name	Street	City	State
0434	7/5/96	Merwin	Adda	947 S. Forest St.	Fort Myers	FL
0728	3/15/97	Roman	Anita	2348 S. Bala Dr.	Fort Myers	FL
0839	8/4/97	Ruiz	Enrique	358 Maple Dr.	Cypress Lake	FL
1115	10/14/97	Sutton	Lisa	4389 Hayden Rd.	Iona	FL
1228	4/12/98	Delano	Gordon	8943 W. Southern Ave.	Iona	FL
9999	2/12/01	Name				

Primary key

Field

Graphics

Record

First Name	Lisa	Employee ID	1115
Last Name	Sutton	Hire Date	12/14/97
Street	4389 S. Hayden Rd.	Picture	
City	Iona		
State	FL		
Zip Code	33101-3309		
Home Phone	(941) 555-19		
Gender			
Birth Date	6/14/75		

FIGURE 2-12   **Forms**

sales database, you could calculate the total and average commissions earned by the sales force in one part of the country.

Most database management programs include a programming control language for advanced users to create sophisticated applications. In addition, they allow direct communication to specialized mainframe databases through languages like **structured query language (SQL).**

Case

Assume that you have accepted a job as an employment administrator for the Lifestyle Fitness Club. One of your responsibilities is to create a database management system to replace the club's manual system for recording employee information. To learn more about database management system features and how they could assist you as an employee administrator at the Lifestyle Fitness Club, see Figures 2-11 through 2-14.

CONCEPT CHECK

✔ What is a database?

✔ What is a database management system (DBMS)?

✔ What are the three most common database management systems?

Collection of data.
Related
Program that sets up
on structures a database
Access, Paradox, Approach

Reports

FIGURE 2-13

Working Together

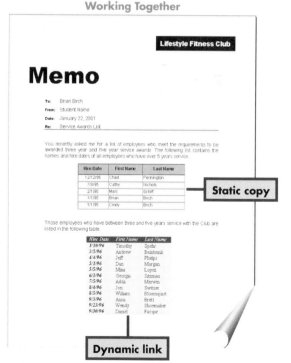

FIGURE 2-14

Presentation Graphics

Presentation graphics software helps you create professional and exciting presentations.

Research shows that people learn better when information is presented visually. A picture is indeed worth a thousand words, or numbers. **Presentation graphics** are used to combine a variety of visual objects to create attractive, visually interesting presentations.

You can use presentation graphics to create class presentations. These programs are excellent tools to communicate a message or to persuade people, such as fellow students, instructors, supervisors, or clients. Presentation graphics programs are often used by marketing or sales people as well as many others.

Three of the most widely used presentation graphics programs are Microsoft PowerPoint, Corel Presentations, and Lotus Freelance Graphics.

Features

Most programs include features that help you organize the content of your presentation. Commonly, an outline feature is included that helps you enter

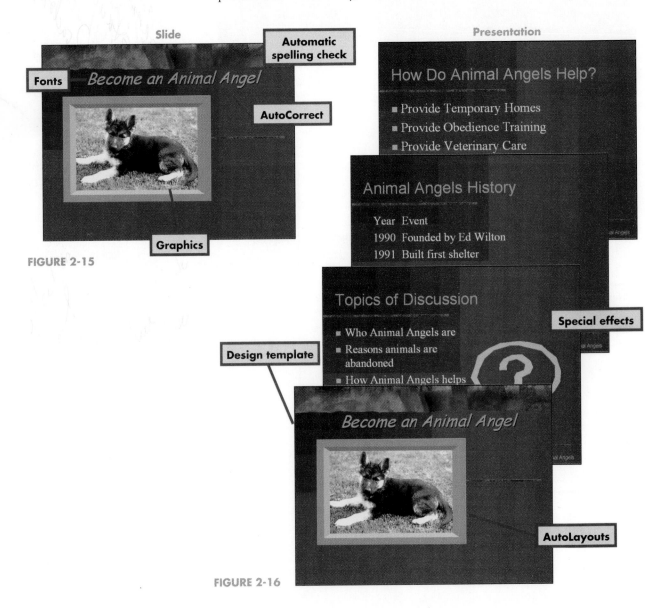

Slide

FIGURE 2-15

FIGURE 2-16

and organize the topics of your talk. Most programs also provide presentation **layout files** that include sample text for many different types of presentations—from selling a product to reporting on progress.

Professionally designed **templates** or model presentations are provided. These can help take the worry out of the design and layout decisions. They include selected combinations of text layouts with features such as title placement and size. Additionally, various bullet styles, background colors, patterns, borders, and other enhancements are provided.

Animations include special visual and sound effects. These effects include blinking text and transitions between topics. You can insert audio and video clips that play automatically or when selected. You can even record your own voice to provide a narration to accompany a slide show.

Case

Assume that you have volunteered to help out at a local animal shelter called Animal Angels. You have been asked to create a powerful and persuasive presentation to encourage other members from your community to volunteer. To learn more about presentation software features and how they could assist you to create a presentation for Animal Angels, see Figures 2-15 through 2-18.

CONCEPT CHECK

✔ What are the three most widely used presentation graphics programs?

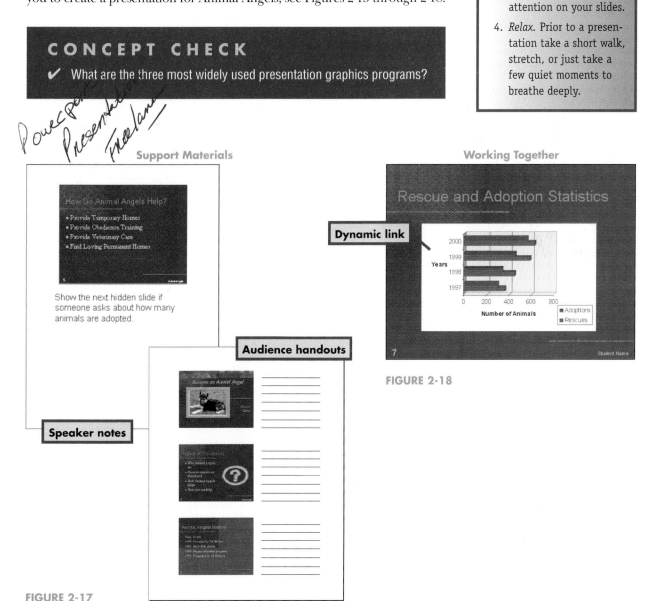

Support Materials

Show the next hidden slide if someone asks about how many animals are adopted.

Audience handouts

Speaker notes

FIGURE 2-17

Working Together

Rescue and Adoption Statistics

Dynamic link

FIGURE 2-18

Software Suites

A software suite is a group of application programs. OLE allows you to share information between applications.

A software suite is a collection of separate applications bundled together and sold as a group. While the applications function *exactly* the same whether purchased in a suite or separately, it is significantly less expensive to buy a suite of applications than to buy each application separately.

Microsoft Office is the most popular software suite. It comes in several different versions. One of the most recent is Microsoft Office 2002 Professional Edition, which includes Word, Excel, Access, and PowerPoint. Other popular software suites are Corel WordPerfect Suite and Lotus SmartSuite.

Object Linking and Embedding

Object linking and embedding (OLE) is a powerful feature of many application programs. Using OLE, you can share information or page objects between files created in different applications. For example, you could create a chart in Excel and then use it in a Word document.

With object linking, a copy of the object from the source file is sent to the destination file along with a *link* or tie back to the source file. If the source file changes, the object in the destination file is updated automatically.

For example, if a chart created in Excel is linked to a Word document, the chart appears in the Word document. Later, if the Excel worksheet changes the chart, the Word document will be automatically updated. Object linking is useful if you want the destination document to always contain the most up-to-date information.

With object embedding, the object from the source file is *embedded* or added in the destination document. While the files are not linked, the embedded file can be run from the destination document.

For example, if a presentation created in PowerPoint is embedded in a Word document, the presentation can be run from the Word document. Object embedding is useful for providing activity and flexibility to a document.

Case

Assume that you are a marketing analyst for The Sports Company and have been working on a presentation for the annual sales meeting. You have analyzed recent sales trends using Excel and have drafted a presentation using PowerPoint. To present your work to your supervisor, you have just completed a Word document. To learn more about OLE and how you could use it, see Figure 2-19.

CONCEPT CHECK

✔ What is a software suite?

✔ What are some popular software suites?

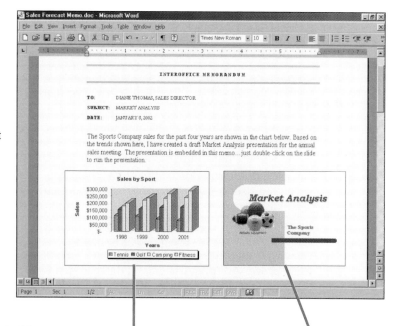

Word Document
This document contains text and two objects. The first object is linked to an Excel file. The second is an embedded PowerPoint presentation.

Linked Object
The chart from the Excel file appears in the word document. Any changes in the Excel file that affect the chart will automatically be updated in the Word document.

Embedded Object
The first slide of the embedded presentation appears in the Word document. Double clicking the slide from the Word document plays the entire presentation.

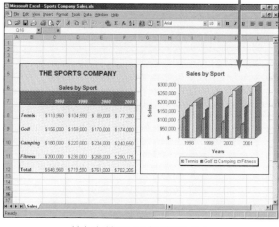

Linked objects can be charts, text, tables, slides, sounds, and video.

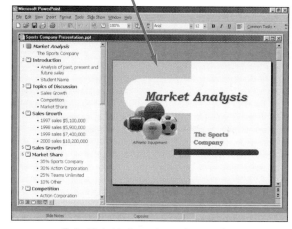

Embedded objects can be word processing, spreadsheet, presentation, and database files.

FIGURE 2-19
Object linking and embedding

Integrated Packages

Integrated software is an all-in-one application program.

An **integrated package** is a single program that provides the functionality of a word processor, spreadsheet, database manager, and more. For example, to create a report on the growth of sales for a sporting goods store, you could use all parts of an integrated package. You could use the database to search and retrieve yearly sales data. The spreadsheet could be used to analyze the data and graphics to visually present the data. You could use the word processor to write the report that includes tables from the spreadsheet and visuals from the graphics program. (See Figure 2-20.)

The primary disadvantage of an integrated package is that the capabilities of each function (such as word processing) are not as extensive as in the specialized programs (such as Microsoft Word). The primary advantage is that the cost of an integrated package is much less than the cost of purchasing a word processor, spreadsheet, and database manager. The most widely used integrated packages are Microsoft Works and AppleWorks.

For a summary of the basic application software, see Figure 2-21.

CONCEPT CHECK

✔ What is an integrated package?

[handwritten margin note: Single program that provides functionality of All.]

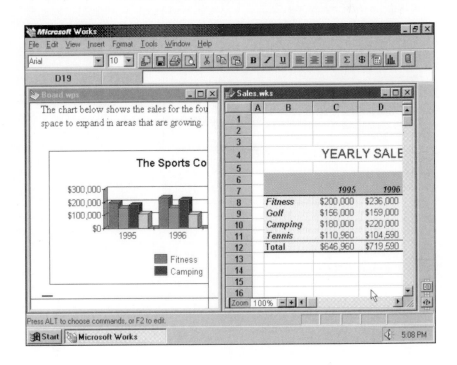

FIGURE 2-20
Integrated package (Microsoft Works for Windows 95)

Word processors	Microsoft Word, Corel WordPerfect, Lotus Word Pro
Spreadsheets	Microsoft Excel, Corel Quattro Pro, Lotus 1-2-3
Database management systems	Microsoft Access, Corel Paradox, Lotus Approach
Presentation graphics	Microsoft PowerPoint, Corel Presentations, Lotus Freelance
Software suites	Microsoft Office, Corel WordPerfect Suite, Lotus SmartSuite
Integrated packages	Microsoft Works, AppleWorks

FIGURE 2-21
Software programs

A LOOK TO THE FUTURE

New software versions will offer more capabilities, freeing your creativity and enhancing the quality and quantity of your work.

New versions of basic application software are being released all the time. One way these programs change is in the way you interact with them. Another way is in the software's capabilities.

Interacting with them may not be as difficult as you think. That's because almost all new software today has a similar command and menu structure. When a new version comes out, it looks and feels quite similar to the previous version. This frees you to focus on the new capabilities.

Basic applications will continue to become more and more powerful by adding breadth to their capabilities. They are no longer limited by the machines that they were designed to replace. Word processors, for example, do much more than typewriters ever could. Recent versions have added desktop publishing and Web page design capabilities.

Some experts predict that our days of buying, installing, and upgrading software will some day be a thing of the past. These activities will be done by specialized Web sites that provide Web-based applications. When you want to run the most recent and powerful applications, you will connect to the appropriate site, pay a fee, and run the application.

What does all this mean to you? You will have access to more powerful applications, which will free your creativity and enhance the quality and quantity of your work. Additionally, you will be challenged to learn how and when to use these more powerful tools.

VISUAL SUMMARY Application Software

GENERAL-PURPOSE APPLICATIONS

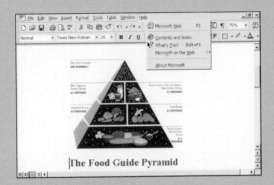

General-purpose applications are for common kinds of tasks. They include word processors, spreadsheets, database management systems, and presentation programs. Common features include the following.

Windows

Windows are rectangular areas that can contain a document, program, or message.

Menus

Menus are lists of optional commands, typically displayed in a menu bar at the top of the screen.

Toolbars

Toolbars contain buttons and menus to provide quick access to commands. Two common types are the **standard toolbar** and the **formatting toolbar.**

Help

Help typically includes a table of contents, an index, and a find or search feature.

WORD PROCESSORS

Word processors allow you to create, edit, save, and print text-based **documents** including brochures, letters, and reports. They are also used to create Web pages.

Features

Principal word processing features include the following:

- **Word wrap**—automatically moves the insertions pointer to the next line.
- **Enter key**—inserts a new line.
- **Spelling checkers**—identify incorrectly spelled words and present alternative spellings.
- **Grammar checkers**—identify poor wording, faulty grammar, and long sentences.
- **Search**—quickly locates characters, words, or phrases.
- **Replace**—replaces the located text with new text.
- **Hypertext links**—provide a connection to cross-referenced information within a document to other documents.

To be a competent end user, you need to know the functionality of systems and application software. Additionally, you need to understand the capabilities of general-purpose applications including word processors, spreadsheets, database management systems, and presentation programs.

SPREADSHEETS

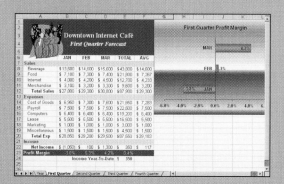

Spreadsheets are used to organize, manipulate, and graph numeric data. Also known as a **worksheet,** a spreadsheet consists of **rows** and **columns** forming **cells.** Individual cells are identified by their **cell address.** A block of adjacent cells is called a **range.**

Features
Principal spreadsheet features include the following:
- **Rows** and **columns** form **cells** in worksheet.
- **Labels** typically identify information in the spreadsheet.
- **Values** include numbers.
- **Formulas** are instructions for calculations.
- **Functions** are prewritten formulas.
- **What-if analysis** is the result of changing one or more values and observing the effect on related cells in the spreadsheet.
- **Analytical** graphs or charts are used to help visualize data in a spreadsheet.

DATABASE MANAGEMENT SYSTEMS

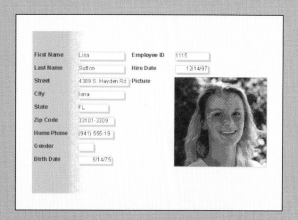

Database management systems are used to create and use databases. A **relational database** organizes data into related **tables** that are linked by **key fields.** In the tables, rows are called **records** and columns are called **fields.**

Features
Principal database management system features include the following:
- **Locate and display**—finding and displaying records.
- **Sort and analyze**—rearranging records in a database. Built-in math formulas may be used to manipulate and analyze data.
- **Program control languages**—like SQL (Structured Query Language) are programming languages for advanced users to create sophisticated database applications.

PRESENTATION GRAPHICS

Presentation graphics are used to create professional and exciting presentations.

Features
Principal presentation graphics features include the following:
- **Content development assistance**—Most provide organizational assistance using an outline feature. **Layout files** are provided to offer content assistance. These files include sample text for a variety of different types of presentations.
- **Professional design**—Sample **templates** or model presentations are provided. They include selected combinations of text layouts, bullet styles, background colors, patterns, borders, and other enhancements.
- **Animations** include special visual and sound effects including blinking text and transitions between topics. Additionally, audio and video clips can be inserted. These features add interest and keep audience attention.

SOFTWARE SUITES

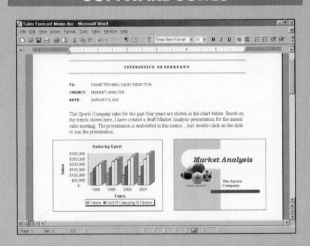

A **software suite** is a collection of individual application packages sold together. While functionally identical, application packages purchased in a suite are significantly less expensive than those purchased separately.

OLE
OLE, or **object linking and embedding,** allows sharing of information (objects) between applications.
- Object linking—linked objects are automatically updated whenever a change in the source file is changed.
- Embedded linking—the object from the source is embedded or added and can be run from the destination file.

INTEGRATED PACKAGES

An **integrated package** is a single program providing the functionality of a word processor, spreadsheet, database manager, and more. Although not as powerful, integrated packages are much less expensive than individual packages.

Key Terms

analytical graphs or charts (35)
animation (39)
basic applications (28)
database management system
 (DBMS) (36)
database manager (36)
documents (32)
Enter key (32)
field (36)
find (32)
formatting toolbar (28)
formula (34)
function (34)
general-purpose
 applications (28)

grammar checker (32)
Help (28)
hyperlinks (33)
hypertext link (33)
integrated package (42)
label (34)
layout files (38)
menu (28)
object linking and embedding
 (OLE) (40)
presentation graphics (38)
record (36)
relational database (36)
replace (33)
search (32)

software suite (40)
sorting (36)
spelling checker (32)
spreadsheet (34)
standard toolbar (40)
structured query language
 (SQL) (37)
table (in database) (36)
templates (39)
toolbar (28)
value (34)
what-if analysis (35)
window (28)
word processing (32)
word wrap (32)

Chapter Review

LEVEL 1

Reviewing Facts and Terms
Matching

Match each numbered item with the most closely related lettered item. Write your answers in the spaces provided.

1. A rectangular area that can contain a document, program, or message. _s_

2. Programs that organize, manipulate, and graph numeric information. _p_

3. Prewritten formulas that perform calculations automatically. _k_

4. Separate software application programs sold as a group. _n_

5. Common feature of database managers that allows you to rearrange information. _o_

6. A feature that contains buttons and menus to provide access to commonly used commands. _f_

7. Software that creates text-based documents such as reports, letters, and memos. _t_

8. Program that sets up, or structures, a database. _D_

9. A single program that provides the functionality of a word processor, spreadsheet, database manager, and more. _b_

10. Command that allows you to quickly locate any character, word, or phrase in your document. _m_

11. In a database, the columns are called _i_ .

12. Most widely used database structure. _c_

13. Word processors, spreadsheets, database management systems, and presentation graphics. _g_

14. Software used to combine a variety of visual objects to create attractive, visually interesting presentations. _l_

a. Word wrap
b. Integrated package
c. Relational database
d. DBMS
e. Help
f. Toolbar
g. General-purpose applications
h. Hypertext link
i. Fields
j. Menu
k. Functions
l. Presentation graphics
m. Search
n. Software suites
o. Sorting
p. Spreadsheets
q. Grammar checker
r. SQL
s. Window
t. Word processor

15. Can be created to cross-reference information within the current document and between documents. _H_

16. Basic feature of a word processor that causes the words to "wrap around" to the next line as you type. _N_

17. The word processing feature that identifies poor wording. _Q_

18. A feature common to most general-purpose applications for displaying commands. ____

19. A language that allows direct communication to specialized mainframe databases. _R_

20. A command on the menu bar of most application packages that provides options such as access to special learning features. _E_

True/False

In the spaces provided, write T or F to indicate whether the statement is true or false.

1. The term *window* is the short way of referring to Microsoft's Windows operating systems. _F_

2. Microsoft, Lotus, and Corel are applications. _T_

3. Quattro Pro is a widely used word processor. _F_

4. Spreadsheet programs are typically used to store and retrieve records quickly. ____

5. Presentation graphics programs provide you with professionally designed templates to simplify design and layout issues. _T_

Multiple Choice

Circle the letter of the correct answer.

1. The following would not be considered a general-purpose application:
   a. word processor
   b. spreadsheet
   c. presentation graphics program
   d. music interface program
   e. database management system

2. The word processing feature that identifies excessively long sentences is a:
   a. spell checker
   b. paste
   c. copy
   d. merge
   e. grammar checker

3. In spreadsheets, the feature that specifies instructions for calculation is:
   a. formulas
   b. format
   c. recalculation
   d. consolidation
   e. value

4. A tool used frequently by marketing people to communicate a message or to persuade clients:
   a. word processors
   b. spreadsheets
   c. browser
   d. database managers
   e. presentation graphics

5. A collection of separate Windows applications sold as a group:
   a. combined
   b. suite
   c. browser
   d. integrated
   e. spreadsheet

Completion

Complete each statement in the spaces provided.

1. A _____ is a rectangular area that can contain a document, program, or message.

2. The _____ is an electronic worksheet.

3. _____ are used to create professional and exciting presentations.

4. _____ software is used to create newsletters.

5. A basic feature of all database programs is the capability to quickly _____ records in the database.

Reviewing Concepts

LEVEL 2

Open-Ended

On a separate sheet of paper, respond to each question and statement.

1. What are general-purpose applications?

2. Describe the concept of what-if analysis.

3. Explain the purpose of presentation graphics.

4. Name three well-known spreadsheet packages.

5. What is the difference between an integrated package and a software suite? What are the advantages and disadvantages to each?

Concept Mapping

On a separate sheet of paper, draw a concept map or a flowchart showing how the following terms are related. Show all relationships. Include any additional terms you can think of.

database management system	analytical graphs	software suite
database manager	column	spreadsheet
documents	key field	structured query language
field	label	toolbar
formula	layout files	value
function	menu	what-if analysis
general-purpose applications	presentation graphics	window
hypertext link	range	word processing
integrated package	record	worksheet
	row	
	search	

Critical Thinking Questions and Projects

LEVEL

3

Read each exercise and answer the related questions on a separate sheet of paper.

1. *New versions and releases*: Software companies seem to be offering new versions and releases of their application packages all the time.

   a. Take one application package (like Microsoft Word, Excel, or PowerPoint) and find out how many variations have been released in the past three years.

   b. If you own one variation of an application package and a new one is released, does yours become obsolete? Discuss and defend your position.

   c. If you own one variation of a software package and a new version is released, should you upgrade to the newer version? What factors should be considered? If you decide to upgrade, how would you go about doing it?

   d. Why do you suppose software manufacturers offer new variations of their software?

2. *Three ways to acquire application software*: The following exercise can be extremely useful. Concentrate on a category of software of personal interest to you—say, word processing or spreadsheets. Go to the library or use the Web to find information on the following three methods of acquiring software. Which route seems to be the best for you? Why?

   a. *Public domain software* is software you can get for free. Someone writes a program and offers to share it with everyone without charge. Generally, you find these programs by belonging to a microcomputer users group. Or, you find them by accessing an electronic bulletin board using your telephone-linked computer. Be aware that the quality of the software can vary widely. Some programs may be excellent and some may be poor.

   b. *Shareware* is inexpensive. It is distributed free, in the same way as public domain software. After you have used it for a while and decide you like it, you're supposed to pay the author for it. Again, the quality varies. Some shareware is excellent.

   c. *Commercial software* consists of brand-name packages, such as those we mentioned in this chapter. Prices and features vary. Fortunately, there are several periodicals that provide ratings and guides. *PC World* polls its readers to find out the best software brands in various categories. The magazine releases the results in its October issue. Other periodicals (for example, *PC* magazine, *MacWorld*, *Infoworld*) also have surveys, reviews, and ratings.

3. *Ethics*: In addition to the three recommended ways to acquire software listed in question 2, two other ways are (1) to copy programs from a friend and (2) to purchase unauthorized copies of programs.

   a. Have you ever had a friend ask to copy one of your programs? Or have you asked someone to make a copy of one of his or her programs? Do you think there is anything wrong with sharing programs in this manner? Identify and discuss the key issues.

   b. Have you ever purchased a bootleg copy of a music CD? As you likely know, bootleg copies are much less expensive and often nearly as good as the originals. Have you ever purchased a bootleg copy of an application program? Do you think there is anything wrong with purchasing unauthorized copies of programs? Identify and discuss the key issues.

4. *Web-based Applications:* As mentioned earlier, some experts predict that today's application software will evolve into Web-based applications. (See Making IT Work for You: Web-based Applications on pages 30–31.)

   a. From your perspective as a user of application programs, discuss the advantages and disadvantages of Web-based applications. Identify key issues and concerns.

b. From the perspective of software manufacturers, such as Microsoft Corporation, discuss the advantages and disadvantages of Web-based applications. Identify their key issues and concerns.

c. Do you think Web-based applications will become widely used? Why or why not? Be specific and defend your answer.

On the Web Exercises

1. Corel Office 2000 Suites

Microsoft's major competitor in the office suite market is Corel. To learn more about Corel and their products, visit the Corel Web site at http://www.corel.com. Once connected to that site, explore and learn more about Corel's office suites. Print out the most informative Web page and write a paragraph describing the different versions of Corel office suites.

2. Presentation Software

Microsoft's PowerPoint can help you create effective and professional-looking presentations. Learn more about this basic application by visiting our site at http://www.mhhe.com/it/oleary/exercise. mhtml, which provides a link to Microsoft's Web site. Once at the site, read about PowerPoint's capabilities and uses. Write a paragraph summarizing some of PowerPoint's most powerful features, and describe how you could use this program to effectively convey ideas to an audience.

3. Computer Games

Computers are not all work and no fun. Learn more about computer games by visiting our site at http://www.mhhe.com/it/oleary/exercise. mhtml, where you'll find a link to a Web site that describes some of the newest and best games. Check out the demos, or go to the pressroom to read reviews on the latest in computer entertainment. Choose a game, and print out the information presented. Write a summary of the game, describe who will likely use the game, and define the system requirements needed to run the game.

4. Web-based PIMs

PIMs (Personal Information Managers) are a popular Web-based application. (See Making IT Work for You: Web-based Applications on pages 30–31.) These programs provide an electronic alternative to the traditional schedule book. Go to www.calendar.yahoo.com and sign up for a free PIM account. Experiment with it by adding yourself and some of your classmates or friends to the address book. Use the calendar to record your class times and other upcoming events. Print out your schedule for the next month. Write a paragraph discussing the advantages and disadvantages of a Web-based PIM as compared to (a) a traditional schedule book such as a day-planner and (b) a PIM stored on your hard disk.

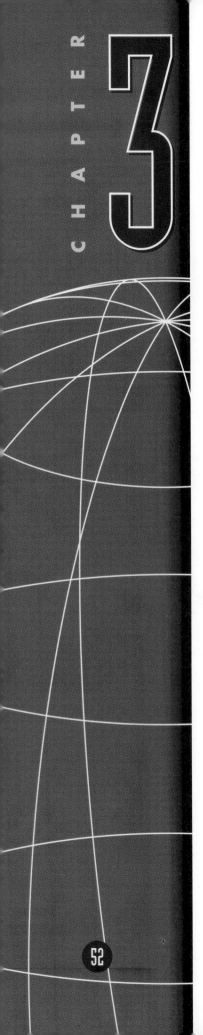

Browsers, Personal Information Managers, Operating Systems, and Utilities

COMPETENCIES

After you have read this chapter, you should be able to:

1. **Describe how browsers are used to navigate, find information, and communicate.**

2. **Discuss the functionality of personal information managers.**

3. **Describe the four kinds of system software.**

4. **Describe the most widely used microcomputer operating systems.**

5. **Describe the five essential utility programs.**

Time—it's a precious thing. Some say it is the most precious of all commodities. Do you have enough of it? Are you balancing school, work, friends, and family? Almost all of us are looking for ways to become more efficient—to do more in less time. We want to meet all of our responsibilities and still have *some* time left for ourselves. That's where this chapter comes in.

With the right kinds of software, you can use a microcomputer as a personal productivity tool. In the previous chapter, we discussed some

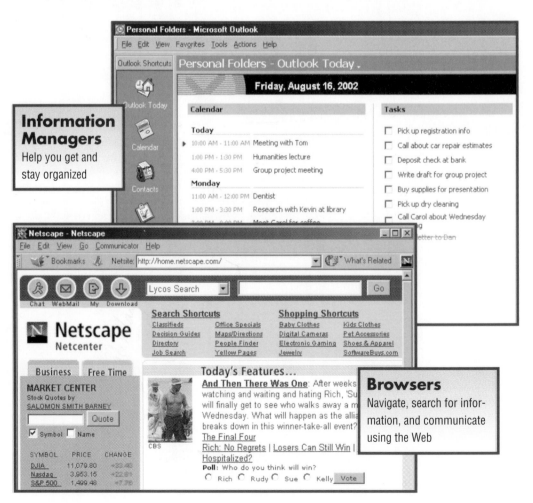

Information Managers
Help you get and stay organized

Browsers
Navigate, search for information, and communicate using the Web

of these programs. Word processors, spreadsheets, database managers, and presentation programs are general-purpose tools that can improve your efficiency and productivity. In this chapter we'll discuss some more specialized tools.

You can surf the Web, locate information, and communicate with friends and family using browser software. You can have active live Web content appear automatically on your computer screen. You can use personal information management software to maintain electronic organizers with calendars, address books, and to-do lists. System software programs help control many of the behind-the-scenes computer operations. In particular, operating systems and utility programs provide tools you can use to maximize your computer's efficiency.

To manage time effectively, competent end users need to understand the functionality and the capabilities of browsers, personal information managers, and system software, including operating systems and utilities.

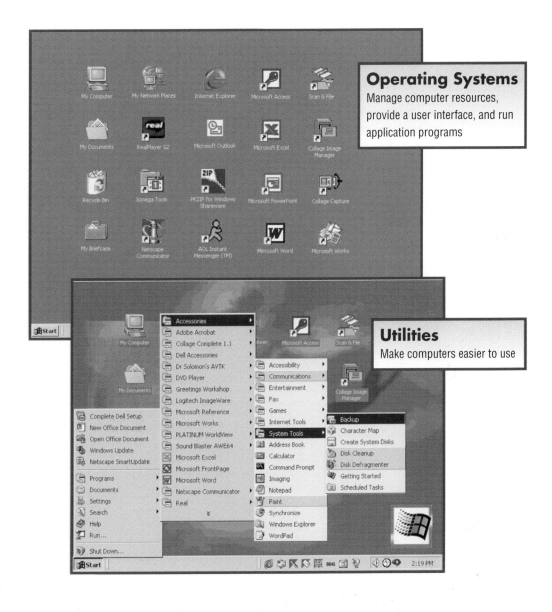

Operating Systems
Manage computer resources, provide a user interface, and run application programs

Utilities
Make computers easier to use

Browsers

Browsers connect to the Web, allowing navigation, searching for information, and communicating with others.

rowsers are used to connect you to remote computers, open and transfer files, display text and images, and provide in one tool an uncomplicated interface to the Internet and the Web. At one time, browsers were considered specialized programs used primarily by computer professionals. Now, browsers are widely used by almost everyone who uses a computer.

The two most widely used browsers are Microsoft's Internet Explorer and Netscape's Navigator. (See Figure 3-1.)

The things you can do on the Internet and the Web are almost endless. The most common activities include navigating or surfing the Web, finding or searching for information, and communicating with others. (A thorough and detailed discussion of the Internet and the Web is presented in Chapter 8.)

Navigating the Web

Navigating the Web means to move from one Web site to another. Often called **surfing** or **browsing** the Web, this activity is like reading a magazine and jumping not only from one article to another within the magazine, but also from one magazine to different magazines, books, movies, and so on. There are millions of different Web sites. See Figure 3-2 for a short list of some popular sites.

There are a number of different ways to navigate from one site or one Web page to another. One of the most common ways is to directly enter a Web site address. Another common way is to use connections provided on a Web page. These connections are called **hypertext links** and are typically represented by colored and underlined text.

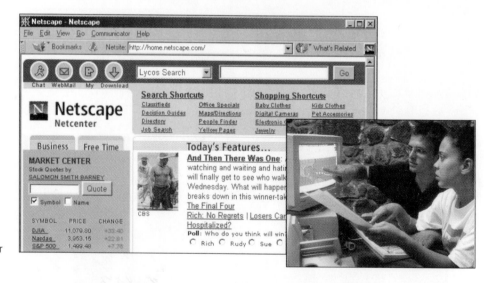

FIGURE 3-1
Netscape Navigator

Description	Site
NBC programming	www.nbc.com
New and used car sales	www.autobytel.com
Medical information	www.mayohealth.org
Active vacations	www.adventurequest.com

FIGURE 3-2
Popular Web sites

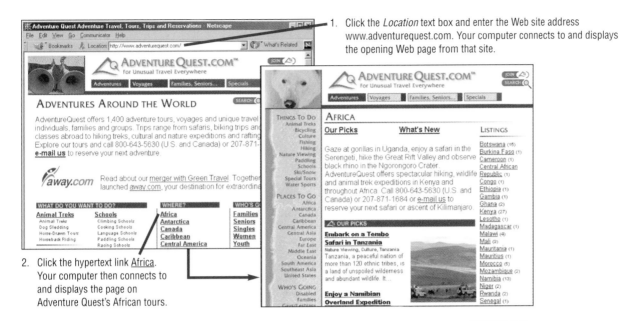

1. Click the *Location* text box and enter the Web site address www.adventurequest.com. Your computer connects to and displays the opening Web page from that site.

2. Click the hypertext link <u>Africa</u>. Your computer then connects to and displays the page on Adventure Quest's African tours.

FIGURE 3-3
Navigating the Web

For example, you could use Netscape to connect to the Adventure Quest Web site and then to find information on Adventure Quest's African tours by taking the steps shown in Figure 3-3.

Finding Information

Do you want to catch up on your favorite television program? Are you buying a new or used car? Planning on doing some research for a school paper? You can do all of these things and much more on the Web. The challenge is finding or searching for the information you want.

Most browsers include a search facility that connects you to Web sites that specialize in finding information. For example, to find information about aerobic exercise, you could use Netscape by following the steps shown in Figure 3-4.

Communicating

Communicating is the most popular Internet activity. Using a browser, you can send and receive messages to and from friends and family located almost anywhere in the world. These messages are called **e-mail.**

FIGURE 3-4
Finding information

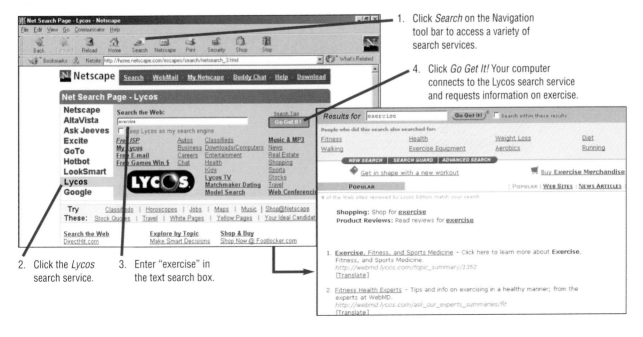

1. Click *Search* on the Navigation tool bar to access a variety of search services.

4. Click *Go Get It!* Your computer connects to the Lycos search service and requests information on exercise.

2. Click the *Lycos* search service.

3. Enter "exercise" in the text search box.

1. Click *Communicator* on the menu bar. This causes a drop-down menu to appear.

2. Click *Messenger* from the drop-down menu to open the Messenger window, which displays the contents of your mailbox.

3. Click *Hello from Dan* in the Message List Window to display Dan's message.

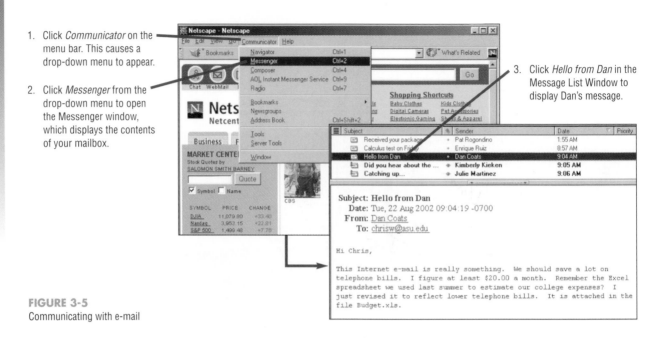

FIGURE 3-5
Communicating with e-mail

For example, you could use Netscape to receive a message from Dan, a friend attending the University of Southern California by taking the steps shown in Figure 3-5.

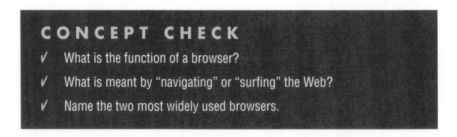

CONCEPT CHECK

✓ What is the function of a browser?

✓ What is meant by "navigating" or "surfing" the Web?

✓ Name the two most widely used browsers.

Personal Information Managers

A personal information manager is a program that helps you get organized and stay organized. It also helps you communicate with others.

FIGURE 3-6
Personal folder—Outlook Today

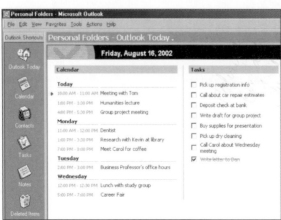

Stop and think for a minute about what you do in a typical day, week, month, and year. Professionals in all kinds of jobs do similar things. They schedule meetings, make to-do lists, and record important names, addresses, and telephone numbers. They jot down notes, make future plans, and record important dates like anniversaries and birthdays.

You may use some of the same tools that many professionals use to keep track of all these things. Such tools include calendars, Rolodex files, address books, index cards, wall charts, notepads, binders, and Post-it notes. **Personal information managers (PIMs),** also known as **desktop managers,** are programs that provide electronic alternatives for these tools.

Two widely used PIMs are Microsoft Outlook and Lotus Organizer. (See Figure 3-6.)

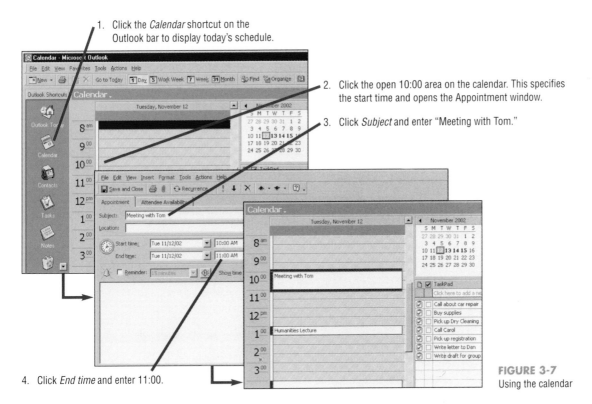

1. Click the *Calendar* shortcut on the Outlook bar to display today's schedule.

2. Click the open 10:00 area on the calendar. This specifies the start time and opens the Appointment window.

3. Click *Subject* and enter "Meeting with Tom."

4. Click *End time* and enter 11:00.

FIGURE 3-7
Using the calendar

Calendar

One of the most important features is the calendar. It operates like an electronic appointment book that keeps track of events, holidays, assignments, and project schedules. Additionally, most calendars provide assistance in scheduling meetings and organizing projects.

For example, you could use Outlook to schedule a meeting with Tom for November 12 that starts at 10:00 and ends at 11:00, as shown in Figure 3-7.

Contacts

Electronic contacts or address books are an essential feature for all PIMs. Like a traditional address book, they are used to record names, addresses, and telephone numbers. In addition, electronic address books are linked to the other parts of a PIM to save you time and energy.

For example, you could use Outlook to locate Lisa Cantrel's postal and e-mail addresses by taking the steps shown in Figure 3-8.

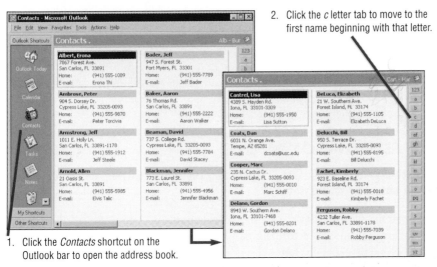

2. Click the *c* letter tab to move to the first name beginning with that letter.

1. Click the *Contacts* shortcut on the Outlook bar to open the address book.

FIGURE 3-8
Using contacts

1. Click the *Tasks* shortcut on the Outlook bar to open the task window.

2. Click the *Complete* box for the task to mark it as completed.

3. Click the *New Task* box to create a new task.

4. Enter the description of the new task.

5. Enter the task due date and press *Enter* to add the task to the list.

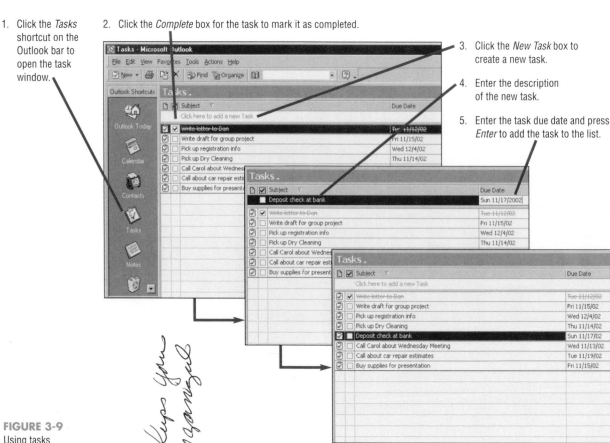

FIGURE 3-9
Using tasks

Tasks

For busy people, organization is often the key to success. A basic feature of all PIMs is a task organizer, or to-do list. It provides two basic functions. First, it records, displays, and reminds you of tasks you need to complete. For example, a task might be a weekly meeting with a project group. Second, it works to record, display, coordinate, and communicate tasks assigned to a group.

You could use Outlook to update your daily task list. For example, you have just completed one of your tasks, writing a letter to Dan. Additionally, you need to add a new task to remind you to deposit a check at the bank. You can update your task list by taking the steps shown in Figure 3-9.

ON THE WEB EXPLORATIONS

Palm is one of the leaders in Web-based PIMs. To learn more about their AnyDay online PIM, visit our Web site at
http://www.mhhe.com/it/oleary/ explore.mhtml

CONCEPT CHECK
✓ What is a personal information manager (PIM)?
✓ What are some features of a typical PIM?

System Software

System software consists of operating systems, utilities, device drivers, and language translators.

System software deals with the physical complexities of computer hardware. It consists of four kinds of programs: operating systems, utilities, device drivers, and language translators.

A program that sets your or ganized I Keep you Monguel calendar contacts Tasks

- **Operating systems** are programs that operate your microcomputer. They coordinate computer resources, provide an interface between users and the computer, and run applications.

- **Utilities,** also known as **service programs,** perform specific tasks related to managing computer resources or files. They include programs to help users identify hardware problems, locate lost files, and back up data.

- **Device drivers** are specialized programs designed to allow particular input or output devices to communicate with the rest of the computer system.

- **Language translators** convert the programming instructions written by programmers into a language that computers understand and process.

CONCEPT CHECK

✓ What is system software?

✓ What kinds of programs can be found in system software?

Operating Systems

Operating systems manage resources, provide a user interface, and run applications. Windows, Mac OS, and Unix are widely used operating systems.

Every computer has an operating system and every operating system performs three basic functions: managing resources, providing a user interface, and running applications.

- **Resources:** These programs coordinate all the computer's **resources** including keyboard, mouse, printer, monitor, storage devices, and memory.

- **User interface:** Users interact with application programs and computer hardware through a **user interface.** Almost all operating systems today provide a windows-like **graphical user interface** (**GUI**) in which graphic objects called **icons** are used to represent commonly used features.

- **Applications:** These programs load and run **applications** such as word processors and spreadsheets. Most operating systems support **multitasking,** or the ability to run more than one application at a time.

For end users, the most important operating systems are those for microcomputers. These include Windows, Mac OS, and Unix.

Windows

By far the most popular microcomputer operating system today is Microsoft's **Windows** with over 80 percent of the market. Windows is designed to run with Intel and Intel-compatible microprocessors such as the Pentium III. It comes in a variety of different versions including Windows 95, Windows 98, and Windows 2000.

Windows gets its name from its use of rectangular boxes called windows. These boxes are extensively used to display information and run programs. Multiple windows can be open at the same time, making it easy to multitask, or work with different programs simultaneously.

Classic view   Web-style view

 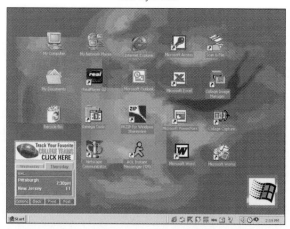

FIGURE 3-10
Desktop views

Windows provides a user interface called the **desktop.** This desktop has two basic views: *classic* and *Web-style.* (See Figure 3-10.) The Web-style view gives Windows a look and feel of the Internet Explorer browser. In the Web-style view, you can use Microsoft's **Active Desktop** to display "active content" from Web pages on your desktop. For example, you could have constantly updating news, weather, sports, and stock prices. (See Making IT Work for You: Active Desktop on pages 62–63.)

One of the most common ways for users to interact with the Windows operating system is by selecting icons. For example, you could use icons to list the contents of your *My Documents* folder by taking the steps shown in Figure 3-11.

The **Start menu** displays a list of commands that can be used to gain access to information, change hardware settings, find information, get online help, run programs, log off a network, and shut down your computer system. For example, you could use the Start menu to run the Netscape Navigator program as shown in Figure 3-12.

Mac OS

The **Mac OS** is designed to run on Macintosh computers. While its market share is much less than that of Windows, it is a very powerful and easy-to-use operating system. It comes in a variety of different versions. Mac OS 8.5 includes **Sherlock,** an innovative search feature for locating information on the Web or on your hard drive. (See Figure 3-13.)

FIGURE 3-11
Using icons

1. Click the *My Computer* icon on the desktop to open a window providing access to information about your computer system.

2. Click the *Local Disk (C:)* icon to open a window providing access to data and programs stored on your internal hard disk.

3. Click the *My Documents* folder to open a window displaying the contents of that folder.

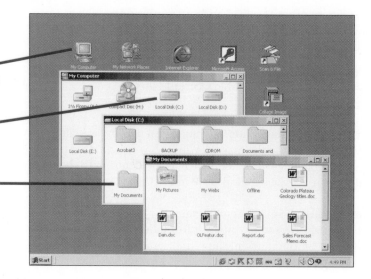

2. Select the *Programs* folder on the Start menu to display program files and folders.

3. Click the *Netscape Communicator* folder to display the contents of that folder.

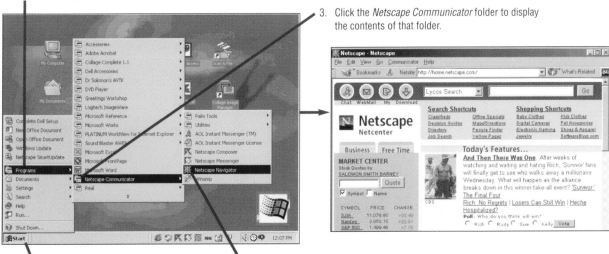

1. Click *Start* on the task bar to open the Start menu.

4. Click the *Netscape Navigator* icon to run the program.

FIGURE 3-12
Using the Start menu

Unix

The **Unix** operating system was originally designed to run on minicomputers in network environments. Now, it is also used by powerful microcomputers and by servers on the Web. There are a large number of different versions of Unix. One receiving a great deal of attention today is **Linux**.

While Windows and the Mac OS are proprietary operating systems (that is, they are owned and licensed by a company), Linux is not. As a graduate student at the University of Helsinki, Linus Torvalds developed Linux in 1991. He has provided the operating system free to others and has encouraged further development. (See Figure 3-14.)

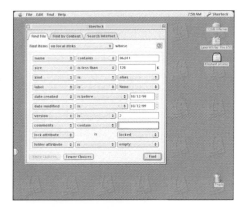

FIGURE 3-13
Mac OS

CONCEPT CHECK

✔ List the three basic functions of an operating system.

✔ Name some common operating systems for microcomputers.

FIGURE 3-14
Recent version of Linux and Linus Torvalds, the founder of Linux

Active Desktop

Want to add some interest to your desktop? Would you like to see the most recent sports scores, news, or stock market updates? Using the Web, you can customize your desktop to provide that information and much more.

How It Works

Specialized Web sites called channels provide access to their continually updated or active content. For example, a sports channel would continually update the scores of ongoing baseball games. Using the Web, you select or add active content to your desktop. Throughout the day, your active desktop automatically connects to the channel site, receives updated information, and displays that information on your desktop.

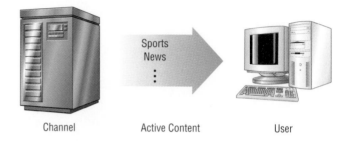

Channel   Active Content   User

Adding Active Content

While you can have any Web page displayed on your desktop, Web pages from specialized channel sites provide the most current information. For example, you can add a dynamic weather map to your desktop using Microsoft's Active Desktop Gallery by:

1. While connected to the Internet, right-click in any open area of the Desktop.

2. Click *Active Desktop*, then *New Desktop Item*.

3. Click the *Visit Gallery* button.

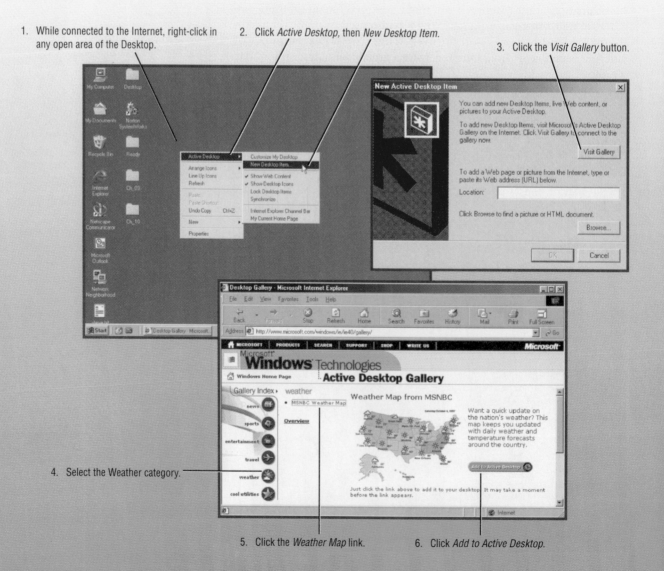

4. Select the Weather category.

5. Click the *Weather Map* link.

6. Click *Add to Active Desktop*.

Refreshing Content

The content of your active desktop is updated or refreshed automatically by periodically connecting to the channel site and downloading the current information from that site. You can specify how frequently this should occur by:

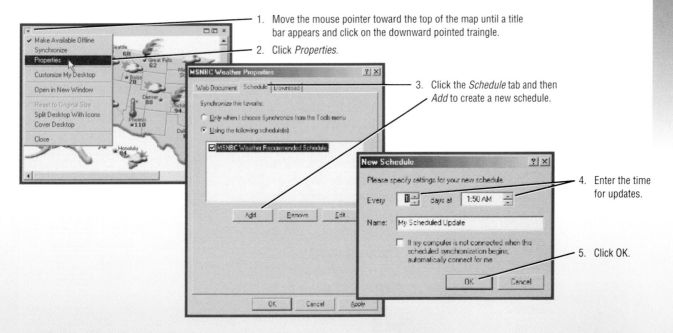

1. Move the mouse pointer toward the top of the map until a title bar appears and click on the downward pointed traingle.

2. Click *Properties*.

3. Click the *Schedule* tab and then *Add* to create a new schedule.

4. Enter the time for updates.

5. Click OK.

Customizing Your Desktop

You can customize the appearance of your active desktop in a variety of ways. For example, you can resize and reposition the weather map and lock its new setting by:

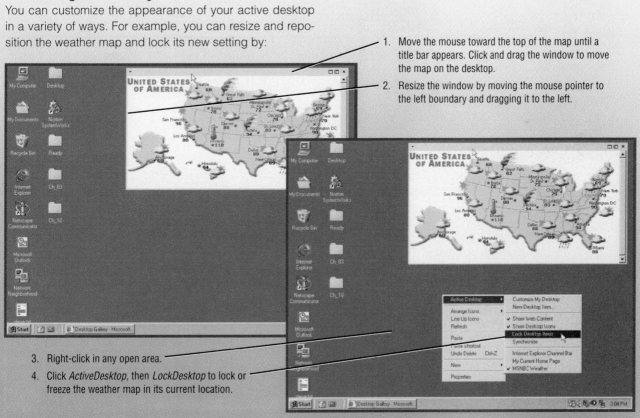

1. Move the mouse toward the top of the map until a title bar appears. Click and drag the window to move the map on the desktop.

2. Resize the window by moving the mouse pointer to the left boundary and dragging it to the left.

3. Right-click in any open area.

4. Click *ActiveDesktop*, then *LockDesktop* to lock or freeze the weather map in its current location.

The Web is continually changing and some of the specifics presented in this Making IT Work for You may have changed. See our Web site at http://www.mhhe.com/it/oleary/IT.mhtml for possible changes and to learn more about this application of technology.

Utilities

Utilities are programs that make computing easier. Operating systems often include utility programs. Norton SystemWorks and McAfee Office are utility suites.

Ideally, microcomputers should run and run and run without problems. However, that simply is not the case. All kinds of things can happen—internal hard disks can crash, destructive programs called viruses can invade a system, computers can freeze up, operations can slow down, and so on. These events can make computing very frustrating. That's where utilities come in. Utilities are specialized programs designed to make computing easier.

There are hundreds of different utility programs. The five most essential utilities are:

- **Troubleshooting programs** that recognize and correct problems, ideally before they become serious problems.

- **Antivirus programs** that guard your computer system against viruses or other damaging programs that can invade your computer system.

- **Uninstall programs** that allow you to safely and completely remove unneeded programs and related files from your hard disk.

- **Backup programs** that make copies of files to be used in case the originals are lost or damaged.

- **File compression programs** that reduce the size of files so they require less storage space and can be sent more efficiently over the Internet.

Most operating system programs provide some utility programs. Even more powerful utility programs can be purchased separately or in utility suites.

Windows Utilities

The Windows operating systems are accompanied with several utility programs, including Backup, Disk Cleanup, and Disk Defragmenter. These utilities can be accessed from the Systems Tools menu. (See Figure 3-15.)

For example by selecting *Backup* from the Windows 2000 System Tools menu, you can create a backup for your hard disk as shown in Figure 3-16.

FIGURE 3-15
Windows utilities

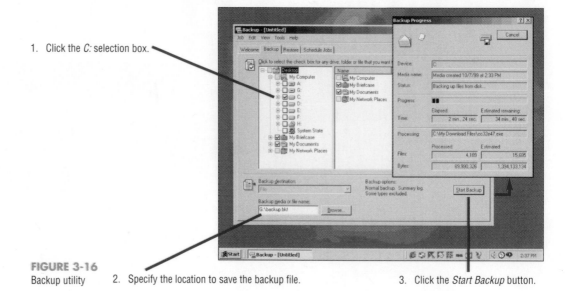

1. Click the *C:* selection box.

FIGURE 3-16
Backup utility   2. Specify the location to save the backup file.   3. Click the *Start Backup* button.

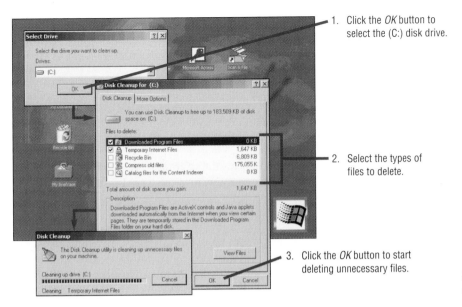

1. Click the *OK* button to select the (C:) disk drive.

2. Select the types of files to delete.

3. Click the *OK* button to start deleting unnecessary files.

FIGURE 3-17
Disk cleanup utility

When you surf the Web a variety of programs and files are saved on your hard disk. Many of these and other files are not essential. **Disk Cleanup** is a trouble shooting utility that identifies and eliminates nonessential files. This frees up valuable disk space and improves system performance.

For example, by selecting Disk Cleanup from the Windows 2000 System Tools menu, you can eliminate unneeded files on your hard disk as shown in Figure 3-17.

Typically, after a hard disk has been used for a period of time, a large file cannot be stored in one location. Rather, the file has to be broken up, or *fragmented,* into small parts and the parts are stored wherever space is available. After a period of time, the hard disk becomes highly fragmented, slowing operations.

Disk Defragmenter is a utility program that locates and eliminates unnecessary fragments and rearranges files and unused disk space to optimize operations. For example, by selecting Disk Defragmenter form the Windows 2000 Systems Tool menu, you can defrag your hard disk as shown in Figure 3-18.

FIGURE 3-18
Disk defragmenter utility

1. Select the disk drive to defrag.

2. Click the *Analyze* button to determine whether defragging is needed.

3. Click the *Defragment* button to begin defragging.

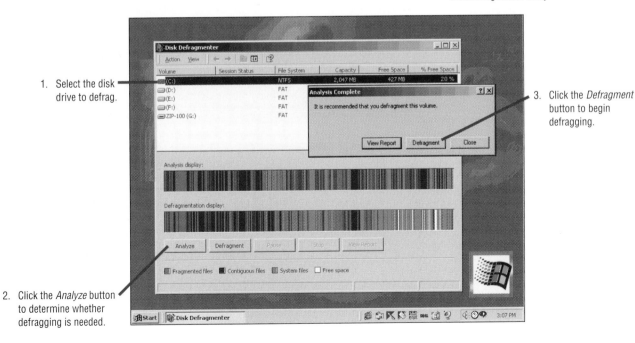

ON THE WEB EXPLORATIONS

Norton is one of the leaders in developing utility software. To learn more about this company, visit our Web site at

http://www.mhhe.com/it/oleary/ explore.mhtml

While these and other utility programs included with Windows are effective, specialty programs offer a wider variety and higher level of support. These programs can be purchased separately or in suites.

Utility Suites

Like application software suites, utility suites combine several programs into one package. Buying the package is less expensive than buying the programs separately. The two best-known utility suites are McAfee Office and Norton SystemWorks. (See Figure 3-19.)

Norton SystemWorks includes a suite of four separate program groups. Each group can be purchased separately or as part of the suite.

- **Norton Utilities** is a collection of 17 separate troubleshooting utilities. These programs can be used to find and fix problems, improve system performance, prevent problems from occurring, and troubleshoot a variety of other problems.

- **Norton AntiVirus** is a collection of antivirus programs that can protect your system from over 21,000 different viruses, quarantine or delete existing viruses, and automatically update its virus list to check for the newest viruses. (See Figure 3-20.)

- **Norton CleanSweep** is a collection of programs that guide you through the process of safely removing programs and files you no longer need. (See Figure 3-21.) Additionally, they will archive, move, and make backups of programs as well as clean up your hard disk. They can also protect your existing files from damage when you install new programs.

- **Norton CrashGuard** is a collection of troubleshooting utilities. These programs can automatically protect you against programs that crash or freeze the display screen. Before an event causes your system to crash or freeze, CrashGuard intervenes, providing you with options to recover your current work.

- **Norton Web Services** monitors your system for out-of-date software and notifies you of available software updates that can be installed automatically from the Internet.

FIGURE 3-19
Norton SystemWorks

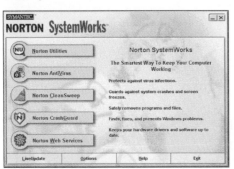

FIGURE 3-20
Norton AntiVirus

CONCEPT CHECK

✓ What are utilities?

✓ List some common utilities.

✓ Name some of the utility suites currently available.

FIGURE 3-21
Norton CleanSweep

A LOOK TO THE FUTURE

New versions of operating systems are being released all the time. While Microsoft's Windows is by far the dominant operating system for today's microcomputers, others are trying to narrow the gap. All are striving toward better hardware support, simpler installations, and greater ease of use.

The next generation of Windows promises to provide a variety of enhanced hardware capabilities, including support for home networks. Using special programs such as Internet Connection Sharing, multiple users on different computers will be able to connect to a home network easily and surf the Web sharing a single connection. Moreover, home networking will be taken to a new level of connectivity. Future Windows versions will support microcomputer interaction and control over smart appliances, including refrigerators, stereos, and heating and cooling systems.

Expect greater challenges to Windows' dominance. Apple will continue to offer newer and more powerful versions of its Macintosh operating system. New versions of Linux will provide user interfaces to match Windows, and new applications written specifically for Linux will enhance its appeal. Finally, BeOS, (pronounced "bee-O-S") is a young operating system gaining popularity with multimedia developers and programmers. It handles complex 3D graphics and animations better than the competitors and could become a serious operating system contender.

In the near future, operating systems will likely retain their current look and feel. Behind the scenes, however, they will continue to become more powerful, more intuitive, and require less user interaction. Increasing competition promises to accelerate this trend.

VISUAL SUMMARY

Browsers, Personal Information Managers, Operating Systems, and Utilities

BROWSERS

Browsers are programs that connect to remote computers, open and transfer files, display text and images, and provide an uncomplicated interface to the Internet and the Web.

Common activities on the Internet and Web are: *navigating* or *surfing* the Web, *finding* or *searching for information*, and *communicating* with others.

Navigating the Web

Navigating the Web (also known as **surfing** or **browsing**) means to move from one Web site to another. Two common ways to navigate:

- Enter Web address in the Location box.
- Use **hyperlinks** or connections between related Web pages.

Finding Information

With so much information available on the Web, the challenge is finding or locating what you want. Most browsers contain search facilities that connect you to Web sites that specialize in finding information.

Communicating

Communicating is the most popular Internet activity. You can communicate with friends and family using **e-mail**.

PERSONAL INFORMATION MANAGERS

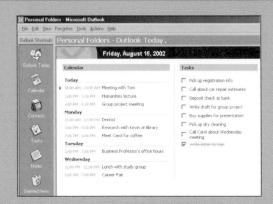

Personal information mangers (also known as **PIMs** and **desktop managers**) are electronic organizers designed to help you get organized and stay organized. PIMs contain a variety of features, including maintaining *calendars, contacts,* and *tasks.*

Calendar

The **calendar** is one of the most important features. Like an electronic appointment book, it tracks events, holidays, assignments, and project schedules.

Contacts

Electronic **contacts** or address books are an essential part of all PIMs. Like a traditional address book, they record names, addresses, and telephone numbers. Linked to other PIM activities, the contacts' lists save time and energy.

Tasks

A basic feature of all PIMs is a **task** organizer, or to-do list. It provides two basic functions: (1) recording, displaying, and reminding you of tasks you need to complete and (2) recording, displaying, coordinating, and communicating tasks assigned to a group.

To manage time effectively, competent end users need to understand the functionality and the capabilities of browsers, personal information managers, and system software, including operating systems and utilities.

OPERATING SYSTEMS

Operating systems perform three basic functions: manage resources, provide a user interface, and run programs. Windows, Mac OS, and Unix are popular operating systems for microcomputers.

Windows

Windows is the most widely used operating system today. The name Windows comes from the rectangular boxes (windows) used to display information and run applications.

Multiple windows can be open to **multitask,** or work with different programs simultaneously. The **desktop** is the user interface provided by Windows. **Icons** are often used to interact with the Windows operating system. Another common way is to use the **Start menu.**

Mac OS

The **Mac OS** runs on Macintosh computers. Although not as widely used as Windows, it is very powerful and easy to use. Mac OS 8.5 includes **Sherlock,** an innovative search feature for locating information on the Web or on your hard drive.

Unix

The **Unix** operating system was originally designed to run on minicomputers in network environments. **Linux** is one version of Unix that is receiving a great deal of attention.

UTILITIES

Utilities are specialized programs designed to make computing easier. While there are hundreds of different utility programs, the most essential are: **troubleshooting, antivirus, uninstall, backup,** and **file compression programs.**

Windows Utilities

The Windows operating systems come with several utility programs. These utilities can be accessed from the Systems Tools menu. Three such utilities are *Backup*, *Disk Cleanup*, and *Disk Defragmenter*.

- **Backup**—to back up your hard disk.
- **Disk Cleanup**—a troubleshooting utility for monitoring storage capacity.
- **Disk Defragmenter**—to locate and eliminate unnecessary fragments, rearranges files and unused disk space.

Utility Suites

Utility suites combine several programs into one package. McAfee Office and Norton SystemWorks are the best-known.

Key Terms

2001
2002

Active Desktop (60)
antivirus programs (64)
applications (59)
backup programs (64)
browser (54)
browsing (54)
desktop (60)
desktop managers (56)
device drivers (59)
Disk Cleanup (65)
Disk Defragmenter (65)
e-mail (55)
file compression programs (64)
graphical user interface (GUI) (59)

hypertext links (54)
icons (59)
language translators (59)
Linux (63)
Mac OS (60)
multitasking (59)
navigating (54)
Norton AntiVirus (66)
Norton CleanSweep (66)
Norton CrashGuard (66)
Norton Utilities (66)
Norton Web Services (66)
operating systems (59)

personal information managers
    (PIMs) (56)
resources (59)
service programs (59)
Sherlock (60)
Start menu (60)
surfing (54)
system software (58)
troubleshooting programs (64)
uninstall programs (64)
Unix (61)
user interface (59)
utilities (59)
Windows (59)

Chapter Review

LEVEL 1

Reviewing Facts and Terms
Matching

Match each numbered item with the most closely related lettered item. Write your answers in the spaces provided.

1. One term for moving between different Web sites. _C_

2. Connections within one Web page that take you to a related Web page. _H_

3. Program that helps you get organized and stay organized. _M_

4. Perform specific tasks related to managing computer resources or files. _S_

5. Allow particular input or output devices to communicate with the rest of the computer system. _E_

6. Operating system originally designed to run on minicomputers in network environments. _Q_

7. Through this, users interact with application programs and computer hardware. _R_

8. Programs such as word processors and spreadsheets. _A_

9. Deals with the physical complexities of computer hardware. _N_

10. Programs that operate your computer. _L_

11. Allow you to safely and completely remove unneeded programs and related files from your hard disk. _P_

12. User interface in Windows. _D_

13. Operating system designed to run only on Macintosh computers. _K_

14. Convert the programming instructions written by programmers into a language that computers understand and process. _I_

a. Applications
b. Backup programs
c. Browsing
d. Desktop
e. Device drivers
f. Disk Cleanup
g. File compression programs
h. Hypertext links
i. Language translators
j. Linux
k. Mac OS
l. Operating systems
m. PIMs
n. System software
o. Troubleshooting programs
p. Uninstall programs
q. Unix
r. User interface
s. Utilities
t. Windows

15. One popular, and free, version of the Unix operating system. _J_

16. Recognize and correct problems. _O_

17. Operating system that gets its name from its use of rectangular boxes called windows. _T_

18. Make copies of files to be used in case originals are lost or damaged. _B_

19. Reduce the size of files so they require less storage space. _G_

20. Troubleshooting utility for monitoring storage capacity. _F_

True/False

In the spaces provided, write T or F to indicate whether the statement is true or false.

1. A connection on one Web page that directs you to a related Web page is called a Web portal. _F_

2. A basic feature of all PIMs is word processing. _F_

3. System software deals with the physical complexities of computer hardware. _T_

4. The Mac OS is designed to run on Macintosh computers. _T_

5. Disk Defragmenter is a utility program that rearranges files and unused disk space to optimize operations. _T_

Multiple Choice

Circle the letter of the correct answer.

1. An operating system provides a:
   - **a.** user interface
   - **b.** window
   - **c.** database
   - **d.** sort box
   - **e.** grammar checker

2. System software consists of all of the following except:
   - **a.** utilities
   - **b.** device drivers
   - **c.** language translators
   - **d.** operating systems
   - **e.** browsers

3. Moving from one Web site to another is called:
   - **a.** navigating
   - **b.** Web hopping
   - **c.** Web cruising
   - **d.** window shopping
   - **e.** multitasking

4. Programs that help you get organized and communicate with others are:
   - **a.** operating systems
   - **b.** browsers
   - **c.** personal information managers
   - **d.** utilities
   - **e.** desktop organizers

5. Designed to repair or prevent certain frustrating problems, these specialized programs perform specific tasks related to managing computer resources or files:
   - **a.** presentation graphics programs
   - **b.** spreadsheets
   - **c.** utilities
   - **d.** browsers
   - **e.** PIMs

Completion

Complete each statement in the spaces provided.

1. _Browsers_ connect to the Web and create Web pages.

2. Windows provides a user interface called the _Desktop_.

3. _Multitasking_ is the ability that most operating systems have to run more than one application at a time.

4. _Device Drivers_ are specialized programs designed to allow particular input or output devices to communicate with the rest of the computer system.

5. Programs that coordinate computer resources, provide an interface between users and the computer, and run applications are called _Operating Systems_.

Reviewing Concepts

LEVEL 2

Open-Ended

On a separate sheet of paper, respond to each question and statement.

1. Describe some of the different ways to navigate from one Web site to another.

2. Compare the three most common microcomputing operating systems. Where are you likely to find each one being used?

3. Explain how you might go about finding information on the Web about the health benefits of vitamins.

4. Outline the features that PIMs typically provide.

5. Describe the five most essential utilities.

Concept Mapping

On a separate sheet of paper, draw a concept map or a flowchart showing how the following terms are related. Show all relationships. Include any additional terms you can think of.

Active Desktop	file compression programs	operating systems
applications	graphical user interface	PIMs
browser	hypertext links	resources
desktop	Linux	system software
desktop managers	Mac OS	Unix
device drivers	multitasking	user interface
disk cleanup	navigating	Windows
disk defragmenter	Norton Utilities	

Critical Thinking Questions and Projects

LEVEL 3

Read each exercise and answer the related questions on a separate sheet of paper.

1. _Future of the Internet:_ More and more people have access to high-speed Internet connections at home, school computer labs, place of employment, and so on. High-speed Internet connections include cable modems, DSL connections, ISDN lines, and some others. Cable modems, for instance, take only seconds to download files that might take a 56K modem several minutes to download. These high-speed connections make it more time-effective, and certainly more enjoyable, to download large files such as music and video files, animations, or complex Web pages.

   As more people gain high-speed Internet access, how will this shape the future of the Internet? Knowing that you will need to keep the attention of a worldwide audience who will be able download large pieces of information quickly, what kind of Internet services can you imagine developing?

2. _Operating systems and competition:_ New operating systems seem to come out every one to two years, depending on the platform. For example, Microsoft plans to introduce Windows 2002 just a few years after it introduced Windows 2000. The average user may not notice many changes from one operating system to the next. Yet they keep coming out. Is this necessary, or is it just a cash cow for computer companies? Can you stay with an older operating system and still remain compatible as others

change? Is LINUX (an operating system easily found for free) a real threat to Windows, especially in light of the antitrust legal proceedings with Microsoft?

3. *Active Desktop:* With Windows 2000 you can have live active Web content appear automatically on your desktop. You can choose from a gallery of topics including news, sports, entertainment, travel, weather, and cool utilities (See Making IT Work for You: Active Desktop on pages 62–63.)

   a. Have you used any of these active desktop topics? If so, identify the topics; describe why you used it, and whether you found it useful.

   b. From the list of gallery topics listed above, select the two most interesting topics to you. For each, discuss why you chose it and how you might use it.

On the Web Exercises

1. Microsoft's Windows

Microsoft created the famous Windows operating system. To learn more about Microsoft's most recent operating systems, visit their Web site at http://www.microsoft.  com. Once connected to that site, explore and learn more about the newest operating system from Microsoft. Print out the most informative Web page and write a paragraph describing the most important features of this new operating system.

2. Linux

Unix is a powerful, versatile operating system used by many computer professionals. Some observers believe that the Linux version will become popular for home users and computer novices. To  learn about Linux visit our Web site at http://www.mhhe. com/it/oleary/exercise. mhtml for a link to a related site. Once connected to that site, explore and learn more about Linux. Print out the Web page you find most interesting. Write a paragraph or two discussing the strengths and weaknesses of Linux in comparison to Microsoft Windows and identifying applications that are better suited for Windows and those that are better suited for Linux.

3. Web Shopping

Shopping on the Web has become an ever increasing activity. What could be more convenient  than to buy groceries on the Web and have them delivered to your door? Of course, there are some downsides to on-line shopping. Visit our site at http://www.mhhe.com/it/oleary/exercise.mhtml to link to an online grocery store. Once connected to that site, browse through some of the products that are available. Print out the most interesting Web page and write a brief paragraph discussing the advantages and disadvantages to shopping this way.

4. Active Desktop

There are any number of ways you can customize your desktop. In addition to the traditional active  desktop (See Making IT Work for You: Active Desktop on pages 62–63.), several other customization options are available to change the look, feel, and performance of your desktop. To learn more about these options, connect to the Yahoo site at http://www.yahoo.com and look at the subject area "Computers and Internet: Desktop Customization" and/or search using the keywords "custom desktop". Explore and learn more about different customization options. Print out the page you find most informative, then write a brief paragraph describing the customization option that is most interesting to you.

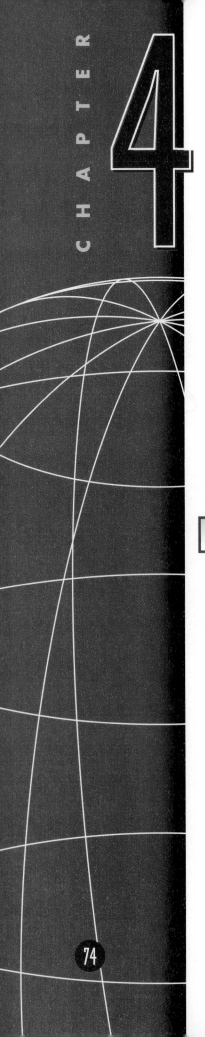

The System Unit

COMPETENCIES

After you have read this chapter, you should be able to:

1. **Describe how a computer uses binary codes to represent data in electrical form.**

2. **Discuss each of the major system unit components.**

3. **Explain the differences among the three types of memory.**

4. **Discuss the three principal types of bus lines.**

5. **Discuss five types of ports.**

How does the system unit work? That is the subject of this chapter. Why are some micro-computers more powerful than others? The answer lies in three words: *speed, capacity,* and *flexibility.* After reading this chapter, you will be able to judge how fast, powerful, and versatile a particular microcomputer is. As you might expect, this knowledge is valuable if you are planning to buy a new microcomputer system or to upgrade an existing system. (The Buyer's Guide and the Upgrader's Guide at the end of this book provide additional information.) It will also help you to evaluate whether or not an existing microcomputer system is powerful enough for today's new and

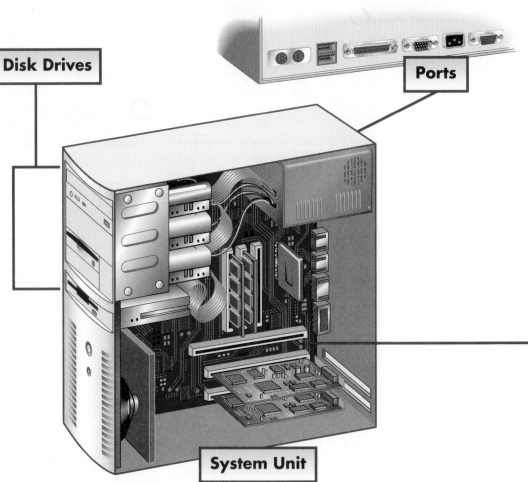

Disk Drives

Ports

System Unit

exciting applications. For example, with the right hardware, you can use your computer to watch TV while you work, and to capture video clips for class presentations.

Sometime you may get the chance to watch when a technician opens up a microcomputer. You will see that it is basically a collection of electronic circuitry. While there is no need to understand how all these components work, it is important to understand the principles. Once you do, you will then be able to determine how powerful a particular microcomputer is. This will help you judge whether it can run particular kinds of programs and can meet your needs as a user.

Competent end users need to understand the functionality of the basic components in the system unit, including the system board, microprocessor, memory, system clock, expansion slots and cards, bus lines, ports, and cables.

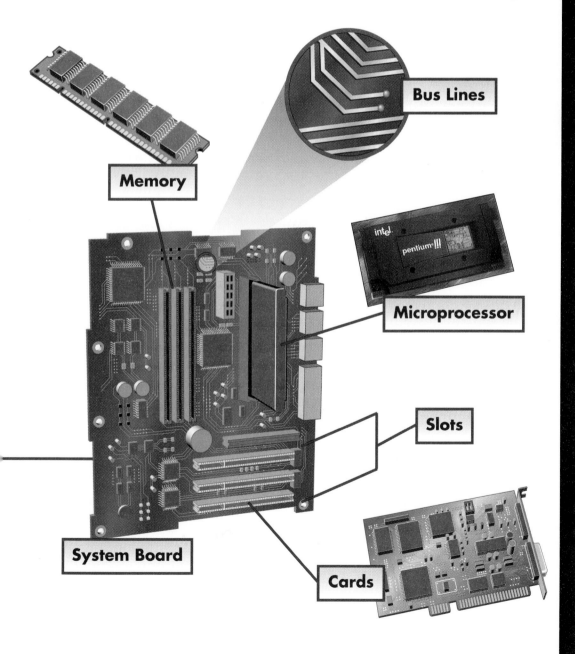

Bus Lines

Memory

Microprocessor

Slots

System Board

Cards

Electronic Data and Instructions

Data and instructions are represented electronically with a binary, or two-state, numbering system. The three principal binary coding schemes are ASCII, EBCDIC, and Unicode.

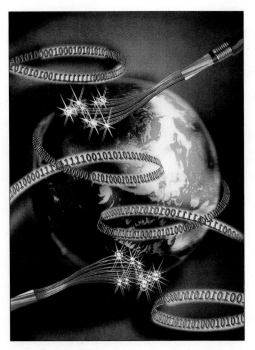

People follow instructions and process data using letters, numbers, and special characters. For example, if we wanted someone to add the numbers 3 and 5 together and record the answer, we might say "please add 3 and 5 and write the sum on a piece of paper." The system unit, however, is electronic circuitry and cannot directly process such a request. Before anything can occur within the system unit, a conversion must occur from what we understand to what the system unit can electronically process.

What is the most fundamental statement you can make about electricity? It is simply this: It can be either *on* or *off*. Indeed, there are many forms of technology that can make use of this two-state on/off, yes/no, present/absent arrangement. For instance, a light switch may be on or off, or an electric circuit open or closed. A magnetized spot on a tape or disk may have a positive charge or a negative charge. This is the reason, then, that the binary system is used to represent data and instructions.

The decimal system that we are all familiar with has 10 digits (0, 1, 2, 3, 4, 5, 6, 7, 8, 9). The **binary system,** however, consists of only two digits—0 and 1. Each 0 or 1 is called a **bit**—short for *bi*nary digi*t*. In the system unit, the 0 can be represented by electricity being off, and the 1 by electricity being on. In order to represent numbers, letters, and special characters, bits are combined into groups of eight bits called **bytes.** Each byte typically represents one character.

Binary Coding Schemes

Now let us consider an important question. How are characters represented as 0s and 1s ("off" and "on" electrical states) in the computer? The answer is in the use of *binary coding schemes*. (See Figure 4-1.)

Two of the most popular binary coding schemes use eight bits to form each byte. These two codes are *ASCII* and *EBCDIC*. (See Figure 4-2.) A recently developed code, *Unicode*, uses sixteen bits.

- **ASCII,** pronounced "*as*-key," stands for *A*merican *S*tandard *C*ode for *I*nformation *I*nterchange. This is the most widely used binary code for microcomputers.

- **EBCDIC,** pronounced "*eb*-see-dick," stands for *E*xtended *B*inary *C*oded *D*ecimal *I*nterchange *C*ode. It was developed by IBM and is used primarily for large computers.

- **Unicode** is a 16-bit code designed to support international languages like Chinese and Japanese. These languages have too many characters to be represented by the eight-bit ASCII and EBCDIC codes. Unicode was developed by Unicode, Inc., with support from Apple, IBM, and Microsoft.

Code	Uses
ASCII	Microcomputers
EBCDIC	Larger computers
Unicode	International languages

FIGURE 4-1
Binary codes

Symbol	ASCII	EBCDIC	Symbol	ASCII	EBCDIC
A	0100 0001	1100 0001	!	0010 0001	0101 1010
B	0100 0010	1100 0010	"	0010 0010	0111 1111
C	0100 0011	1100 0011	#	0010 0011	0111 1011
D	0100 0100	1100 0100	$	0010 0100	0101 1011
E	0100 0101	1100 0101	%	0010 0101	0110 1100
F	0100 0110	1100 0110	&	0010 0110	0101 0000
G	0100 0111	1100 0111	(	0010 1000	0100 1101
H	0100 1000	1100 1000	)	0010 1001	0101 1101
I	0100 1001	1100 1001	*	0010 1010	0101 1100
J	0100 1010	1101 0001	+	0010 1011	0100 1110
K	0100 1011	1101 0010	0	0011 0000	1111 0000
L	0100 1100	1101 0011	1	0011 0001	1111 0001
M	0100 1101	1101 0100	2	0011 0010	1111 0010
N	0100 1110	1101 0101	3	0011 0011	1111 0011
O	0100 1111	1101 0110	4	0011 0100	1111 0100
P	0101 0000	1101 0111	5	0011 0101	1111 0101
Q	0101 0001	1101 1000	6	0011 0110	1111 0110
R	0101 0010	1101 1001	7	0011 0111	1111 0111
S	0101 0011	1110 0010	8	0011 1000	1111 1000
T	0101 0100	1110 0011	9	0011 1001	1111 1001
U	0101 0101	1110 0100			
V	0101 0110	1110 0101			
W	0101 0111	1110 0110			
X	0101 1000	1110 0111			
Y	0101 1001	1110 1000			
Z	0101 1010	1110 1001			

FIGURE 4-2
ASCII and EBCDIC binary coding schemes

When you press a key on the keyboard, a character is automatically converted into a series of electronic pulses that the system can recognize. For example, pressing the letter *A* on a keyboard causes an electronic signal to be sent to the microcomputer's system unit. The system unit then converts it to the ASCII code of 0100 0001. (See Figure 4-3.)

FIGURE 4-3
How the letter *A* is represented in ASCII code

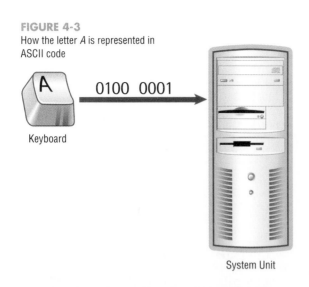

Keyboard

0100 0001

System Unit

All instructions and data have to be converted into binary data before they can be executed. For example, the instructions 3 + 5 requires 24 bits using the ASCII coding scheme. (See Figure 4-4.)

Why are coding schemes important? Whenever files are used or shared by different computers or applications, the same coding scheme must be used. Generally, this is not a problem if both computers are microcomputers since both would most likely use ASCII code. And most microcomputer applications store files using this code. However, problems occur when files are shared between microcomputers and larger computers that use EBCDIC code. The files must be translated from one coding scheme to the other before processing can begin. Fortunately, special conversion programs are available to help with this translation.

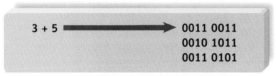

3 + 5 → 0011 0011
0010 1011
0011 0101

FIGURE 4-4
ASCII code for 3 + 5

CONCEPT CHECK

✔ The _____ system is used to represent data and instructions.

✔ Name the three binary coding schemes.

✔ All instructions and data must be converted into _____ _____ before they can be executed.

System Board

The system board connects all system components and allows input and output devices to communicate with the system unit.

FIGURE 4-5
System board

The **system board** is also known as the **main board** or **motherboard.** (See Figure 4-5.) The system board is the communications web for the entire computer system. Every component of the system unit connects directly to the system board. It acts as a data path allowing the various components to communicate with one another. External devices such as keyboard, mouse, and monitor could not communicate with the system unit without the system board.

On a desktop computer, the system board is located at the bottom of the systems unit. On a tower unit, the system board is located on one of the sides. It is a large flat circuit board covered with sockets and other electronic parts, including a variety of chips. A **chip** consists of a tiny circuit board etched on a postage-stamp-sized square of sandlike material called silicon. (See Figure 4-6.) A chip is also called a **silicon chip, semiconductor,** or **integrated circuit.** Chips are mounted on carrier packages, which then plug into sockets on the system board.

Microprocessor

The CPU is located on the microprocessor chip and has two components—the control unit and the arithmetic-logic unit. RISC and CISC are two types of microprocessors.

In a microcomputer system, the **central processing unit (CPU)** or **processor** is contained on a single chip called the microprocessor. Typically, this microprocessor is contained within a cartridge that plugs in to the system board. (See Figure 4-7.) The microprocessor is the "brains" of the system. It has two basic components: the control unit and the arithmetic-logic unit.

Control Unit

The **control unit** tells the rest of the computer system how to carry out a program's instructions. It directs the movement of electronic signals between memory—which temporarily holds data, instructions, and processed information—and the arithmetic-logic unit. It also directs these control signals between the CPU and input and output devices.

Arithmetic-Logic Unit

The **arithmetic-logic unit,** usually called the **ALU,** performs two types of operations—arithmetic and logical. *Arithmetic* operations are, as you might expect, the fundamental math operations: addition, subtraction, multiplication, and division. *Logical* operations consist of comparisons. That is, two pieces of data are compared to see whether one is equal to (=), less than (<), or greater than (>) the other.

Microprocessor Chips

Chip capacities are often expressed in word sizes. A **word** is the number of bits (such as 16, 32, or 64) that can be accessed at one time by the CPU. The more bits in a word, the more powerful—and the faster—the computer is. A 32-bit-word computer can access 4 bytes at a time. A 64-bit-word computer can access 8 bytes at a time. Therefore, the 64-bit computer is faster.

Microcomputers process data and instructions in millionths of a second, or **microseconds.** Supercomputers, by contrast, operate at speeds measured in nanoseconds and even picoseconds—1,000 to 1 million times as fast as microcomputers. (See Figure 4-8.)

There are two types of microprocessor chips.

- **CISC chips:** The most common type of microprocessor is the **complex instruction set computer (CISC)** chip. This design was popularized by Intel and is the basis for their line of microprocessors. It is the most widely used chip design with thousands of programs written specifically for it. Intel's Pentium II and Pentium III are recent CISC chips. Two other manufacturers of CISC chips are AMD and Cyrix.

FIGURE 4-6
Chip

FIGURE 4-7
Microprocessor cartridge

ON THE WEB EXPLORATIONS

Motorola is a leader in RISC research and development. To learn more about the company, visit our Web site at

http://www.mhhe.com/it/oleary/ explore.mhtml

Unit	Speed
Millisecond	Thousandth of a second
Microsecond	Millionth of a second
Nanosecond	Billionth of a second
Picosecond	Trillionth of a second

FIGURE 4-8
Processing speeds

Microprocessor	Type	Typical Use
Pentium	CISC	Microcomputers
PowerPC	RISC	Apple Computers
ALPHA	RISC	Supercomputers, workstations
MIPS	RISC	Workstations, video games

FIGURE 4-9
Popular microprocessors

- **RISC chips: reduced instruction set computer (RISC)** chips use fewer instructions. This design is simpler and less costly than CISC chips. A recent Motorola chip developed with IBM and Apple is the PowerPC chip. Two other recent RISC chips are Digital Equipment Corporation's (DEC) Alpha chip and Silicon Graph's MIPS chip. These chips are used in many of today's most powerful microcomputers. See Figure 4-9 for a table of popular microprocessors.

Some specialized processor chips are available. One example is the tiny built-in microprocessor used in **smart cards.** They can be used to hold health insurance information, driver's license information, credit records, and so on. American Express and Visa have recently introduced smart credit cards. In addition to traditional credit card uses, these cards can be used to connect to selected Web sites, complete online purchases, conduct online banking, and more.

CONCEPT CHECK

✔ What role does the system board play in a computer?

✔ Which two basic components make up the CPU?

✔ Name two types of microprocessor chips.

Memory

Three types of memory are RAM, ROM, and CMOS.

There are three well-known types of memory chips: random-access memory (RAM), read-only memory (ROM), and complementary metal-oxide semiconductor (CMOS).

RAM

Random-access memory (RAM) chips hold the program and data that the CPU is presently processing. (See Figure 4-10.) That is, it is *temporary* or volatile storage. (Secondary storage, which we shall describe in Chapter 6, is *permanent* storage, such as the data stored on diskettes. Data from this kind of storage must be loaded into RAM before it can be used.)

RAM is called temporary because as soon as the microcomputer is turned off, everything in RAM is lost. It is also lost if there is a power failure that

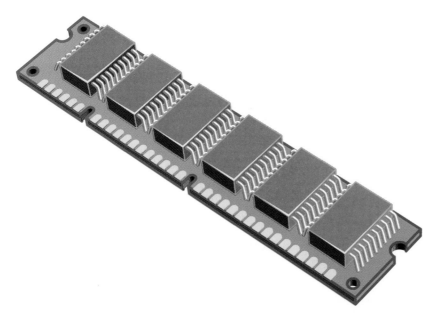

FIGURE 4-10
RAM chips

disrupts the electric current going to the microcomputer. For this reason, as we mentioned earlier, it is a good idea to save your work in progress. That is, if you are working on a document or a spreadsheet, every few minutes you should save, or store, the material.

There is a relatively new type of RAM, however, that is not temporary. **Flash RAM** or **flash memory** chips can retain data even if power is disrupted. This type of memory is more expensive and used primarily in high-end portable computers.

Having enough RAM is important! Some programs may require more memory than a particular microcomputer offers. For instance, Excel 2000 requires 20MB of RAM. Additional RAM is needed to hold any data or other applications. However, many microcomputers—particularly older ones—may not have enough memory to hold the program or to run the program. The capacity or amount of RAM is expressed in bytes. There are four commonly used units of measurement to describe memory capacity. (See Figure 4-11.)

Even if your computer does not have enough RAM to hold a program, it might be able to run the program using **virtual memory.** Most of today's operating systems support virtual memory. With virtual memory, large programs are divided into parts and the parts stored on a secondary device, usually a hard disk. Each part is then read into RAM only when needed. In this way, computer systems are able to run very large programs.

Unit	Capacity
Kilobyte (KB)	1,000 bytes
Megabyte (MB)	1 million bytes
Gigabyte (GB)	1 billion bytes
Terabyte (TB)	1 trillion bytes

FIGURE 4-11
Memory capacity

TIPS Does your computer seem to be getting slower and slower? Perhaps it's so slow you are thinking about buying a new one. Before doing that, consider the following suggestions that might add a little zip to your current system.

1. *Uninstall programs you no longer need.* Explore the contents of your hard disk and identify programs that you no longer need. If you have Windows 95 or Windows 98, use Start/Settings/Control Panel/Add/Remove Programs to access the Uninstall feature.

2. *Remove unneeded fonts.* If you have Windows 95 or 98, use Start/Settings/Control Panel/Fonts to determine the different font types stored on your system. To see a sample of any font type, double-click it. Review the fonts and delete those you will not need.

3. *Empty the Recycle Bin.* If you have Windows 95 or Windows 98, files are not removed from your hard disk when you delete them. Rather, they are moved to the Recycle Bin. To empty or remove files from the Recycle Bin, open the Recycle Bin and use File/Empty Recycle Bin.

Another term you are apt to hear about in conjunction with RAM is **cache memory** or **RAM cache.** Cache (pronounced "cash") memory is used to store the most frequently accessed information stored in RAM. The cache acts as a temporary high-speed holding area between the memory and the CPU. In a computer with a cache (not all machines have one), the computer detects which information in RAM is most frequently used. It then copies that information into the cache. When needed, the CPU can quickly access the information from the cache. Most newer microprocessors have cache memory built in.

ROM

Read-only memory (ROM) chips have programs built into them at the factory. Unlike RAM chips, ROM chips are not volatile and cannot be changed by the user. "Read only" means that the CPU can read, or retrieve, the programs written on the ROM chip. However, the computer cannot write—encode or change—the information or instructions in ROM.

ROM chips typically contain special instructions for detailed computer operations. For example, ROM instructions may start the computer, give keyboard keys their special control capabilities, and put characters on the screen. ROMs are also called **firmware.**

CMOS

A **complementary metal-oxide semiconductor (CMOS)** chip provides flexibility and expandability for a computer system. It contains essential information that is required every time the computer system is turned on. The chip supplies such information as the amount of RAM, type of keyboard, mouse, monitor, and disk drives. Unlike RAM, it is powered by a battery and does not lose its contents when the power is turned off. Unlike ROM, its contents can be changed to reflect changes in the computer system such as increased RAM and new hardware devices.

See Figure 4-12 for a summary of the three types of memory.

CONCEPT CHECK

✔ Name three well-known types of memory chips.

✔ What makes RAM different from ROM?

Type	Use
RAM	Programs and data
ROM	Fixed start-up instructions
CMOS	Flexible start-up instructions

FIGURE 4-12
Memory

System Clock

Speed of computer operations is measured in megahertz, or millionths of a second.

The **system clock** controls the speed of operations within a computer. This speed is expressed in **megahertz** (abbreviated **MHz**). One megahertz equals 1 million cycles (beats) per second. The faster the clock speed, the faster the computer can process information.

Expansion Slots and Cards

Expansion slots provide an open architecture. Expansion cards provide network connections, TV tuner cards, and more.

Computers are known for having different kinds of "architectures." Machines that have **closed architecture** are manufactured in such a way that users cannot easily add new devices. Most microcomputers have **open architecture.** They allow users to expand their systems by providing **slots** on the system board. Users can insert optional devices known as **expansion cards** into these slots. (See Figure 4-13.) Expansion cards are also called **plug-in boards, controller cards, adapter cards,** and **interface cards.**

Expansion cards plug into slots inside the system unit. Ports on the cards allow cables to be connected from the expansion cards to devices outside the system unit. Among the kinds of expansion cards available are:

- **Network adapter cards:** These cards are used to connect a computer to one or more other computers. This forms a communication network whereby users can share data, programs, and hardware. The **network adapter card** typically connects the system unit to a cable that connects to the other devices on the network. The network adapter card plugs into a slot inside the system unit.

- **Small computer system interface (SCSI—pronounced "scuzzy") card:** Most computers have only a limited number of expansion slots. A **SCSI card** uses only one slot and can connect as many as seven devices to the system unit. These cards are used to connect such devices as printers, hard-disk drives, and CD-ROM drives to the system unit.

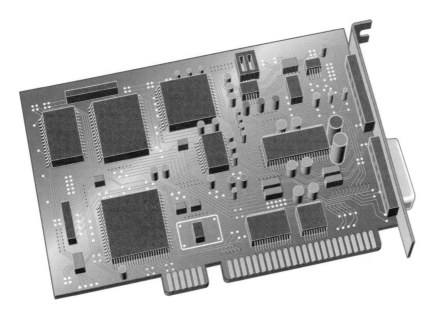

FIGURE 4-13
Expansion card

FIGURE 4-14
TV tuner card

- **TV tuner cards:** Now you can watch television and surf the Internet at the same time. **Television boards** (see Figure 4-14) contain a TV tuner and a video converter that changes the TV signal into one that can be displayed on your monitor. (See Making IT Work for You: TV Tuner Cards and Video Clips, shown below.)

- **PC cards:** To meet the size and constraints of portable computers, credit card–sized expansion boards have been developed. These cards can be easily inserted and replaced from the outside of a portable computer. They are called **PC cards** or **Personal Computer Memory Card International Association (PCMCIA) cards.** (See Figure 4-15.) These cards can be used for a variety of purposes, including increasing memory and connecting to other computers.

MAKING **IT** WORK FOR YOU

TV Tuner Cards and Video Clips

Want to watch your favorite television program while you work? Perhaps you would like to include a video clip from television and include it in a class presentation. It's easy using a video TV card.

TV Signal

(Analog)

Computer Signal

(Digital)

How It Works

A video capture card converts analog signals from a television or VCR into digital signals that your computer can process. Once the card has been installed, you can view, capture, and use television video clips in a variety of ways.

FIGURE A

Viewing

You can be running an application such as PowerPoint and view your favorite TV shows, by taking the steps shown in Figure A.

1. Click the TV icon on the desktop.

2. Size and move the television window and control box window.

3. Select the channel.

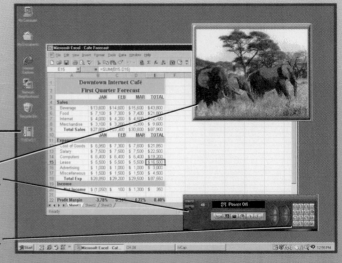

A wide variety of other expansion boards exist. Some of the most widely used are the following: Video adapter cards are used to adapt a variety of color video display monitors to a computer. CD-ROM cards connect optical disk drives (which we discuss in Chapter 6), and sound boards can record and play back digital sound.

FIGURE 4-15
A PC card

Capturing

You can capture the video playing in the TV window into a digital file by taking the steps shown in Figure B.

FIGURE B

2. Click the *Record* button to start recording.

1. Specify where to save the video clip (e.g., on your internal hard disk).

3. Click the *Stop* button to stop recording.

FIGURE C

Using

Once captured in a file, a video can be used in any number of ways. It can be added to a Web page, attached to an e-mail, or added to a class presentation.

For example, you could include a video clip into a PowerPoint presentation by taking the steps shown in Figure C.

TV tuner cards are relatively inexpensive and easy to install. Some factors limiting their performance on your computer are the speed of your processor, the amount of memory, and secondary storage capacity.

TV tuner cards are continually changing and some of the specifics presented in this Making IT Work for You may have changed. See our Web site at http://www.mhhe.com/it/oleary/IT.mhtml for possible changes and to learn more about this application of technology.

1. Insert the video clip into a page in the presentation.

2. Click the image anytime during your presentation to play the video.

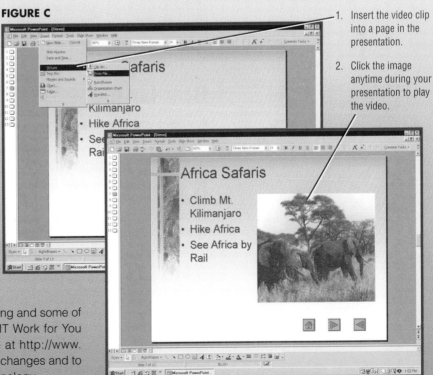

TIPS Having problems or want to upgrade your system and would like professional help? Here are a few suggestions.

1. *Select a reputable computer store.* Consider local as well as national chain stores. Check them out with the Better Business Bureau.

2. *Visit the store with your computer.* Ideally, have a knowledgeable friend accompany you. Describe the problem and get a written estimate. Ask about the company's warranty.

4. *Tag your system.* If you leave the system, attach a tag with your name, address, and telephone number.

5. *Pay by credit card.* If a dispute occurs, many credit card companies will intervene on your side.

To access the capabilities of an expansion board, the board must be inserted into a slot in the system unit and the system reconfigured to recognize the new board. Reconfiguration may require setting special switches on the expansion board and creating special configuration files. This can be a complex and difficult task. A recent development known as **Plug and Play** promises to eliminate this task.

Plug and Play is a set of hardware and software standards recently developed by Intel, Microsoft, and others. It is an effort by hardware and software vendors to create operating systems, processing units, and expansion boards, as well as other devices, that are able to configure themselves. Ideally, to install a new expansion board all you have to do is insert the board and turn on the computer. As the computer starts up, it will search for these Plug and Play devices and automatically configure the devices and the computer system.

Plug and Play is an evolving capability. A limited number of completely Plug and Play–ready systems exist today. However, observers predict that within the next few years this will become a widely adopted standard, and adding expansion boards will be a simple task.

CONCEPT CHECK

✔ What is the difference between a closed architecture and an open architecture?

✔ There are several kinds of expansion cards available. In general, these cards allow users to do what for their system?

✔ What role does a SCSI card play in expansion?

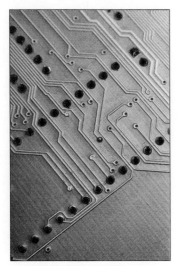

FIGURE 4-16
Bus lines

Bus Lines

Bus lines provide data pathways that connect various system components.

A bus line—also known as **data bus** or simply **bus**—connects the parts of the CPU to each other. It also links the CPU to various other components on the system board. A bus is a data roadway along which bits travel. (See Figure 4-16.) Such data pathways resemble a multilane highway. The more lanes there are, the faster traffic can go through. Similarly, the greater the capacity of a bus, the more powerful and faster the operation. A 64-bit bus has greater capacity than a 32-bit bus, for example.

Why should you even have to care about what a bus line is? The answer is that, as microprocessor chips have changed, so have bus lines. Many devices, such as expansion boards, will work with only one type of bus line.

A system unit has more than one type of bus line. The three principal bus lines (or "architectures") are:

- **Industry Standard Architecture (ISA)** was developed for the IBM Personal Computer. First it was an 8-bit-wide data path; then it was 16 bits wide. Although too slow for many of today's applications, the ISA bus is still widely used.

- **Peripheral Component Interconnect (PCI)** was originally developed to meet the video demands of graphical user interfaces. PCI is a high-speed 32-bit or 64-bit bus that is over 20 times faster than ISA buses. The PCI is expected to replace the ISA bus in the near future. PCI buses are widely used to connect the CPU, memory, and expansion boards.

- **Accelerated Graphics Port (AGP)** is the newest bus and over twice as fast as the PCI bus. While the PCI bus is used for a variety of purposes, the AGP bus is dedicated to the acceleration of graphics performance. Widely used for graphics and 3-D animations, the AGP is replacing the PCI bus for the transfer of video data.

CONCEPT CHECK

✔ What is a bus?

✔ What is the function of a bus line?

✔ List the three principal bus lines. Which is fastest?

Ports and Cables

Ports are connecting sockets. Cables connect input and output devices to ports.

A **port** is a connecting socket on the outside of the system unit. (See Figure 4-17.) Some ports, like the mouse and keyboard ports, are for specific devices. Others, like those listed below, can be used for a variety of different devices.

- **Serial ports** are used for a wide variety of purposes. They are used to connect a mouse, keyboard, modem, and many other devices to the system unit. Serial ports send data one bit at a time and are very good for sending information over a long distance.

- **Parallel ports** are used to connect external devices that need to send or receive a lot of data over a short distance. These ports typically send eight bits of data simultaneously across eight parallel wires. Parallel ports are mostly used to connect printers to the system unit.

FIGURE 4-17
Ports

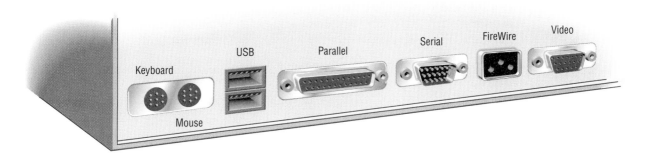

- **Accelerated graphics ports (AGPs)** are used to connect monitors.
  They are able to support high-speed graphics and other video input.
- **Universal serial bus (USB) ports** are gradually replacing serial and
  parallel ports. They are faster, and one USB port can be used to
  connect several devices to the system unit.
- **High perfomance serial bus (HPSB),** also know as **FireWire ports,** are
  the newest type. They are 33 times faster than USB ports and are used to
  connect high-speed printers and even video cameras to the system unit.

Cables are used to connect input and output devices to the system unit
via the ports. See Figure 4-18 for some of the most common types of cables.
For a summary of the major system unit components, see Figure 4-19.

CONCEPT CHECK

✔ There are several different kinds of ports. What is the general function
of a port?

✔ Cables connect _____ and _____ devices to the system unit
via the ports.

Cable	Port	Devices
	Mouse	Mouse
	Keyboard	Keyboard
	Serial	Mouse, modem, keyboard
	Parallel	Printer, CD-ROM drive, Zip drive
	AGP Video	Monitor
	USB	Modem, joy stick, scanner
	HPSB	Video camera

FIGURE 4-18
Cables

Component	Function
System board	Holds the various other system components
Microprocessor	Contains the CPU on a single chip
Memory	Holds programs and instructions
System clock	Controls the speed of computer operations
Expansion slots and boards	Connect to system board to allow expansion of system capabilities
Bus lines	Connect various internal system components
Ports	Connect outside devices to system unit

FIGURE 4-19
Major components of a system unit

A LOOK TO THE FUTURE

Intel's Mobil Module and Xircom's PC cards could lead the way to desktop computing power in notebook computers.

Wouldn't it be nice if you could conveniently carry one of today's most powerful microcomputers? One that was light, compact, and would run on batteries? You could take it to class, to a group project meeting, or even to the beach.

Of course that's what notebook computers are all about. Unfortunately, the power of today's notebook microprocessors lags behind their desktop counterparts. Additionally, notebooks are priced higher than desktops with comparable performance. But all that may change in the next few years.

Intel, a leader in microprocessor technology, has recently taken steps to reduce the performance lag between notebooks and desktops. One step is the recent development of the Intel Mobil Module. This module combines multiple functions including the microprocessor onto a single circuit board. It will allow notebook manufacturers to readily switch to newer processors as they become available. Another step is Intel's purchase of 12.5 percent of Xircom, a maker of PC cards used to connect notebooks to corporate networks.

Observers note that future Xircom products will likely encourage managers to replace their corporate desktop systems with notebooks.

Will your next computer be a notebook? There is a clear trend. Today more than one in four microcomputer buyers purchase a notebook computer and that number is expected to increase in the next few years.

VISUAL SUMMARY The System Unit

ELECTRONIC REPRESENTATION

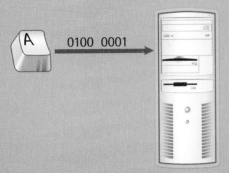

Data and instructions are represented **electronically** with a two-state **binary system** of numbers (0 and 1). Each 0 or 1 is called a **bit**. A **byte** consists of eight bits and represents one character.

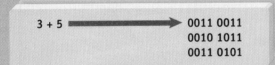

3 + 5 → 0011 0011
0010 1011
0011 0101

Binary Coding Schemes
Binary coding schemes convert binary data into characters. Three such schemes are:

- **ASCII**—the most widely used for microcomputers
- **EBCDIC**—developed by IBM and used primarily by large computers
- **Unicode**—16-bit code to support international languages like Chinese and Japanese.

SYSTEM BOARD

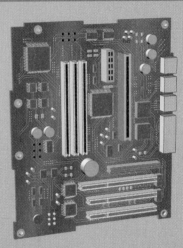

The **system board**, also known as the **main board** or the **motherboard**, connects all system components. It is a flat circuit board covered with sockets and other electronic parts, including a variety of chips.

A **chip**, also known as a **silicon chip, semiconductor,** or **integrated circuit,** is a postage-stamp-sized circuit board.

To be a competent end user, you need to understand how data and programs are represented electronically. Additionally, you need to understand the functionality of the basic components in the system unit: system board, microprocessor, memory, system clock, expansion slots and cards, bus lines, and ports and cables.

MICROPROCESSOR

Unit	Speed
Millisecond	Thousandth of a second
Microsecond	Millionth of a second
Nanosecond	Billionth of a second
Picosecond	Trillionth of a second

The **microprocessor** plugs into the system board, contains the **CPU,** and is the "brains" of the system unit. Two basic components of the microprocessor are the *control unit* and the *arithmetic-logic* unit.

Control Unit
The **control unit** executes programs by directing the other system components. It directs electronic signals between memory, the arithmetic-logic unit, and input/output devices.

Arithmetic-Logic Unit
The **arithmetic-logic unit,** commonly referred to as the **ALU,** performs arithmetic (math) and logical (comparisons) operations.

Microprocessor Chips
A **word** is the number of bits (such as 16, 32, or 64) that can be accessed by the microprocessor at one time. The more bits in a word, the more powerful the microprocessor.

Two types of microprocessor chips are:
- **Complex instruction set computer (CISC)** chips are the basis for Intel's Pentium II and Pentium III microprocessors.
- **Reduced instruction set computer (RISC)** chips use fewer instructions. They are the basis for IBM and Motorola's PowerPC microprocessor.

Smart cards contain built-in microprocessor chips.

MEMORY

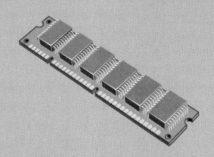

There are three types of **memory** chips: RAM, ROM, and CMOS.

RAM
RAM chips are called **temporary** or **volatile** because their contents are lost if power is disrupted.
- **Flash RAM** or **flash memory** is a special type of RAM that does not lose its contents when power is disrupted.
- **Virtual memory** uses the hard disk to run large programs on systems with limited memory.
- **Memory cache** or **RAM cache** is a high-speed holding area for frequently used data and information.

ROM
ROM chips are permanent and control essential system operations.

CMOS
CMOS chips provide flexibility and expandability to computer systems.

Unit	Capacity
Kilobyte (KB)	1,000 bytes
Megabyte (MB)	1 million bytes
Gigabyte (GB)	1 billion bytes
Terabyte (TB)	1 trillion bytes

SYSTEM CLOCK

The **system clock** controls the speed of computer operations. It is measured in **megahertz (MHz).**

EXPANSION SLOTS AND CARDS

Expansion slots and **cards** allow additional devices to be added to a computer system.

Expansion Slots
Expansion slots connect the system board to expansion cards.

Expansion Cards
Expansion cards provide network connections, SCSI connections, PC/TV or combined computer and TV operations, PC cards for expanding portable computer capabilities, and more.

    Plug and Play is an evolving set of hardware and software designed to assist with the installation of expansion cards.

BUS LINES

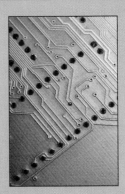

Bus lines provide data pathways that connect various system components. Three principal types are:
- **ISA**—older and slower but still widely used.
- **PCI**—high-speed; used to connect CPU, memory, and expansion boards.
- **AGP**—fastest; used for video data.

PORTS AND CABLES

Ports and **cables** allow external devices to connect to the system unit.

Ports
Ports are connecting sockets on the outside of the system unit. They are used to connect keyboards, mouse, monitors, modems, and printers. The five most common types are **serial, parallel, AGP, USB,** and **HPSB.**

Cables
Cables are used to connect external devices to the system unit via ports.

Key Terms

accelerated graphics port (AGP) (87)

adapter card (83)

arithmetic-logic unit (ALU) (79)

ASCII (76)

binary system (76)

bit (76)

bus (86)

bus line (86)

byte (76)

cable (87)

cache memory (82)

central processing unit (CPU) (79)

chip (78)

closed architecture (83)

complementary metal-oxide semiconductor (CMOS) (82)

complex instruction set computer (CISC) (79)

control unit (79)

controller card (83)

data bus (86)

EBCDIC (76)

expansion card (83)

FireWire ports (87)

firmware (82)

flash memory (81)

flash RAM (81)

High Performance Serial Bus (HPSB) (88)

Industry Standard Architecture (ISA) (86)

integrated circuit (78)

interface card (83)

main board (78)

megahertz (MHz) (83)

microseconds (79)

motherboard (78)

network adapter card (83)

open architecture (83)

parallel port (87)

PC card (84)

Peripheral Component Interconnect (PCI) (87)

Personal Computer Memory Card International Association (PCMCIA) card (84)

Plug and Play (86)

plug-in board (83)

port (87)

processor (79)

RAM cache (82)

random-access memory (RAM) (80)

read-only memory (ROM) (82)

reduced instruction set computer (RISC) (82)

semiconductor (78)

serial port (87)

silicon chip (78)

slots (83)

small computer system interface (SCSI) card (83)

smart card (80)

system board (78)

system clock (83)

television board (84)

Unicode (76)

universal serial bus (USB) port (88)

virtual memory (81)

word (79)

Chapter Review

LEVEL 1

Reviewing Facts and Terms
Matching

Match each numbered item with the most closely related lettered item. Write your answers in the spaces provided.

1. This system consists of only two digits—0 and 1. _C_

2. This is the number of bits that can be accessed at one time by the CPU. _√_

3. This feature allows your computer to run a large program even if there is not enough RAM to hold the program. _X_

4. Cards with built-in, specialized microprocessors that can hold information such as health insurance or frequent flier records. _w_

5. A CPU can only read, or retrieve, programs written on this type of memory chip. _√_

6. With reduced instruction sets, these chips are used in many of today's most powerful microcomputers. _u_

7. This type of memory is volatile, and all data stored in this fashion are lost when the computer is turned off. _√_

8. Credit card–sized expansion boards that can be used for a variety of functions, including increasing memory and connecting to other computers. _5_

a. AGP
b. ALU
c. Binary system
d. Bus
e. Byte
f. Cache memory
g. CISC
h. Closed architecture
i. CMOS
j. Control unit
k. CPU
l. EBCDIC
m. Expansion card
n. Flash RAM
o. Megahertz (MHz)

9. Used to connect external devices that need to send or receive a lot of data over a short distance. _R_

10. Most microcomputers have this kind of design that allows users to expand their system via slots on the system board. _M_

11. Another name for the system board—the part that connects all system components and allows input and output devices to communicate with the system unit. _P_

12. The speed of operations within a computer is measured in these units. _O_

13. Type of memory that can retain data even if power is disrupted. _N_

14. These plug into slots inside the system unit and allow the user to expand the computer's capabilities. _I_

15. Binary coding scheme used primarily for large computers. _L_

16. The "brains" of the system, it contains the control unit and the arithmetic-logic unit. _K_

17. Tells the rest of the computer how to carry out a program's instructions. _J_

18. Contains essential information that is required every time the computer system is turned on. _____

19. Computers with this design are manufactured in such a way that users cannot easily add new devices. _H_

20. The most common type of microprocessor. _G_

21. Memory used to store the most frequently accessed information stored in RAM. _F_

22. A group of eight bits. _E_

23. A data roadway between various system components along which bits travel. _D_

24. The newest bus, more than twice as fast as the PCI bus, dedicated to the acceleration of graphics performance. _A_

25. Performs two types of operations—arithmetic and logical. _B_

p. Motherboard
q. Open architecture
r. Parallel port
s. PCMCIA
t. RAM
u. RISC
v. ROM
w. Smart card
x. Virtual memory
y. Word

True/False

In the space provided, write T or F to indicate whether the statement is true or false.

1. In a microcomputer, the CPU is located on a single chip called the microprocessor. _A_

2. A grouping of eight bytes is called a bit. _F_

3. Another name for the system board is the processor board. _F_

4. CISC chips use fewer instructions than RISC chips. _F_

5. A SCSI card can be used to connect several different devices to the system unit. _T_

6. ROM is volatile memory that cannot be changed by the user. _F_

Multiple Choice

Circle the letter of the correct answer.

1. The ALU performs arithmetic operations and:
   - **a.** stores data
   - **b.** logical operations
   - **c.** binary calculations
   - **d.** reduced instruction calculations
   - **e.** parity checks

2. The binary code that is widely used with microcomputers is:
   a. ASCII
   b. EBCDIC
   c. BCD
   d. DEC/MVS
   e. Unicode

3. The number of bits that can be processed at one time is a:
   a. register
   b. cycle
   c. byte
   d. word
   e. PROM

4. Three types of memory chips are RAM, ROM, and:
   a. RISC
   b. main
   c. MCA
   d. CD-ROM
   e. CMOS

5. The bus originally designed to meet video demands is:
   a. EISA
   b. ISA
   c. PCI
   d. MCA
   e. PCMCIA

Completion

Complete each statement in the spaces provided.

1. Bit is short for _Binary digit_
2. The most common type of microprocessor chip is _CISC_.
3. _RAM_ memory allows computers with limited memory to run large programs.
4. _ROM_ memory chips retain data even if the power is disrupted.
5. Data and instructions move along the system board using _BUS_ lines.

Reviewing Concepts

Open-Ended

On a separate sheet of paper, respond to each question and statement.

1. Describe two basic components of the CPU.
2. What are the four commonly used units of measurements to describe memory capacity? Define each.
3. What is the difference between RISC and CISC chips?
4. Name four expansion cards and describe the function of each.
5. Name the three most common bus lines and describe their differences.
6. Name and describe describe five ports.

Concept Mapping

On a separate sheet of paper, draw a concept map or a flowchart showing how the following terms are related. Show all relationships. Include any additional terms you can think of.

arithmetic-logic unit	CPU	RAM
ASCII	EBCDIC	RAM cache
binary system	expansion card	RISC
bus	flash RAM	ROM
byte	ISA	SCSI
cache memory	motherboard	semiconductor
CISC	open architecture	serial port
closed architecture	parallel port	system board
CMOS	PC card	system clock
control unit	PCMCIA	virtual memory

Critical Thinking Questions and Projects

LEVEL 3

Read each exercise and answer the related questions on a separate sheet of paper.

1. *An inexpensive microcomputer system:* The prices of microcomputer systems have been decreasing for some time. You can buy them from computer stores, pay to have one built for you, or purchase them via mail order or the Internet. Look through print and Web sources to find three low-priced systems. Read each advertisement, make a list of all the computer terms used, and write down their definitions. Prepare a table comparing the three systems. If you were to purchase one of these systems, which would you select, and why?

2. *The mobile office:* The mobile office does not have a fixed location, but like the traditional office it has support facilities such as computers, printers, and fax and copy machines. The central piece of equipment in a mobile office is the notebook computer. Since many jobs now require frequent travel, a functional mobile office is essential.

   Contact a major hotel chain to determine its support for mobile computing. Specifically, find out the following:

   a. Are in-room microcomputers and fax machines available? What are the charges?

   b. Is there a business center on site? What support is provided? What are the hours? What are the charges?

   c. If a business center is not available, will the front desk provide basic services such as faxing, copying, and printing? What are the charges?

3. *Security:* Expensive equipment has a way of getting lost. Most organizations report that theft of their computer equipment is an ever-increasing problem. Police report that one of the most common items taken in home burglaries is computer hardware. People who use and travel with notebook computers need to be particularly careful.

   a. What security measures are in place at your school or business to ensure against theft of keyboards, mice, printers, and entire systems? Can you suggest additional safeguards?

   b. Do you have any insurance policies that cover theft of a computer system from your home, your dorm, or your car? How much coverage, if any, do you have?

   c. In the recent past, the FBI issued a warning to travelers who use notebook computers. The warning described a popular scam that involves a team of three members and takes place at typical airport security checkpoints with a conveyor belt and a walk-through metal detector. Can you figure out how this scam might work and what you can do to avoid being a victim?

4. *TV Tuner Cards and Video Clips:* TV tuners convert traditional analog TV signals into digital signals that can be captured and edited by computers. (See Making IT Work for You: TV Tuner Cards and Video Clips on pages 84–85.)

a. Discuss the relative advantages and disadvantages of using a computer rather than a standard television set to access television programming. How might you use this technology? Provide specific examples.

b. It is projected that within the next five years, all television broadcasts will be digital. Will that make today's traditional analog television sets obsolete? Will that make today's TV tuner cards obsolete? Discuss the potential impact of digital broadcasting on the television and the computer industries.

On the Web Exercises

1. Microprocessors

The Intel Corporation is one of the leaders in microprocessor design. To learn more about their most recent microprocessors, visit our Web site at http://www.

mhhe.com/it/oleary/exercise.mhtml to link to Intel's Web site. Once connected to that site, select two current microprocessors. Print out the pages that best describe each microprocessor. Write a paragraph comparing the two.

2. Universal Serial Bus

The Universal serial bus (USB) is the next evolution of computer ports. It was developed through the collaboration of seven leading-edge companies. To learn more about this technology, visit the Yahoo site at http://www.yahoo.com

and search using the keywords "USB" and "universal serial bus". Follow several of the links to learn more and print the Web page you find the most informative. Write a brief summary of what you learned about USB.

3. Desktops and Notebooks

There are several sites on the Web from which you can learn more about the newest desktops and notebooks on the market today. Visit our Web site at http://www.

mhhe.com/it/oleary/exercise.mhtml to link to one of these sites. Once connected to that site, check out the different models and select the one that best meets your needs. Print out the specifications for your chosen computer and write a paragraph describing how this computer best suits your needs.

4. TV Tuner Cards and Video Clips

Video capture technology is used to transform traditional video images into digital files that can be played and modified on a computer. (See Making

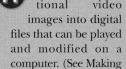

IT Work for You: TV Tuner Cards and Video Clips on pages 84–85.) This technology allows distribution of movies and movie clips over the Internet. Visit our Web site at http://www. mhhe.com/it/oleary/exercise.mhtml to link to a site that provides digital video. Once connected to that site, print out the first page of the site. Choose a movie or movie clip that looks interesting to you and watch it. Write a paragraph describing the video and the quality of your viewing experience. How did it compare to watching a video on television?

CHAPTER 5

Input and Output

COMPETENCIES

After you have read this chapter, you should be able to:

1. Explain the difference between keyboard and direct-entry input devices.

2. Describe the features of keyboards and the four types of terminals.

3. Describe direct-entry devices used with microcomputers.

4. Discuss voice recognition systems.

5. Describe monitors and monitor standards.

6. Describe printers (ink-jet, laser, thermal) and plotters (pen, ink-jet, electrostatic, and direct-imaging).

7. Describe voice-output devices.

How do you get data to the CPU? How do you get information out? Here we describe the two most important places where the computer interfaces with people. The first half of the chapter covers input devices; the second half covers output devices.

People understand language, which is constructed of letters, numbers, and punctuation marks. However, computers can understand only the binary machine language of 0s and 1s. Input and output devices are essentially translators. Input devices translate symbols that people understand into

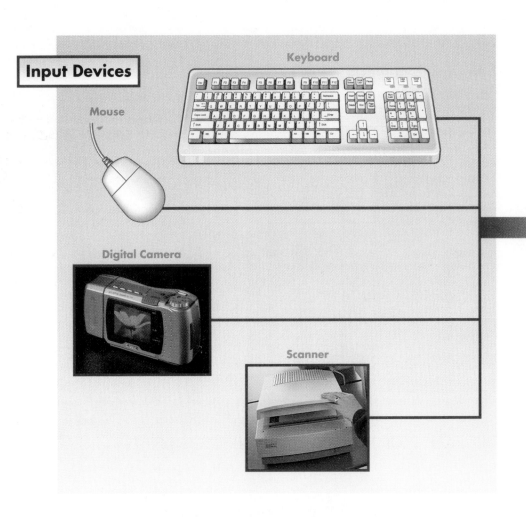

Input Devices

Mouse

Keyboard

Digital Camera

Scanner

symbols that computers can process. Output devices do the reverse: They translate machine output to output people can comprehend.

How would you like talking to your computer and having it respond to you? You can! You can create e-mail, dictate term papers, and control computer operations with the appropriate input and output devices.

Competent end users need to know about the most commonly used input devices, including keyboards, mice, digital cameras, scanners, and voice recognition devices. Additionally, they need to know about the most commonly used output devices, including monitors, printers, plotters, and speakers.

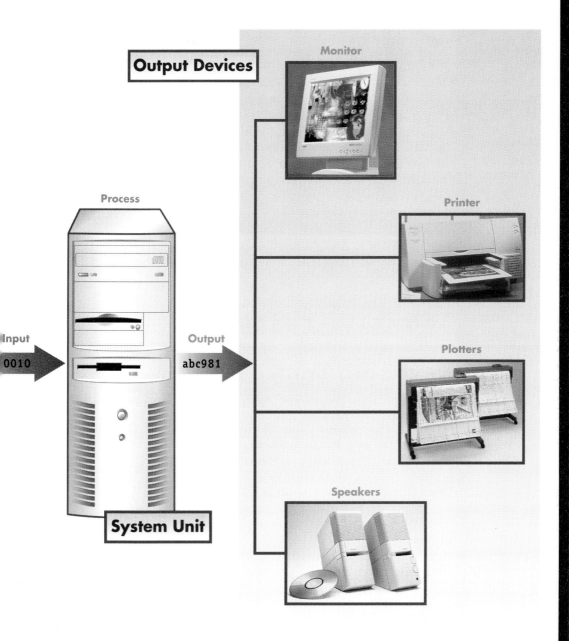

Output Devices

Monitor

Printer

Plotters

Speakers

Process

Input

0010

Output

abc981

System Unit

Input: Keyboard versus Direct Entry

Input devices convert people-readable data into machine-readable form. Input may be by keyboard or direct entry.

Input devices take data and programs people can read or understand and convert them to a form the computer can process. This form consists of machine-readable electronic signals of 0s and 1s that we described in Chapter 4. Input devices are of two kinds: keyboard entry and direct entry.

FIGURE 5-1
A point-of-sale terminal

- **Keyboard entry:** Data is input to the computer through a *keyboard* that looks like a typewriter keyboard but has additional keys. In this method, the user typically reads from an original document called the **source document.** The user enters that document by typing on the keyboard.
- **Direct entry:** Data is made into machine-readable form as it is entered into the computer; no keyboard is used.

An example of an input device that uses both keyboard and direct entry is a **point-of-sale (POS) terminal.** This is the sort of electronic cash register you see in many stores. (See Figure 5-1.) When clerks sell a can of paint, for example, they can record the sale by typing in the information (product code, purchase amount, tax) on the keyboard. Or they can use a handheld **wand reader** or **platform scanner** to read special characters on price tags as direct entry. The wand reflects light on the characters. The reflection is then changed by photoelectric cells to machine-readable code. Whether by keyboard entry or direct entry, the results will appear on the POS terminal's digital display.

CONCEPT CHECK

✔ What are input devices?

✔ How does keyboard entry differ from direct entry?

Keyboard Entry

In keyboard entry, people type input. There are four types of terminals: dumb, intelligent, network, and Internet.

Probably the most common way in which you will input data, at least at the beginning, is by using a keyboard. Keyboards convert numbers, letters, and special characters that people understand into electrical signals. These signals are sent to and processed by the system unit.

Keyboards

A computer keyboard combines a typewriter keyboard with a numeric keypad. Additionally, it has many special-purpose keys. Some keys such as the *Caps Lock* key are **toggle keys.** These keys turn a feature on or off. Others such as the *Ctrl* key are **combination keys** that perform an action when held down in combination with another key. To learn more about keyboard features, see Figure 5-2.

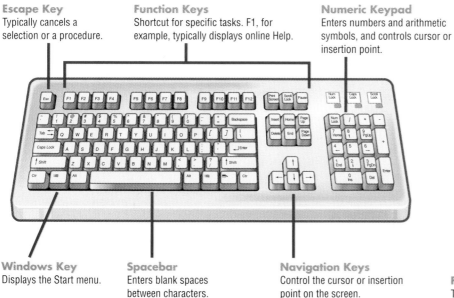

Escape Key
Typically cancels a selection or a procedure.

Function Keys
Shortcut for specific tasks. F1, for example, typically displays online Help.

Numeric Keypad
Enters numbers and arithmetic symbols, and controls cursor or insertion point.

Windows Key
Displays the Start menu.

Spacebar
Enters blank spaces between characters.

Navigation Keys
Control the cursor or insertion point on the screen.

FIGURE 5-2
Traditional keyboard

There are two basic keyboard designs. The traditional design or straight design is shown in Figure 5-2. The contour design splits and slopes the keyboard. (See Figure 5-3.) Many people prefer the contour design because it is more natural and comfortable.

Terminals

A **terminal** is a form of input (and output) device that connects you to a mainframe or other type of computer called a *host computer* or *server*. There are four kinds of terminals:

FIGURE 5-3
Microsoft's Natural keyboard

- A **dumb terminal** can be used to input and receive data, but it cannot process data independently. It is used only to gain access to information from a computer. Such a terminal may be used by an airline reservations clerk to access a mainframe computer for flight information.

- An **intelligent terminal** includes a processing unit, memory, and secondary storage such as a magnetic disk. Essentially, an intelligent terminal is a microcomputer with communications software and a telephone hookup (modem) or other communications link. These connect the terminal to the larger computer or to the Internet. An increasingly popular type is the **Net PC**, also known as the **Net Personal Computer.** These low-cost and limited microcomputers typically have only one type of secondary storage (an internal hard disk drive), a sealed system unit, and no expansion slots.

- A **network terminal,** also known as a **thin client** or **network computer,** is a low-cost alternative to an intelligent terminal. Most network terminals do not have a hard-disk drive and must rely on a host computer or server for application and system software. These devices are becoming increasingly popular in many organizations.

TIPS Is your keyboard looking tired and dirty? Are the keys sticking? Then it may be time to clean your keyboard. Follow the guidelines below to make it shine again.

1. *Disconnect.* Turn off the computer and disconnect the keyboard.
2. *Vacuum.* Use a vacuum or a blow-dryer to remove dust and particles from the keyboard.
3. *Clean keys.* Using a soft cloth moistened with 90 percent isopropyl alcohol, gently clean the key surfaces. Carefully dry with another soft cloth.
4. *Clean case.* Using a soft cloth moistened with liquid cleaning solution, clean the keyboard case. Dry with another soft cloth.
5. *Reconnect.* When the keys and case are completely dry, plug in the keyboard and turn on the computer.

FIGURE 5-4
Internet terminal

- An **Internet terminal,** also known as a **Web terminal** or **Web appliance,** provides access to the Internet and typically displays Web pages on a standard television set. (See Figure 5-4.) These special-purpose terminals offer Internet access without a microcomputer. Unlike the other types of terminals, Internet terminals are used almost exclusively in the home.

CONCEPT CHECK

✔ Keyboards convert numbers, letters, and special characters into _____ _____ to be processed by the system unit.

✔ List the four types of keyboard entry terminals.

Direct Entry

Direct entry creates machine-readable data that can go directly to the CPU. Direct entry includes pointing, scanning, and voice-input devices.

Direct entry is a form of input that does not require data to be keyed by someone sitting at a keyboard. Direct-entry devices create machine-readable data on paper or magnetic media, or feed it directly into the computer's CPU. This reduces the possibility of human error being introduced (as often happens when data is being entered through a keyboard). It is also an economical means of data entry. Direct-entry devices may be categorized in three ways: pointing devices, scanning devices, and voice-input devices.

Pointing Devices

Pointing, of course, is one of the most natural of all human gestures. There are a number of devices that use this method as a form of direct-entry input:

- **Mouse:** A **mouse** controls a **pointer** that is displayed on the monitor. The pointer usually appears in the shape of an arrow. It frequently, however, changes shape depending on the application. The standard mouse has a ball on the bottom and is attached with a cord to the system unit. (See Figure 5-5.) The standard mouse has evolved over time with many predecessors. (See Figure 5-6.)

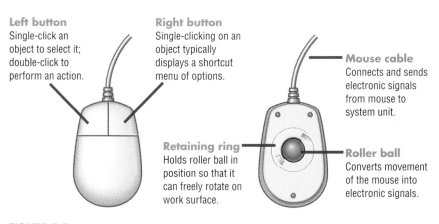

Left button
Single-click an object to select it; double-click to perform an action.

Right button
Single-clicking on an object typically displays a shortcut menu of options.

Mouse cable
Connects and sends electronic signals from mouse to system unit.

Retaining ring
Holds roller ball in position so that it can freely rotate on work surface.

Roller ball
Converts movement of the mouse into electronic signals.

FIGURE 5-5
Standard mouse

FIGURE 5-6
The evolving mouse

Three devices similar to a mouse are trackballs, touch surfaces, and pointing sticks. You can use the **trackball,** also known as the **roller ball,** to control the pointer by rotating a ball with your thumb. You can use **touch surfaces** to control the pointer by moving and tapping your finger on the surface of a pad. (See Figure 5-7.) You can use a **pointing stick,** located in the middle of the keyboard, to control the pointer by directing the stick with your finger. (See Figure 5-8.)

- **Joystick:** The **joystick** is the most popular input device for computer games. You control game actions by varying the pressure, speed, and direction of the joystick. Additional controls such as buttons and triggers are used to specify commands or initiate specific actions. (See Figure 5-9.)

- **Touch screen:** A **touch screen** is a particular kind of monitor screen covered with a plastic layer. (See Figure 5-10.) Behind this layer are crisscrossed invisible beams of infrared light. This arrangement enables someone to select actions or commands by touching the screen with a finger. Touch screens are easy to use, especially when people need information quickly. They are commonly used at automated teller machines (ATMs), information centers, restaurants, and department stores.

touch surface

FIGURE 5-7
A touch surface: typically part of a portable computer

pointing stick

FIGURE 5-8
A pointing stick: typically part of a portable computer

FIGURE 5-9
A joystick: used for computer games

FIGURE 5-10
A touch screen: a consumer application

TIPS

Does your mouse pointer move smoothly across your screen? Or does it jerk, start, and stop occasionally? If it does, then it may be time for a cleaning. To clean a standard mouse, follow these steps:

1. *Disconnect.* Turn off the computer and disconnect the mouse from the computer.

2. *Take out ball.* Turn the mouse upside down, gently unscrew the cover to the compartment containing the mouse ball with your thumbs, and allow the ball to drop into your hand.

3. *Clean ball.* Clean the ball with a mild detergent or alcohol and a soft cloth; then dry it with a clean, lint-free cloth.

4. *Clean compartment.* Blow into the ball compartment to remove any dust or lint from the inside. Remove any oil on the rollers in the compartment by cleaning with a cotton swab and tape-head cleaner.

5. *Reassemble.* Put the ball back into the compartment and replace the cover.

6. *Reconnect.* Connect the mouse and you're ready to go.

If you do not have a standard mouse, consult the manufacturer's Web page for cleaning instructions.

- **Light pen:** A **light pen** is a light-sensitive penlike device. (See Figure 5-11.) The light pen is placed against the monitor. This closes a photoelectric circuit and identifies the spot for entering or modifying data. Light pens are used by engineers, for example, in designing anything from microprocessor chips to airplane parts.

- **Digitizer:** A **digitizer** is a device that can be used to trace or copy a drawing or photograph. The item to be copied is placed on a flat **digitizing tablet.** A special **stylus** connected to a computer is then used to trace the item. (See Figure 5-12.) As the stylus moves from one position to another, the computer records its position on the digitizing tablet. After the item has been traced, its image can be displayed on the screen, printed out on paper, or stored on the computer system for later use. Digitizers are often used by designers, architects, and engineers.

- **Digital camera:** **Digital cameras** are similar to traditional cameras except that images are recorded digitally on a disk or in the camera's memory rather than on film. (See Figure 5-13.) You can take a picture, view it immediately, and even place it on your own Web page, within minutes. Although faster than traditional photography, digital camera prices are typically higher, and image quality is not quite as good. Digital cameras are particularly popular with real estate agents, who use them to capture images of homes to be displayed on their Web pages.

FIGURE 5-11
A light pen: a hospital application

FIGURE 5-12
A digitizer: an engineering application

stylus

FIGURE 5-13
A digital camera and a digitial video camera

- **Digital video cameras:** Unlike traditional video cameras, **digital video cameras** record motion digitally on a disk or in the camera's memory. Compared to traditional video cameras, the image quality is better and the price is higher. **Webcams** are specialized digital video cameras that capture images and send them to a computer for broadcast over the Internet.

- **Digital notebook:** Taking class notes may never be the same again. **Digital notebooks** use a regular notepad on top of an electronic pad. (See Figure 5-14.) Whenever you move a special pen across the notepad, a signal tracing your movements is sent to and stored in the electronic pad. These notebooks can store up to 50 pages of notes before they reach their capacity. The notes are transferred to a microcomputer where they can be viewed, organized, and edited as needed.

Scanning Devices

Direct-entry scanning devices record images of text, drawings, or special symbols. The images are converted to digital data that can be processed by a computer or displayed on a monitor. Scanning devices include the following:

- **Image scanner:** An **image scanner,** also known simply as a **scanner,** copies or reproduces images. There are two basic types. One, the

FIGURE 5-14
A digital notebook: a student application

FIGURE 5-15
Portable scanner: capturing text and graphics

flatbed scanner, is much like a copy machine. (See Figure 5-15.) The image to be scanned is placed on a glass surface and the scanner records the image from below. The other type, the **portable scanner,** is typically a handheld device that slides across the image, making direct contact. (See Figure 5-15.) A **pen scanner** is a highly portable device for capturing text. (See Figure 5-16.) With these devices, the light and dark areas as well as color of the image are recorded as points or dots. After the image has been scanned, it can be displayed, printed on paper, and stored for later use. Image scanners are commonly used in desktop publishing to scan graphic images and in research to scan text.

- **Fax machine: Facsimile transmission machines,** commonly called **fax machines,** scan the image of a document to be sent. The light and dark areas of the image are converted into a format that can be sent

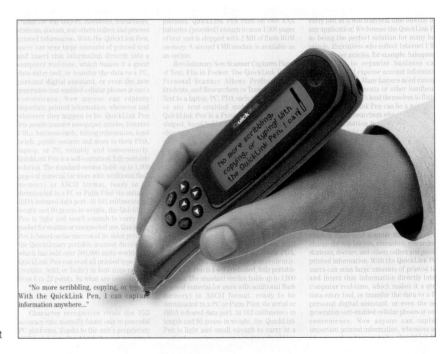

FIGURE 5-16
Pen scanner: capturing text

electronically over telephone lines. The receiving fax machine converts the signals back to an image and recreates it on paper.

Although dedicated fax machines are popular, many people use their microcomputers with a **fax/modem board** that provides the independent capabilities of a fax and a modem.

- **Bar-code readers:** You are probably familiar with **bar-code readers** from grocery stores. (See Figure 5-17.) Bar-code readers are photo-electric scanners that read the **bar codes,** or vertical zebra-striped marks, printed on product containers. Supermarkets use a bar-code system called the Universal Product Code (UPC). The bar code identifies the product to the supermarket's computer, which has a description and the latest price for the product. The computer automatically tells the POS terminal what the price is. And it prints the price and the product name on the customer's receipt.

FIGURE 5-17
A bar-code reader: recording product codes

- **Character and mark recognition devices:** There are three kinds of scanning devices—formerly used only with mainframes—now found in connection with the more powerful microcomputers.

  Magnetic-ink character recognition (MICR) is a direct-entry method used in banks. This technology is used to automatically read those futuristic-looking numbers on the bottom of checks. A special-purpose machine known as a **reader/sorter** reads characters made of ink containing magnetized particles.

  Optical-character recognition (OCR) uses special preprinted characters, such as those printed on utility and telephone bills. They can be read by a light source and changed into machine-readable code. A common OCR device is the handheld *wand reader* discussed earlier in this chapter. (See Figure 5-18.) These are used in department stores to read retail price tags by reflecting light on the printed characters.

  Optical-mark recognition (OMR) is also called **mark sensing.** An OMR device senses the presence or absence of a mark, such as a pencil mark. OMR is often used to score multiple-choice tests such as the College Board's Scholastic Aptitude Test and the Graduate Record Examination.

FIGURE 5-18
A wand reader: recording product codes

Voice-Input Devices

Voice-input devices convert a person's speech into a digital code. By far the most widely used voice-input device is the microphone. This input device, when combined with a sound card and appropriate software, forms a **voice recognition system.** (See Making IT Work for You: Voice Recognition Systems and Dictating a Paper shown below.) These systems enable users to operate microcomputers and to create documents using voice commands. (See Figure 5-19.) Recently, portable voice recognition systems have been introduced. (See Figure 5-20.) These units typically are able to record for up to one hour. They can also connect to a computer system through the system unit's serial port to save files and print documents.

Most voice recognition systems must be "trained" to the particular user's voice. This is done by matching the user's spoken words to patterns previously stored in the computer. More advanced systems that can recognize the same

MAKING IT WORK FOR YOU

Voice Recognition Systems and Dictating a Paper

Tired of using your keyboard to type term papers and to control programs? Perhaps voice recognition is just what you're looking for. It's easy with the right software and hardware.

How It Works

A voice recognition system consists of a combination of standard hardware components and some specialized software like NaturallySpeaking from Dragon Systems, Inc. Once the software has been trained, you can dictate papers and surf the Web using only voice commands.

FIGURE A

Training the Software

The first step is to train your system to recognize your voice. For example, you can train NaturallySpeaking as shown in Figure A.

1. Create your personal user profile, which will store your speech patterns.

2. Read prearranged standard text to teach the software your unique speech patterns.

3. Refine your profile to include specialized vocabulary.

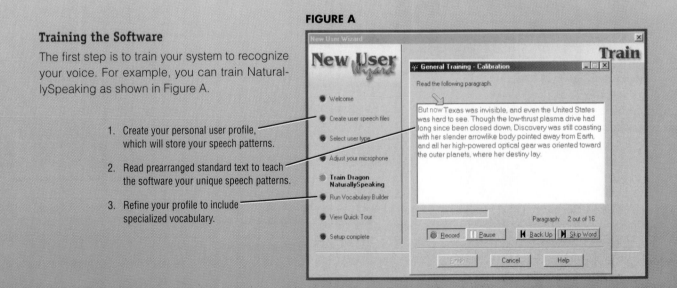

Dictating

Once your system has been trained, you can use it to interact with a variety of applications.

For example, you can create a text document using voice recognition software and Word. As you speak into the microphone, your words will appear on your monitor and in the word processing document. (See Figure B.)

FIGURE B

"...combine many different outdoor activities while exploring the world."

1. Using a microphone, dictate the text you want to appear in the Word document.

2. As you speak, the voice recognition system interprets your spoken words.

3. Once interpreted, text is inserted into the Word document.

Surfing

You can control many computer operations with just your voice. You can load, run, and interact with programs without using your mouse or keyboard.

For example, you could connect to the Web through your browser, enter Web addresses, and use hyperlinks as shown in Figure C.

1. From the desktop, instruct your computer to load your browser.

"Start Internet Explorer."

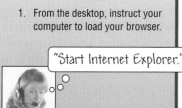

2. Instruct your computer to go to the location box, dictate the Web address and instruct your computer to connect to that Web address.

"Go to address www.yahoo.com. Go there."

3. Surf by selecting and connecting to hypertext links.

"Click Auctions."

FIGURE C

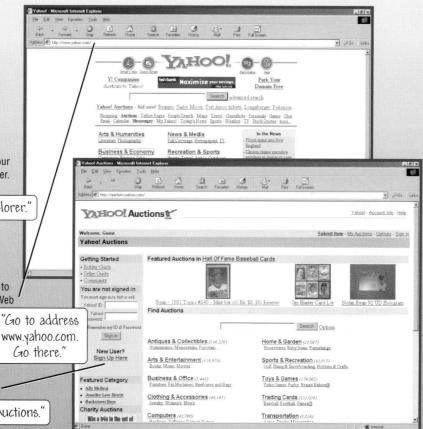

Many voice recognition programs are reasonably priced and relatively easy to install. For effective utilization, however, you must carefully train your system to recognize your speech patterns and vocabulary.

Voice dictation systems are continually changing and some of the specifics presented in this Making IT Work for You may have changed. See our Web site at http://www.mhhe.com/it/oleary/IT.mhtml for possible changes and to learn more about this application of technology.

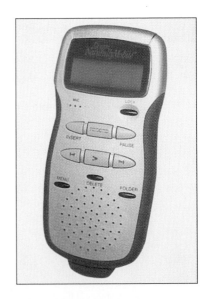

FIGURE 5-20
Portable voice recognition system

FIGURE 5-19
Voice recognition system

word spoken by many different people have been developed. However, until recently the list of words has been limited. One voice recognition system, the Dragon Dictate, identifies over 30,000 words and adapts to individual voices. There are even systems that will translate from one language to another, such as from English to Japanese.

There are two types of voice recognition systems:

- **Discrete speech:** A common activity in business is preparing memos and other written documents. **Discrete-speech recognition systems** have been widely used in the legal, medical, and other professions to directly convert the spoken word into printed material. Users dictate into a microphone that is connected to a microcomputer. The microcomputer takes the audio signal, converts it to a digital signal, and then analyzes it using a special program. The program recognizes individual words and stores them in a file. Using a standard word processing program, the file can be retrieved, edited, and printed out.

- **Continuous speech: Continuous-speech recognition systems** are considered to be one of the key technologies for the 21st century. They are more natural and conversational than discrete word systems. Continuous-speech systems are able to recognize individual words and phrases in context. For example, continuous-speech recognition systems are able to distinguish between same-sounding words like *there, their,* and *they're*. These systems are able to accept dictation as well as spoken commands to operate many of today's applications including Word and Excel. Two well-known systems are NaturallySpeaking from Dragon Systems and ViaVoice from IBM.

ON THE WEB
EXPLORATIONS

Dragon Systems Inc. is a leader in developing continuous-speech systems. To learn more about the company, visit our Web site at
http://www.mhhe.com/it/oleary/ explore.mhtml

CONCEPT CHECK

✔ Define direct entry.

✔ Three devices similar to a mouse are _____, _____, and _____.

✔ What are two basic types of image scanners?

✔ What do voice-input devices do?

Output: Monitors, Printers, Plotters, Voice

Output devices convert machine-readable information into people-readable form.

Data that is input to and then processed by the computer remains in machine-readable form until output devices make it people-readable. The output devices we shall describe for microcomputers are monitors, printers, plotters, and voice-output.

Monitors

Monitor standards indicate screen quality. Some monitors are used on the desktop; others are portable.

The most frequently used output device is the **monitor.** Two important characteristics of monitors are *size* and *clarity*. A monitor's size is indicated by the diagonal length of its viewing area. (See Figure 5-21.) Common sizes are 15, 17, 19, and 21 inches. Larger monitors have the advantage of displaying more information at one time; however, they are more expensive.

A monitor's clarity is indicated by its **resolution,** which is measured in pixels. **Pixels** are individual dots or "picture elements" that form images on a monitor. For a given size monitor, the greater the resolution (the more pixels), the better the clarity of the image. For a given level of clarity, larger monitors require a higher resolution (more pixels).

Standards

To indicate a monitor's resolution capabilities, several standards have evolved. The four most common today are SVGA, XGA, SXGA, and UXGA. (See Figure 5-22.)

- **SVGA** stands for *Super Video Graphics Array*. It has a minimum resolution of 800 by 600 pixels. A few years ago, SVGA was the most popular standard. Today, it is used primarily with 15-inch monitors.
- **XGA** stands for *Extended Graphics Array*. It has a resolution of up to 1,024 by 768 pixels. It is a popular standard today, especially with 17- and 19-inch monitors.
- **SXGA** stand for *Super Extended Graphics Array*. This standard has a resolution of 1,280 by 1,024 pixels. It is popular with 19- and 21-inch monitors.

FIGURE 5-21
A 17-inch monitor

Standard	Pixels
SVGA	800 × 600
XGA	1,024 × 768
SXGA	1,280 × 1,024
UXGA	1,600 × 1,200

FIGURE 5-22
Resolution standards

- **UXGA** stands for *Ultra Extended Graphics Array.* UXGA is the newest and highest standard. Although not as widely used as the XGA and SXGA monitors, its popularity is expected to increase dramatically as 21-inch monitors become more widely used. UXGA monitors are primarily used for high-end engineering design and graphic arts.

Cathode-Ray Tubes

The most common type of monitor for the office and the home is the **cathode-ray tube (CRT).** (See Figure 5-23.) These monitors are typically placed directly on the system unit or on the desktop. CRTs are similar in size and technology to televisions. Compared to other types of monitors, their primary advantages are low cost and excellent resolution. Their primary disadvantage is size.

Flat-Panel Monitors

ON THE WEB EXPLORATIONS

NEC Technologies is a leader in developing flat-panel monitors. To learn more about the company, visit our Web site at

http://www.mhhe.com/it/oleary/ explore.mhtml

Because CRTs are too bulky to be transported, portable monitors (see Figure 5-24) known as **flat-panel monitors** or **liquid crystal display (LCD) monitors** were developed. Unlike the technology used in CRTs, the technology for portable monitors involves liquid crystals. Flat-panel monitors are much thinner than CRTs. Once used exclusively for portable computers, flat-panel monitors are now starting to be used for desktop systems as well.

There are two basic types of flat-panel monitors: *passive-matrix* and *active-matrix.* **Passive-matrix** or **dual-scan monitors** create images by scanning the entire screen. This type requires very little power, but the clarity of the images is not as sharp. **Active-matrix** or **thin film transistor (TFT) monitors** do not scan down the screen; instead, each pixel is independently activated. More colors with better clarity can be displayed. Active-matrix monitors are more expensive and require more power.

FIGURE 5-23
CRT monitors

FIGURE 5-24
A flat-panel monitor

An exciting recent development is the merger of microcomputers and television called **PC/TV.** This is becoming possible through the establishment of all-digital **high-definition television (HDTV).** HDTV delivers a much clearer and more detailed wide-screen picture than regular television. Additionally, because the output is digital, it enables users to readily freeze video sequences to create still images. These images can then be digitized and output as artwork or stored on disks for later use. This technology is very useful to graphic artists, publishers, and educators.

CONCEPT CHECK

✔ What do output devices do?

✔ _____, measured in pixels, indicates a monitor's clarity.

✔ List the four standards of monitor resolution.

Printers

There are three types of printers: ink-jet, laser, and thermal.

The images output on a monitor are often referred to as **soft copy.** Information output on paper—whether by a printer or by a plotter—is called **hard copy.** Three popular kinds of printers used with microcomputers are ink-jet, laser, and thermal.

FIGURE 5-25
An ink-jet printer

Ink-Jet Printer

An **ink-jet printer** sprays small droplets of ink at high speed onto the surface of the paper. This process not only produces a letter-quality image but also permits printing to be done in a variety of colors. (See Figure 5-25.) Ink-jet printers are the most widely used printer. They are reliable, quiet, and inexpensive. Ink-jet printers are used wherever color and appearance are important, as in advertising and public relations.

Laser Printer

The **laser printer** uses a technology similar to that used in a photocopying machine. (See Figure 5-26.) It uses a laser beam to produce images with excellent letter and graphics quality. More expensive than ink-jet printers, laser printers are used in applications requiring high-quality output.

There are two categories of laser printers. **Personal laser printers** are inexpensive and used by many single users. They typically can print four to six pages a minute. **Shared laser printers** are more expensive and are used (shared) by a group of users. Shared laser printers typically print over 30 pages a minute.

TIPS Is your ink-jet printer producing smeared pages? Are the characters looking blurry? Then it could be time to give it a cleaning. Here are a few cleaning suggestions:

1. *Disconnect.* Turn off the printer and unplug the electricity.
2. *Open.* Open the printer and locate the spray nozzles.
3. *Clean.* Wipe the nozzles with a dry lint-free cloth or a swab moistened with distilled water.
4. *Reconnect.* When completely dry, plug in the printer and turn it on.

To prevent clogging in the future, always allow your printer to finish printing the current page and to return the print head to its resting position before removing the page or turning off the power.

FIGURE 5-26
A laser printer

Printer	Characteristics	Typical Use
Ink-jet	High color quality; inexpensive; sprays drops of ink on paper	Internal and external communications, advertising pieces
Laser	Very high quality; uses photo-copying process	Desktop publishing, external documents
Thermal	Very high quality; uses heat elements on special paper	Art and design work

FIGURE 5-27
Three types of printers

ON THE WEB EXPLORATIONS

Hewlett-Packard is a leading manufacturer of printers and of multifunction devices that combine a printer, copier, scanner, and fax into one product. To learn more about the company, visit our Web site at

http://www.mhhe.com/it/oleary/ explore.mhtml

Thermal Printer

A **thermal printer** uses heat elements to produce images on heat-sensitive paper. Originally these printers were used in scientific labs to record data. More recently, color thermal printers have been widely used to produce very high quality color artwork and text.

Color thermal printers are not as popular because of their cost and the requirement of specially treated paper. They are a more special-use printer that produces near-photographic output. They are widely used in professional art and design work where very high quality color is essential.

Some of the important characteristics of the three most widely used microcomputer printers are summarized in Figure 5-27.

Other Printers

There are several other types of printers. Two are the dot-matrix printer and the chain printer. Dot-matrix printers form characters or images using a series of small pins on a print head. Once the most widely used microcomputer printer, dot-matrix printers are inexpensive and reliable but quite noisy. In general, they are used for tasks where a high-quality image is not essential. Thus, they are often used for documents that are circulated within an organization rather than shown to clients and the public. Their sales have declined dramatically in the past few years.

You probably won't find a chain printer standing alone on a desk next to a microcomputer. This is because a chain printer is an expensive, high-speed machine originally designed to serve minicomputers and mainframes. However, you may see one in organizations that link several microcomputers together by a communications network.

CONCEPT CHECK

✔ What are the three printer types commonly used with microcomputers?

✔ What type of printer is typically used for art and design work?

FIGURE 5-28
A pen plotter

Plotters

Plotters are special-purpose drawing devices.

lotters are special-purpose output devices for producing bar charts, maps, architectural drawings, and even three-dimensional illustrations. Plotters can produce high-quality multicolor documents and also documents that are larger than most printers can handle. There are four types of plotters: pen, ink-jet, electrostatic, and direct imaging.

Pen Plotter

Pen plotters (see Figure 5-28) create plots by moving a pen or pencil over drafting paper. (With some pen plotters, the paper moves and the pen remains stationary.) Pen plotters are the least expensive plotters and the easiest to maintain. Their major limitations are slower speed and an inability to produce solid fills and shading. Nevertheless, they have been the most popular type of plotter.

Ink-Jet Plotter

Ink-jet plotters create line drawings and solid-color output by spraying droplets of ink onto paper. Their best features are their speed, high-quality output, and quiet operation. The major disadvantage of ink-jet plotters is that the spray jets can become clogged and thus require more maintenance. Ink-jet plotters are used by a wide variety of workers, including engineers and automotive designers.

Electrostatic Plotter

Whereas pen plotters use pens, **electrostatic plotters** use electrostatic charges to create images made up of tiny dots on specially treated paper. (See Figure 5-29.) The image is produced when the paper is run through a developer. Electrostatic plotters produce high-resolution images and are much faster than either pen or ink-jet plotters. These plotters, unfortunately, use expensive chemicals that are considered hazardous. Electrostatic plotters are used for applications that require high-volume and high-quality outputs such as in advertising and graphic arts design.

FIGURE 5-29
An electrostatic plotter

Direct-Imaging Plotter

Direct-imaging plotters or **thermal plotters** create images using heat-sensitive paper and electrically heated pins. This type of plotter is comparably priced with electrostatic plotters, quite reliable, and good for high-volume work. However, direct-imaging plotters require expensive paper and typically create only two-color output. These plotters are typically used for very specific applications such as creating maps.

CONCEPT CHECK

✔ What are plotters?

✔ When might you need a plotter instead of a printer?

FIGURE 5-30
Stereo speakers

Voice-Output Devices

Voice-output devices vocalize prerecorded sounds.

Voice-output devices make sounds that resemble human speech but actually are prerecorded vocalized sounds. With one Macintosh program, the computer speaks the synthesized words "We'll be right back" if you type in certain letters and numbers. (The characters are *Wiyl biy ray5t bae5k*—the numbers elongate the sounds.) Voice output is not anywhere near as difficult to create as voice input. In fact, there are many occasions when you will hear synthesized speech being used. Examples are found in soft-drink machines, on the telephone, and in cars.

The most widely used voice-output devices are stereo speakers and headphones. (See Figures 5-30 and 5-31.) These devices are connected to a sound card in the system unit. The sound card is used to capture as well as play back recorded sounds.

Voice output is used as a reinforcement tool for learning, such as to help students study a foreign language. It is also used in many supermarkets at the checkout counter to confirm purchases. One of its most powerful capabilities is to assist the physically challenged. Some Web sites use so-called **chatterbots,** or virtual characters, that allow you to ask questions and get answers. (See Figure 5-32.)

FIGURE 5-31
Flat panel speakers

CONCEPT CHECK

✔ What do voice-output devices do?

✔ List some examples of where you might encounter voice-output.

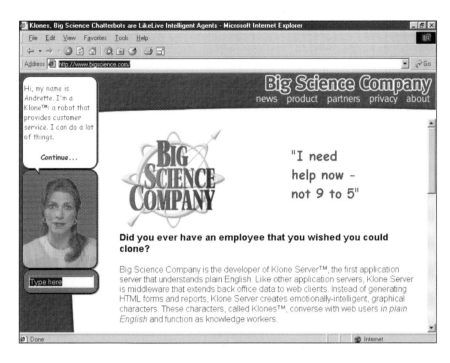

FIGURE 5-32
Chatterbot

A LOOK TO THE FUTURE

Microsoft's Investment in WebTV and introduction of Broadcast Technology could lead the way to the merger of all digital television and microcomputers.

Have you been hearing a lot about digital TV? It promises to combine the power and flexibility of a microcomputer with the entertainment capabilities of a large-screen TV. You and a whole group of friends could play computer games, interact with life-sized figures, surf the Web, capture video, edit it, and paste into electronic presentations.

Of course, that's what PC/TV is all about. These systems contain special devices that convert analog signals required for today's television sets to digital signals required for microcomputers. Unfortunately, this conversion greatly constrains the speed, flexibility, and quality of these systems. But all that's going to change in the next few years.

All-digital television broadcasting is just getting started and promise greater image and sound quality. The Federal Communications Commission is leading the way. Every major network is required to offer digital signals to their top 30 markets this year.

Microsoft, a leader in software technology, has recently taken steps to position itself for this change. One step has been the recent investment of nearly a half billion dollars in WebTV, a company that manufactures Internet terminals for television sets. Another step is the introduction of Broadcast Technology in Microsoft's Windows operating systems. This technology presents a cable TV interface that lets you choose content on screen from the Internet, local TV, cable TV, and other sources.

Will your current television set become useless? It will not be able to display the new digital signals; by the year 2006, all analog broadcasts are expected to be eliminated. However, the transition to digital will occur gradually over the next few years.

VISUAL SUMMARY Input and Output

KEYBOARD ENTRY

Input devices translate data and programs that people understand into a form that a computer can process. Two kinds of input are *keyboard* and *direct entry.*

Keyboard entry can be categorized as *keyboards* and *terminals.*

Keyboards
Both traditional and natural **keyboards** have various types of keys including **toggle keys**, and **combination keys**.

Terminals
Terminals connect to the host computer or server. Four kinds of terminals are:

- **Dumb**—sends and receives only; no processing.
- **Intelligent**—sends, receives, and processes; popular type is **Net PC.**
- **Network**—provides low-cost alternative to intelligent terminal; also known as **thin client** and **network computer.**
- **Internet**—accesses Internet and typically displays on a television set; does not require microcomputer; also known as **Web terminal** and **Web appliance.**

DIRECT ENTRY

Direct entry devices can be categorized as *pointing, scanning,* or *voice-input.*

Pointing
The most common pointing device is the **mouse.** Three similar devices are **trackballs, touch surfaces,** and **pointing sticks.**
Other pointing devices include:
- **Joysticks**—control game action by varying pressure, speed, and direction.
- **Touch screens**—touching finger to screen to control operations.
- **Light pens**—directed at screen to control operations.
- **Digitizers**—convert images to digital data using **stylus** and **digitizing tablet.**
- **Digital cameras**—record still-image photos on disk and in memory. **Webcams** are used for Internet brodcasting.
- **Digital video cameras**—record motion digitally on tape, on disk, or in memory.
- **Digital notebooks**—electronic pads record handwriting as digital data.

To be a competent end user you need to be aware of the most commonly used input and output devices. These devices are translators for information into and out of the system unit. Two kinds of input are keyboard and direct entry. Output devices include monitors, printers, plotters, and voice-output.

DIRECT ENTRY

Scanning
Scanners convert images to digital data. There are two basic types of scanners:

- **Flatbed scanners** are like copy machines.
- **Portable scanners** are handheld devices. **Pen scanners** are highly portable.

Widely used scanning devices include **image scanners, fax machines, bar-code readers, MIRC, OCR,** and **OMR** devices.

Voice-Input Devices
These devices convert a person's spoken words into digital data. **Voice recognition systems** are a combination of hardware and software that allows users to control operations and create documents using voice commands.

Two types of voice recognition systems:

- **Discrete-speech recognition systems** recognize only individual words.
- **Continuous-speech recognition systems** recognize individual words and phrases in context; key technology for the 21st century.

MONITORS

Output devices translate machine output to output that people can understand. Output devices include *monitors, printers, plotters,* and *voice-output.*

    Monitor size is indicated by diagonal length of viewing area. Monitor clarity is indicated by **resolution,** measured in **pixels.**

Standards
Monitor standards indicate resolution capabilities.

Standard	Pixels
SVGA	800 × 600
XGA	1,024 × 768
SXGA	1,280 × 1,024
UXGA	1,600 × 1,200

Cathode-Ray Tubes
CRTs are similar to televisions in size and technology. They are typically placed on a system unit or directly on a desk. Their primary advantages are low cost and high resolution.

Flat-Panel Monitors
Flat-panel monitors are used for portable computers and becoming more popular for desktop systems. Two basic types are:

- **Passive-matrix (dual-scan)**—requires little power but image is not as sharp.
- **Active matrix** or **thin film transistor (TFT)**—requires more power, is more expensive, but produces sharper images.

PRINTERS

Output from monitors is called **soft copy.** Output from a **printer** is called a **hard copy.**

Three widely used types are: *ink-jet, laser,* and *thermal.*

Ink-Jet
Ink-jet printers spray ink onto the surface of the paper. They are reliable, quiet, and inexpensive.

Laser
Laser printers use technology similar to that used in copy machines. They are more expensive than ink-jet printers. Produce excellent quality letter and graphics quality.

Personal lasers are for single users. **Shared lasers** are faster, more expensive and powerful. They are designed for multiple users.

Thermal
Thermal printers use heat elements to produce images on heat-sensitive paper. Thermal printers are the most expensive and produce the highest quality color images.

Other Types
Other types of printers include *dot-matrix printers* and *chain printers.*

PLOTTERS

Plotters produce multicolor bar charts, maps, architectural drawings, and three-dimensional illustrations. Four types are as follows.

Pen
Pen plotters create a drawing by moving a pen or pencil over drafting paper. They are the most popular and least expensive.

Ink-Jet
Ink-jet plotters are fast, quiet, and very good at producing solid-color output. The jets, however, can become clogged and require maintenance.

Electrostatic
These plotters use **electrostatic** charges to create high-quality and high-volume output on specially treated paper. Costly chemicals considered dangerous to the environment, however, are required for the development process.

Direct-Imaging
Direct-imaging plotters use electrically heated pins to create two-color output on special heat-sensitive paper. Due to expensive requirements and limited color output, these plotters are specialty devices.

VOICE-OUTPUT DEVICES

Voice-output devices make sounds resembling human speech. The most widely used devices are stereo speakers and headphones.

Key Terms

active-matrix monitor (112)
bar codes (107)
bar-code reader (107)
cathode-ray tube (CRT) (112)
chatterbots (116)
combination key (100)
continuous-speech recognition
   system (110)
digital camera (104)
digital notebook (105)
digital video camera (105)
digitizer (104)
digitizing tablet (104)
direct entry (102)
direct-imaging plotter (116)
discrete-speech recognition
   systems (110)
dual-scan monitor (112)
dumb terminal (105)
electrostatic plotter (115)
facsimile transmission (fax)
   machine (106)
fax/modem board (107)
flatbed scanner (106)
flat-panel monitor (112)
hard copy (113)
high-definition television
   (HDTV) (113)
image scanner (105)
ink-jet plotter (115)
ink-jet printer (113)

intelligent terminal (111)
Internet terminal (102)
joystick (103)
keyboard entry (100)
laser printer (113)
light pen (104)
liquid crystal display (LCD)
   monitor (112)
magnetic-ink character
   recognition (MICR) (107)
mark sensing (107)
monitor (111)
mouse (102)
Net Personal Computer
   (Net PC) (101)
network computer (101)
network terminal (101)
optical-character recognition
   (OCR) (107)
optical-mark recognition
   (OMR) (107)
passive-matrix monitor (112)
PC/TV (113)
pen plotter (115)
pen scanner (119)
personal laser printer (113)
pixels (111)
platform scanner (100)
plotters (115)
pointer (102)
pointing stick (97)

point-of-sale (POS) terminal (100)
portable scanner (106)
reader/sorter (107)
resolution (111)
roller ball (103)
scanner (105)
shared laser printer (113)
soft copy (113)
source document (100)
stylus (104)
SVGA (111)
SXGA (111)
terminal (101)
thermal plotter (114)
thermal printer (114)
thin client (101)
thin film transistor (TFT)
   monitor (112)
toggle key (100)
touch screen (103)
touch surface (103)
trackball (103)
UXGA (111)
voice-input device (108)
voice-output device (116)
voice recognition systems (108)
wand reader (100)
Web appliance (118)
Web terminal (102)
Webcam (118)
XGA (111)

Chapter Review

LEVEL 1

Reviewing Facts and Terms
Matching

Match each numbered item with the most closely related lettered item. Write your answers in the spaces provided.

1. Vocalizes prerecorded sounds. _I_

2. Controls a pointer that is displayed on the monitor. _J_

3. Inexpensive laser printers used by many single users. _M_

4. Considered to be one of the key technologies for the 21st century. _C_

5. All-digital, this delivers a much clearer and more detailed wide-screen picture. _G_

6. Special-purpose output device that creates line drawings and solid-color output by spraying droplets of ink onto paper. _H_

7. Flat-panel monitor where each pixel is independently activated, allowing more colors and better clarity to be displayed. _A_

a. Active-matrix
   monitor
b. Bar-code
   reader
c. Continuous-
   speech
   recognition
   system
d. Direct entry
e. Dumb
   terminal
f. Electrostatic
   plotter
g. HDTV

8. Input that does not require data to be keyed—machine-readable data can go directly to the CPU. _D_

9. Can be used to input and receive data, but it cannot process data independently. _E_

10. Uses electrostatic charges to create images made up of tiny dots on specially treated paper. _F_

11. Printer that sprays small droplets of ink at high speed onto the surface of the paper. _i_

12. Scanning device that uses special preprinted characters. _J_

13. Special-purpose drawing devices. _N_

14. Describes a monitor's clarity, measured in pixels. _q_

15. A specialized pointing device, connected to a computer, used to trace an item placed on a digitizing tablet. _r_

16. Photoelectric scanners that read bar codes. _B_

17. Low-cost and limited microcomputers that typically have only one type of secondary storage, a sealed system unit, and no expansion slots. _k_

18. Usually appears in the shape of an arrow on the monitor. _o_

19. Super Extended Graphics Array. _S_

20. Input device that uses both keyboard and direct entry—a sort of electronic cash register. _p_

h. Ink-jet plotter
i. Ink-jet printer
j. Mouse
k. Net PC
l. OCR
m. Personal laser printer
n. Plotters
o. Pointer
p. POS terminal
q. Resolution
r. Stylus
s. SXGA
t. Voice-output device

True/False

In the spaces provided, write T or F to indicate whether the statement is true or false.

1. Input devices translate symbols that people understand into symbols that computers can process. _T_

2. A plotter is a device that can be used to trace or copy a drawing or photograph. _F_

3. Banks use a method called magnetic-ink character recognition (MICR) to automatically read and sort checks. _T_

4. Laser printers are highly reliable, but the quality of their output limits their use to rough drafts and in-house communications. _F_

5. Plotters are special-purpose drawing devices. _T_

6. The most common type of home or office monitor is an UXGA. _F_

Multiple Choice

Circle the letter of the correct answer.

1. The *Caps Lock* key is a _____ c _____ key.
   a. function
   b. numeric
   c. toggle
   d. cursor control
   e. combination

2. A device that converts images on a page to electronic signals that can be stored in a computer:
   a. monitor
   b. scanner
   c. plotter
   d. MICR
   e. POS

3. The type of intelligent terminal that typically has only one type of secondary storage, a sealed system unit, and no expansion slots:
   a. PC/TV
   b. Web terminal
   c. digital modem
   d. Net PC
   e. network device

4. The printer that can produce very high quality images using heat elements on heat-sensitive paper:
   a. dot-matrix
   b. laser
   c. ink-jet
   d. plotter
   e. thermal

5. The plotter that creates images using heat-sensitive paper and electrically heated pins:
   a. pen
   b. ink-jet
   c. direct-imaging
   d. scanner
   e. electrostatic

6. Which of the following is NOT a scanning device:
   a. bar-code reader
   b. digitizing tablet
   c. fax machine
   d. optical-mark recognition
   e. portable scanner

Completion

Complete each statement in the spaces provided.

1. A device similar to a mouse that controls the pointer by rotating a ball controlled with the thumb is _Track Ball_

2. _FAX_ machines are popular office machines that scan the image of a document to be sent.

3. _SVGA_ is a monitor standard used primarily with 15-inch monitors.

4. The _InkJet_ printer is the most widely used.

5. _Voice Output_ devices make sounds that resemble human speech.

6. _Touch Screens_ control the pointer by moving and tapping your finger on the surface of a pad.

Reviewing Concepts

LEVEL 2

Open-Ended

On a separate sheet of paper, respond to each question and statement.

1. What are the differences between keyboard entry and direct entry?

2. What is a POS terminal? Describe the processing capabilities of a POS terminal and examples of where you might find POS terminals in use.

3. Distinguish among the four kinds of terminals: dumb, intelligent, network, and Internet.

4. What are pixels? What do they have to do with screen resolution?

5. Compare the various kinds of direct-entry devices. For the ones you have never used personally, give an example of where you might encounter them.

6. Distinguish between discrete-speech and continuous-speech recognition systems.

Concept Mapping

On a separate sheet of paper, draw a concept map or a flowchart showing how the following terms are related. Show all relationships. Include any additional terms you can think of.

active-matrix monitor	laser printer	portable scanner
bar codes	LCD monitor	stylus
bar-code reader	monitor	SVGA
digital camera	mouse	SXGA
digital notebook	network computer	terminal
digitizer	optical-character	thermal printer
direct entry	recognition	touch screen
dual-scan monitor	passive-matrix monitor	UXGA
electrostatic plotter	pen plotter	video display
fax machine	plotters	voice recognition systems
ink-jet plotter	pointer	voice-input/output device
ink-jet printer	POS terminal	wand reader
Internet terminal	resolution	Webcam
joy stick	scanner	XGA

Critical Thinking Questions and Projects

LEVEL 3

Read each exercise and answer the related questions on a separate sheet of paper.

1. *Multifunction devices*: Multifunction devices are also known as information appliances. These all-in-one devices provide a variety of input and output capabilities in one piece of equipment. Most offer scanning, faxing, copying, and printing capabilities. To learn more about them, visit a local computer store and/or use the Web. Determine what advantages and disadvantages they offer compared to purchasing separate, more specialized devices.

2. *Security and privacy*: Security cameras are becoming more widely used by retail stores to monitor customers as they select items, as well as to monitor parking structures, elevators, dressing rooms, and restrooms. Many homes have security monitors for the front door and for a child's room. Computer-enhanced videotapes from security cameras have been successfully used to catch and prosecute shoplifters and abusive care providers.

   a. Does your school, home, or place of work use security cameras? If it does, describe where and why they are used.

    b. Some people would point out that while security devices may be effective for security, they represent a threat to personal privacy. Discuss your position on this potential trade-off.

    c. Do you think the use of security cameras should be controlled? If so, how and under what conditions?

3. *Voice Recognition Systems and Dictating a Paper:* Voice recognition systems offer an alternative to the keyboard and mouse use for many applications. (See Making IT Work for You: Voice Recognition Systems and Dictating a Paper on pages 108–109.) Although far from perfect, this technology is expected to be widely used in the near future.

    a. Discuss the relative advantages and disadvantages of using voice input over other types of input. Be specific.

    b. Identify and describe three specific jobs or job functions where voice input would be a particularly effective tool.

    c. Identify and describe three specific jobs or job functions where voice input would not be an effective tool.

On the Web Exercises

1. SoundBlaster

Creative Lab is one of the pioneers in sound input and output technologies. Its SoundBlaster family of sound cards is an industry standard for affordability, quality, and reliability.  Visit our Web site at http://www.mhhe.com/it/oleary/exercise.mhtml to link to the SoundBlaster site. Once connected to that site, check out the newest sound card. Print out the Web page describing the specifications of this card and write a paragraph comparing the card to its predecessors.

2. Movie Magic

A variety of input and output devices have been used to create dramatic special effects in movies. To learn how these devices are used to create movie magic, visit our Web site at  http://www.mhhe. com/it/oleary/exercise.mhtml to link to a Web site on special effects. Once connected to that site, print out the Web page that you find most interesting. Write a brief paragraph on one movie magic technique that most interests you and how you found it on the Web site.

3. WebTV

In order to provide Internet access to a wider audience, the company WebTV Networks, Inc. has eliminated the need for a traditional computer when surfing  the Web. Using a Web terminal connected to a TV set, anyone can access the Internet without the expense of a computer. To learn more about WebTV's Web terminal, visit our Web site at http://www.mhhe.com/it/oleary/exercise.mhtml to link to WebTV's site. Once connected to that site, learn how to use your TV to access the Web. Print out the Web page you find most informative. Write a paragraph discussing the relative advantages and disadvantages of TV versus computer access to the Internet.

4. Voice Recognition

Voice activation and control of computers may someday be as common as using a mouse and keyboard is today. (See Making IT Work for You: Voice  Recognition Systems and Dictating a Paper on pages 108–109.) Visit our Web site at http://www.mhhe.com/it/oleary/exercise.mhtml to link to a site on speech recognition. Once connected to that site, explore it and print out the Web page that you found the most informative. Write a paragraph describing how you might use this technology.

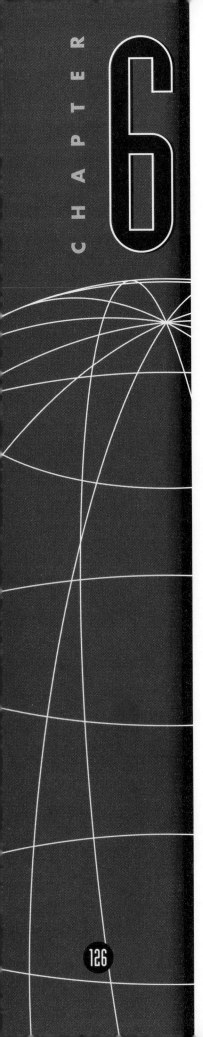

Secondary Storage

COMPETENCIES

After you have read this chapter, you should be able to:

1. Describe today's standard floppy disk and compare it to Zip, SuperDisks, and HiFD disks.

2. Describe the following kinds of disks: internal hard disks, hard-disk cartridges, and hard-disk packs.

3. Describe ways to improve hard-disk operations: disk caching, redundant arrays of inexpensive disks, and data compression.

4. Compare the CD and DVD formats.

5. Describe the different types of optical disks.

6. Describe magnetic tape streamers and magnetic tape reels.

An essential feature of every computer is the ability to save—that is, store—information. Computers can save information permanently, after you turn them off. This way, you can save your work for future use, share information with others, or modify information already available. Secondary storage holds information external from the CPU. Secondary storage allows you to store programs, such as Word and Excel. It also allows you to store the data processed by programs, such as text or the numbers in a spreadsheet.

We described random-access memory (RAM) in Chapter 4. This is

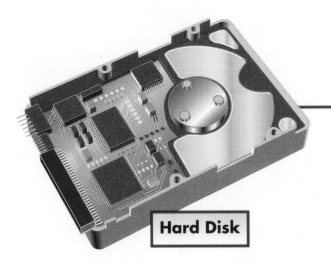

Hard Disk

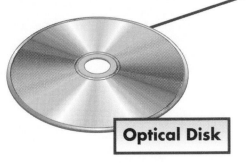

Optical Disk

the *internal* and *temporary* storage of data and programs in the computer's memory. Once the power is turned off or interrupted, everything in internal storage disappears. Such storage is therefore said to be **volatile storage.** Thus, we need *external, more permanent,* or **nonvolatile storage** for data and programs. We also need external storage because users need much more capacity than is possessed by a computer's primary memory.

CDs, for example, store data that can be used over and over again. You can download music from the Internet, play it on your computer, and create custom CDs with the right hardware and software.

Competent end users need to be aware of the different types of secondary storage. They need to know the capabilities, limitations, and uses of floppy disks, hard disks, optical disks, and magnetic tapes.

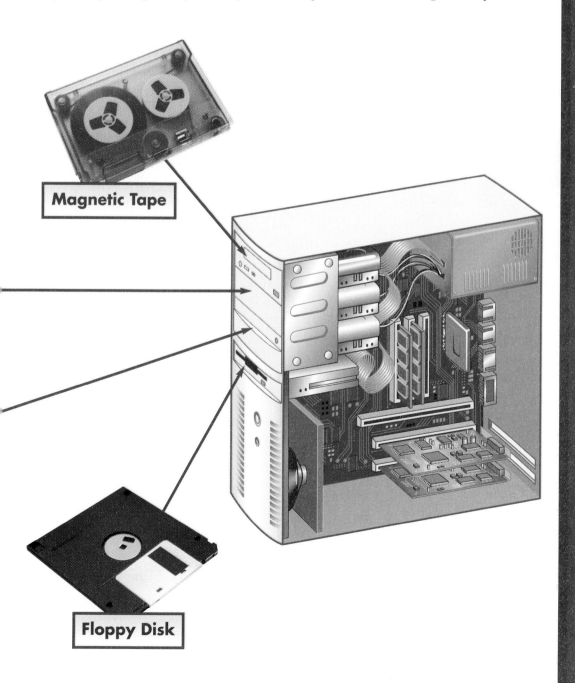

Magnetic Tape

Floppy Disk

Floppy Disks

Floppy disks are removable storage media. Today's standard is 1.44 MB. Tomorrow's might be a Zip, SuperDisk, or HiFD. Data is recorded on tracks and sectors.

Floppy disks, often called **diskettes** or simply **disks,** are a portable or removable storage media. They are used to store and to transport word processing, spreadsheet, and other types of files. They use flat circular pieces of Mylar plastic that rotate within a jacket. Data is stored as electromagnetic charges on a metal oxide film coating the Mylar plastic. Data and programs are represented by the presence or absence of these charges, using the ASCII or EBCDIC data representation codes.

Floppy disks are also called **flexible disks,** and **floppies.** This is because the plastic disk inside the diskette covers is flexible, not rigid. Although there are several different types of floppy disks, there is just one standard and several others that are competing to become the next standard.

Today's Standard

The most widely used floppy disk is the **1.44 MB 3½-inch disk.** (See Figure 6-1.) These disks are typically labeled **2HD,** which means "two-sided, high-density." This disk can store 1.44 megabytes—the equivalent of 400 typewritten pages. It has a thin exterior jacket made of hard plastic to protect the flexible disk inside. When the disk is inserted into a floppy disk drive, the shutter slides open to expose the 3½-inch flexible disk. A read-write head from the disk drive moves across the exposed disk to store and retrieve data.

These disks also have a **write-protect notch.** When the notch is open, data cannot be added on the disk. This is to provide protection against accidentally writing over information on the disk that you want to keep.

Tomorrow's Standard Floppy Disk

Today's standard floppy disk is very reliable and widely used to store data. Its major limitation is its capacity. While 1.44 MB is fine for many text and spreadsheet files, it is not sufficient to hold larger files. For example, many presentations far exceed 1.44 MB. Most multimedia applications require greater capacity.

Several other floppy disks, known as **floppy-disk cartridges,** are competing to become the next higher capacity floppy disk standard. These contenders are also 3½-inch; however, they are thicker and require special disk drives. The three best known are Zip disks, SuperDisks, and HiFD disks.

- **Zip disks** are produced by Iomega and typically have a 100 MB or 250 MB capacity—over 170 times as much as today's standard floppy disk. Zip drives are a standard feature on many of today's system units. They are widely used to store multimedia, database, large text, and spreadsheet files. Many observers predict that Zip disks will become the next floppy disk standard. (See Figure 6-2.)

- **SuperDisks** are produced by Imation and have a 120 MB capacity. They have one major advantage over Zip disks. SuperDisk drives are able to read and store data on today's 1.44 MB standard disk. Zip disk drives cannot. For this reason, SuperDisks are popular for use with notebook computers.

- **HiFD disks** from the Sony Corporation are the newest challenger. They have a capacity of 200 MB, and, like SuperDisk drives, HiFD disk drives are able to use today's 1.44 MB standard disks.

ON THE WEB EXPLORATIONS

Iomega and Imation are leaders in developing high-capacity floppy disks. To learn more about the companies, visit our Web site at

http://www.mhhe.com/it/oleary/
explore.mhtml

FIGURE 6-1
A 3½-inch floppy disk

FIGURE 6-2
Zip disk

Description	Capacity
2HD	1.44 MB
Zip	100 MB/250 MB
SuperDisk	120 MB
HiFD	200 MB

FIGURE 6-3
Typical floppy-disk capacities

Each of these will likely improve its capacity and speed in the near future. Will one of them become the next standard floppy disk? While most observers believe one *will* become the next standard, they are less certain *when* this will occur. For a summary of floppy disks, see Figure 6-3.

TIPS Are you concerned about losing the data stored on your floppy disks? Actually, floppy disks are quite durable, and taking care of them boils down to just a few basic rules.

1. *Don't bend.* Don't bend, flex, or put heavy weights on them.
2. *Don't touch.* Don't touch anything visible through the protective jacket.
3. *Avoid extreme conditions.* Keep disks away from strong magnetic fields, extreme heat, and any chemicals such as alcohol and other solvents.
4. *Use storage boxes.* Store disks in a sturdy plastic storage box.

Of course the best protection is to make a backup or duplicate copy of your disk.

The Parts of a Floppy Disk

Data is recorded on a disk in rings called **tracks.** (See Figure 6-4.) These tracks are closed concentric circles, not a single spiral as on a phonograph record. Unlike a phonograph record, these tracks have no visible grooves. Looking at an exposed floppy disk, you would see just a smooth surface. Each track is divided into invisible wedge-shaped sections known as **sectors.**

Some disks are manufactured without tracks and sectors in place. They must be adapted to the type of microcomputer and disk drive you are using. You do this using a process called *formatting,* or *initializing.*

CONCEPT CHECK

✔ How do floppy disks store information?

✔ What is today's standard floppy disk?

✔ What are likely to be tomorrow's standard floppy disks?

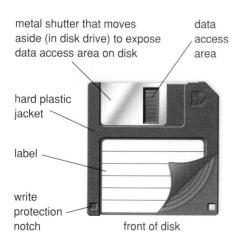

metal shutter that moves aside (in disk drive) to expose data access area on disk

data access area

hard plastic jacket

label

write protection notch

front of disk

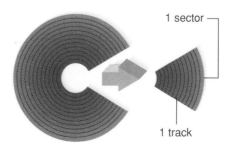

1 sector

1 track

FIGURE 6-4
The parts of a 3½-inch floppy disk

Hard Disks

Hard disks are of three types: internal hard disk, hard-disk cartridge, and hard-disk pack.

While floppy disks use thin flexible plastic disks, **hard disks** use thicker, ridged metallic platters. (See Figure 6-5.) Hard disks are able to store and retrieve information much faster and have a greater capacity. They are sensitive instruments. The read-write head rides on a cushion of air about 0.000001 inch thick. It is so thin that a smoke particle, fingerprint, dust, or human hair could cause what is known as a head crash. (See Figure 6-6.)

A **head crash** happens when the surface of the read-write head or particles on its surface contact the magnetic disk surface. A head crash is a disaster for a hard disk. It means that some or all of the data on the disk is destroyed.

There are three types of hard disks: *internal hard disk, hard-disk cartridge,* and *hard-disk pack.*

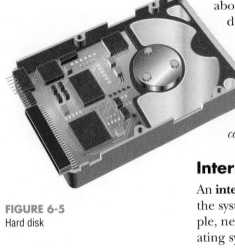

FIGURE 6-5
Hard disk

Internal Hard Disk

An **internal hard disk** is also known as a **fixed disk** because it is located inside the system unit. It is used to store programs and large data files. For example, nearly every microcomputer uses its internal hard disk to store its operating system and major applications like Word and Excel.

An internal hard disk consists of one or more metallic platters sealed inside a container. The container includes the motor for rotating the disks. It also contains an access arm and read-write heads for writing data to and reading data from the disks. From the outside of a microcomputer, an internal hard disk looks like part of a front panel on the system cabinet. Typically, inside are four 3½-inch metallic platters with access arms that move back and forth. (See Figure 6-7.)

Internal hard disks have two advantages over floppy disks: capacity and speed. A 40-gigabyte internal hard disk, for instance, can hold almost as much information as 28,000 standard floppy disks. Moreover, access is faster. For these reasons, almost all of today's powerful applications are designed to be stored on and run from an internal hard disk. Adequate capacity or size of a microcomputer's internal hard disk is essential.

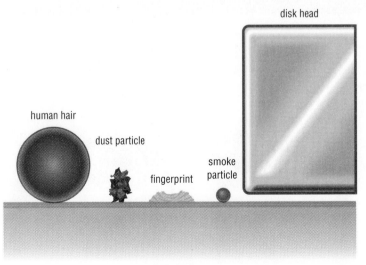

FIGURE 6-6
Materials that can cause a head crash

Hard-Disk Cartridges

While internal hard disks provide fast access, they have a fixed amount of storage and cannot be easily removed. **Hard-disk cartridges** are as easy to remove as a cassette from a videocassette recorder. The amount of storage available to a computer system is limited only by the number of cartridges.

Hard disk cartridges are used primarily to complement an internal hard disk. Because the cartridges are easily removed, they are particularly useful to protect or secure sensitive information. For example, personnel administrators need to have access to highly confidential employee information. When information is stored on a hard-disk cartridge, it is readily available, yet can be easily locked up when not in use. Other uses for hard-disk cartridges include backing up the contents of the internal hard disk and providing additional hard disk capacity.

The typical capacity for a hard-disk cartridge is 2 gigabytes. Two well-known hard-disk cartridges are Jaz from Iomega and SparQ from SyQuest. (See Figure 6-8.)

> **TIPS** Does your internal hard-disk drive run a lot and seem slow? Are you having problems with lost or corrupted files? The problem could be with fragmented files—files that when saved were broken into pieces (fragments) and stored in different locations on your hard disk. To clean up the disk and speed up access, consider defragging. If you are using Windows 95, 98, or 2000:
>
> 1. *Start Disk Defragmenter.* Use Start/Programs/ Accessories/System Tools/Disk Defragmenter to start defragmenting your disk. Defragging rearranges the file parts so that they are stored in adjacent locations.
>
> 2. *Keep working.* You can continue running other applications while your disk is being defragmented. Unfortunately, your computer operates more slowly, and Disk Defragmenter takes longer to finish.
>
> It is recommended that you defrag your disk on a regular basis. If you have Windows 98 or 2000, you can schedule this task to be done automatically for you—use Start/Programs/Accessories/System tools/Scheduled tasks.

Hard-Disk Packs

Hard-disk packs are removable storage devices used to store massive amounts of information. (See Figure 6-9.) Their capacity far exceeds the other types of hard disks. Although you may never have seen one, it is almost certain that you have used them. Microcomputers that have access to the Internet, minicomputers, or mainframes often have access to external hard-disk packs through communication lines. Banks and credit card companies use them to record financial information.

FIGURE 6-7
Inside of an internal hard-disk drive

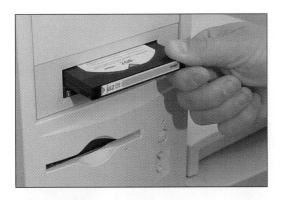

FIGURE 6-8
Removable hard-disk cartridge

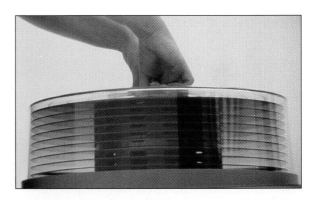

FIGURE 6-9
Hard-disk packs

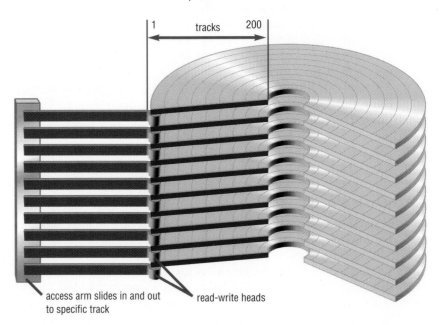

FIGURE 6-10
How a hard-disk pack works

access arm slides in and out
to specific track

read-write heads

1   tracks   200

Hard-disk packs consist of several platters aligned one above the other. They resemble a stack of phonograph records. The difference is that there is space between the disks to allow the access arms to move in and out. (See Figure 6-10.) Each access arm has two read-write heads. One reads the disk surface above it; the other reads the disk surface below it. A disk pack with 11 disks provides 20 recording surfaces. This is because the top and bottom outside surfaces of the pack are not used.

All the access arms move in and out together. However, only one of the read-write heads is activated at a given moment. **Access time** is the time between the computer's request for data from secondary storage and the completion of the data transfer.

You may well use your microcomputer to gain access to information over a telephone or other communications line. (We show this in the next chapter.) Such information is apt to be stored on disk packs. For example, one large information service, named Dialog, has over 300 databases. These databases cover all areas of science, technology, business, medicine, social science, current affairs, and humanities. All of these are available through a telephone link with your desktop computer. There are more than 100 million items of information, including references to books, patents, directories, journals, and newspaper articles. Such an information resource may be of great value to you in your work.

For a summary of the different types of hard disks, see Figure 6-11.

Type	Description
Internal	Fast access to applications, fixed
Cartridge	Complement to internal hard disk, removable
Disk pack	Massive storage capacity, removable

FIGURE 6-11
Types of hard disks

Performance Enhancements

Three ways to improve the performance of hard disks are disk caching, redundant arrays of inexpensive disks, and file compression/decompression.

Disk caching improves hard-disk performance by anticipating data needs. It requires a combination of hardware and software. During idle processing time, frequently used data is read from the hard disk into memory (cache). When needed, the data is then accessed directly from memory. The transfer rate from memory is much faster than from hard disk. As a result, overall system performance is often increased by as much as 30 percent.

Redundant arrays of inexpensive disks (RAIDs) improve performance by expanding external storage. Groups of inexpensive hard-disk drives are related or grouped together using networks and special software. These grouped disks are treated as a single large-capacity hard disk. They can outperform single disks of comparable capacities.

File compression and **file decompression** increase storage capacity by reducing the amount of space required to store data and programs. File compression is not limited to hard-disk systems. It is frequently used to compress files on floppy disks as well. File compression also helps to speed up transmission of files from one computer system to another. Sending and receiving compressed files across the Internet is a common activity.

Special file compression programs scan files for ways to reduce the amount of required storage. One way is to search for repeating patterns. The repeating patterns are replaced with a token, leaving enough so that the original can be rebuilt or decompressed.

File compression programs typically can shrink files to a quarter of their original size. Two well-known file compression programs are WinZip from Nico Mak Computing and PKZip from PKWare.

> **TIPS** Are you running short of hard-disk storage space? Want to send a large file or several files at once over the Internet? You can save both space and valuable connection time by compressing the files first. Here's how, using WinZip, a popular compression program:
>
> 1. *Start*. Start the WinZip program.
> 2. *Create file*. Click the *New* button on the toolbar to create and name a zip file.
> 3. *Select*. Locate and select the file(s) you want to compress.
> 4. *Compress*. Click the *Add* button to compress and add the selected file(s) to the zip file.
>
> You can now replace the selected file(s) with the much smaller zip file. To access the original files at any time, use the *WinZip Extract* button.

CONCEPT CHECK

✔ What are the three types of hard disks?

✔ Internal hard disks have two advantages over floppy disks: _____ and _____.

✔ _____-_____ _____ are removable storage devices used to store massive amounts of information.

✔ List three ways to improve the performance of hard disks.

FIGURE 6-12
Optical disk

Optical Disks

Optical disks use laser technology. CD and DVD are optical disk formats.

Today's optical disks can hold over 4.7 gigabytes of data. (See Figure 6-12.) That is the equivalent of over 1 million typewritten pages or a medium-sized library all on a single disk. Optical disks are having a great impact on storage today, but we are probably only beginning to see their effects.

In optical-disk technology, a laser beam alters the surface of a plastic or metallic disk to represent data. Unlike floppy and hard disks, which use magnetic charges to represent 1s and 0s, optical disks use reflected light. The 1s and 0s are represented by flat areas called **lands** and bumpy areas called **pits** on the disk surface. The disk is read by a laser that projects a tiny beam of light on these areas. The amount of reflected light determines whether the area represents a 1 or a 0. (See Figure 6-13.)

Optical disks come in many different sizes including 3½, 4¾, 5¼, 8, 12, and 14 inches. The most common size is 4¾-inch. Data is stored on these disks in different ways or different formats. The two most common are CD and DVD.

Compact Disc

Compact disc, or as it is better known, **CD,** format is the most widely used today. (See Figure 6-14.) CD drives are standard on many microcomputer systems. Typically, CD drives can store 650 megabytes of data on one side of a CD. One important characteristic of CD drives is their speed. The rotational speed is important because it determines how fast data can be transferred from the CD. For example, a 24X or 24-speed

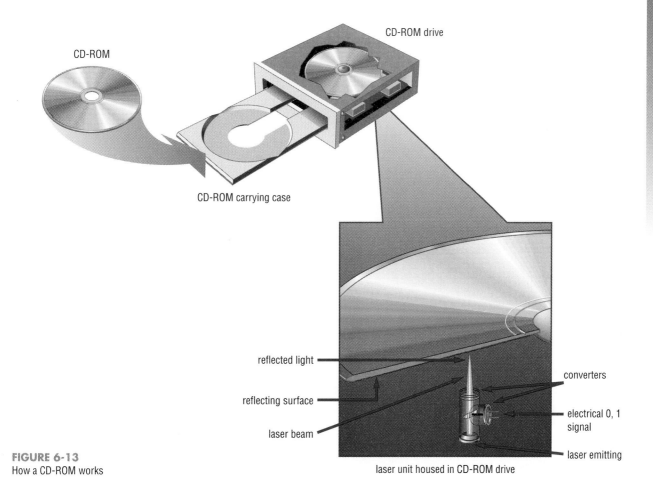

CD-ROM

CD-ROM drive

CD-ROM carrying case

reflected light

reflecting surface

laser beam

converters

electrical 0, 1 signal

laser emitting

laser unit housed in CD-ROM drive

FIGURE 6-13
How a CD-ROM works

CD drive can transfer 3.6 MB per second, while a 32X drive can transfer 4.8 MB per second. The faster the drive, the faster data can be read from the CD and used by the computer system.

There are three basic types of CDs: CD-ROM, CD-R, and CD-RW:

- **CD-ROM,** which stands for *compact disc–read only memory,* is similar to a commercial music CD. *Read-only* means it cannot be written on or erased by the user. Thus, you as a user have access only to the data imprinted by the publisher. CD-ROMs are used to distribute large databases and references. They are also used to distribute large software application packages. For example, Microsoft Office is available on a single CD-ROM or on 38 floppy disks.

- **CD-R** stands for *CD-recordable.* Also known as **WORM,** or *write once, read many,* CD-Rs can be written to once. After that they can be read many times without deterioration but cannot be written on or erased. CD-Rs are used to create custom music CDs (see making IT Work for You: CD-R Drives and Music from the Internet on pages 136–137) and to archive data.

- **CD-RW** stands for *compact disc rewritable.* Also known as **erasable optical disks,** these disks are very similar to other CD-Rs except that the disk surface is not permanently altered when data is recorded. Because they can be changed, CD-RWs are often used to create and edit multimedia presentations.

FIGURE 6-14
CD in drive

Digital Versatile Disc

DVD stands for *digital versatile disc* or *digital video disc*. This is a relatively new and evolving format. DVD and DVD drives are very similar to CDs except that more data can be packed into the same amount of space. The DVD drives can store 4.7 gigabytes on one side of a DVD—more than seven times the capacity of a CD. Furthermore, many DVD drives can store data on both sides of the disk, thereby doubling the capacity. Soon-to-be-released DVDs are likely to have a 17-gigabyte capacity.

MAKING IT WORK FOR YOU

CD-R Drives and Music from the Internet

Did you know that you can use the Internet to locate music, download it to your computer, and create your own compact discs? All it takes is the right software, hardware, and a connection to the Internet.

How It Works

Music is available on the Internet in special compressed music files called MP3s. You can download these music files to your system via your Internet connection. Using your system and some specialized software, you can store and play these files. You can even create a custom CD using a CD-R drive.

Internet   Your System   CD

Downloading Music

There are several sites on the Web that offer free music that you can download to your system. Some Web sites are set up to offer a convenient way to browse and select these files, such as www.mp3.com.

For example, you can download files to your system as shown in Figure A.

FIGURE A

1. Enter www.mp3.com in the Location box of your browser.

2. Select a featured artist.

3. Select a song to download.

While CDs may be the standard optical disk format today, most observers predict that DVDs will soon replace them. Many of today's more powerful microcomputer systems come with DVD drives. Like CDs, there are three basic types of DVDs:

• **DVD-ROM** stands for *digital versatile disc–read only memory.* DVD-ROMs are having a major impact on the video market. While CD-ROMs can store just over an hour of fair-quality video, DVD-ROMs can provide over two hours of very high quality video and sound comparable to that found in motion picture theaters.

FIGURE B

Playing Music

To arrange and play downloaded music files, you need specialized software called a player. WinAmp from Nullsoft (www.winamp.com) is one of the best known.

For example, using WinAmp, you can play music files as shown in Figure B.

2. Click the *Play* button to play the music.

1. Click the *Add* button to select the music files to play from your hard-disk drive.

Creating a Custom CD

If your computer is equipped with a CD-R or CD-RW drive, you can create your own music CDs. You'll need a blank recordable compact disc and special CD-creation software, such as Easy CD Creator by Adaptec. You use this software to organize and save the music files onto your CD.

For example, using Easy CD Creator, you can create a custom CD as shown in Figure C.

2. Click the *Create CD* button to start the process of recording the CD.

FIGURE C

1. Select the music files from your hard drive.

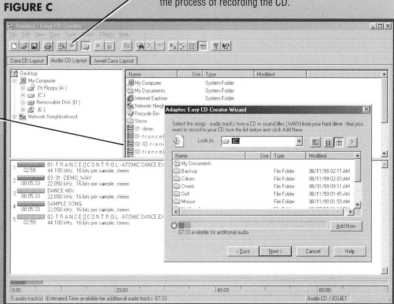

Not all MP3 files can be legally copied. To protect yourself, only download music files from reputable sites.

The Web is continually changing and some of the specifics presented in this Making IT Work for You may have changed. See our Web site at http://www.mhhe.com/it/oleary/IT.mhtml for possible changes and to learn more about this application of technology.

Format	Type	Typical Capacity	Description
CD	CD-ROM	650MB	Fixed content, used to distribute databases, reference books, and software
	CD-R	650 MB	Written-to one time only, used to archive large amounts of data
	CD-RW	650 MB	Reusable, used to create and edit large multimedia presentations
DVD	DVD-ROM	4.7 GB	Fixed content, distribute theater-quality video and sound presentations
	DVD-R	4.7 GB	Written-to one time only, used to archive very large amounts of data
	DVD-RAM (DVD-RW)	2.6 to 5.2 GB	Reusable, used to create and edit large-scale multimedia presentations

FIGURE 6-15
Types of optical disks

- **DVD-R** stands for *DVD-recordable*. DVD-Rs and DVD-R drives are starting to be used to create permanent archives for large amounts of data. Currently, they are not as widely used as CD-Rs because of their higher costs. This is expected to change as the costs for DVD-R become lower.

- **DVD-RAM** (*DVD–random-access memory*) and **DVD-RW** (*DVD-rewritable*) have recently been introduced. They are two different types of reusable DVD disks. Like CD-RW, they can be used over and over again. They will be used for a wide variety of applications including the creation and editing of large-scale multimedia presentations.

For a summary of optical disk storage capacities, see Figure 6-15.

CONCEPT CHECK

✔ Optical disks can hold over _____ gigabytes of data.

✔ How much data can be stored on one side of a CD?

✔ How much data can DVD drives store on one side of a DVD?

Magnetic Tape

Magnetic tape streamers and magnetic tape reels are used primarily for backup purposes.

To find a particular song on an audiotape, you may have to play several inches of tape. Finding a song on an audio compact disc, in contrast, can be much faster. You select the song, and the disc player moves directly to it. That, in brief, represents the two different approaches to external storage. The two approaches are called *direct access* and *sequential access*.

Disks provide fast, direct access. Tapes provide sequential access. With tape, information is stored in sequence, such as alphabetically. You may have to search a tape past all the information from A to P, say, before you get to Q.

FIGURE 6-16
Magnetic tape

This may involve searching several inches or feet, which takes time. (See Figure 6-16.)

Although tape may be slow to access specific information, it is an effective way of making a *backup,* or duplicate, copy of your programs and data. Today, magnetic tape is used almost exclusively for that purpose.

There are two forms of tape storage. These are *magnetic tape streamers,* for use with microcomputers, and *magnetic tape reels,* for use with minicomputers and mainframes.

Magnetic Tape Streamers

Many microcomputer users with hard disks use a device called a **magnetic tape streamer** or a **backup tape cartridge unit.** (See Figure 6-17.) This enables you to duplicate or make a backup of the data on your hard disk onto a tape cartridge. Typical capacities of such tape cartridges are 10 to 20 gigabytes. Advanced forms of backup technology known as **digital audiotape (DAT) drives,** which use 2-inch by 3-inch cassettes, store 4 gigabytes or more. If your internal hard disk fails, you can have it repaired (or get another hard disk). You can restore all your lost data and programs in a matter of minutes from the backup tapes.

Magnetic Tape Reels

The cassette tapes you get for an audiotape recorder are only about 200 feet long. They record 200 characters to the inch. A reel of magnetic tape used with minicomputer and mainframe systems, by contrast, is ½-inch wide and

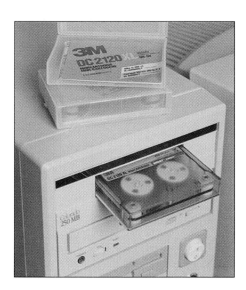

FIGURE 6-17
Magnetic tape streamer

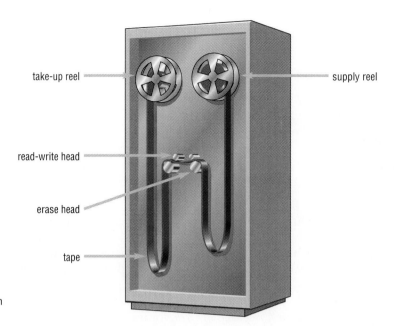

take-up reel

supply reel

read-write head

erase head

tape

FIGURE 6-18
Tape reels for recording data on
magnetic tape

½-mile long. It stores 1,600 to 6,400 characters to the inch. Such tapes are
run on **magnetic tape drives** or **magnetic tape units.** (See Figure 6-18.) You
may never actually see these devices yourself. However, as a microcomputer
user sharing storage devices with others, you may have access to them through
a minicomputer or mainframe.

For the typical microcomputer user, the four storage options—floppy
disk, hard disk, optical disk, and magnetic tape—are complementary, not
competing. Almost all microcomputers today have at least one floppy-disk
drive and one hard-disk drive. For those users who need access to vast
amounts of data, an optical drive is added. Those who need to back up lots
of data and programs may add magnetic tape drives.

A LOOK TO THE FUTURE

Near-field recording devices use lasers and solid immersion lens. Holographic storage systems use three-dimensional holograms.

Have you ever wondered why we need such large-capacity secondary storage devices? As we mentioned earlier, DVD-RWs with a 17-gigabyte capacity are expected. Soon after, we can expect 140-gigabyte FMD-ROM 3D (Flourescent Multilayer Disk) from Constellation 3D. When will this trend of increasing capacity end? Probably not in the near future.

As we use more graphical interfaces like Windows 2000, store images from the Web, and work with more advanced applications like multimedia and virtual reality, the demand for larger and faster secondary storage devices will continue to grow. Fortunately, several new technologies promise to meet this demand.

One is called near-field recording. These devices use a revolutionary new drive head that utilizes lasers and a special focusing lens called a solid immersion lens. This lens precisely directs the laser's path, thereby allowing more information to be stored on a disk. Near-field recording devices are expected to cost the same as today's hard disk systems yet provide ten times more storage capacity.

A bit further on the horizon is holographic storage. Holograms, as you may know, are those shimmering, three-dimensional images often seen on credit cards. Holographic systems can store data equivalent to thousands of books inside a container the size of a sugar cube.

When will we be able to purchase systems using these new technologies? Prototypes have been created and we can expect to see near-field recording devices and holographic systems in the next few years.

VISUAL SUMMARY   Secondary Storage

FLOPPY DISKS	HARD DISKS

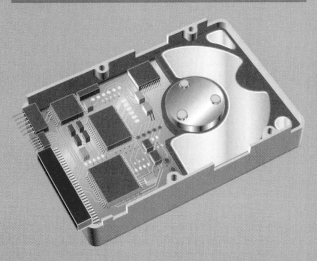

Floppy disks are also known as **diskettes** or simply as **disks.** They are inexpensive, removable storage media. Floppy disks are primarily used to save and transport documents.

Today's Standard
Today's standard floppy disk is the **1.44 MB 3½-inch disk,** or **2HD.** It is by far the most widely used floppy disk today. Due to its relatively low storage capacity, however, it is expected to be replaced by higher capacity floppy disks.

Tomorrow's Standard
Three types of **floppy-disk cartridges** are competing to become the next standard:
- **Zip disks**—considered by many to be the leading contender to replace today's standard; storage capacity 100 MB and 250 MB.
- **SuperDisks**—120 MB capacity; SuperDisk drives are able to read today's standard disks.
- **HiFD**—200 MB capacity; like SuperDisk drives, HiFD drives are able to read today's standard disks.

Parts of a Floppy Disk
Data is recorded in rings called **tracks.** Each track is divided into wedge-shaped sections known as **sectors.** The process of preparing a disk with tracks and sectors is called **formatting** or **initializing.**

Compared to floppy disks, **hard disks** are much faster and provide much greater storage. They are used to store programs and large amounts of data. Three types are *internal, hard disk cartridge,* and *hard-disk pack.*

Internal Hard Disk
An **internal hard disk** or **fixed disk** is located inside the system unit.

Hard-Disk Cartridge
Unlike internal hard disks, **hard-disk cartridges** are removable and only the number of cartridges limits the amount of storage capacity.

Hard-Disk Pack
Hard-disk packs consist of several platters and have a capacity that greatly exceeds both internal and hard-disk cartridges. Hard-disk packs are used primarily by minicomputer and mainframe computer systems.

Performance Enhancements
Three ways to improve hard disk performance are:
- **Disk caching**—reduces time to access data by anticipating data needs.
- **RAIDs**—expand storage capacity by grouping inexpensive hard disk drives.
- **File compression and decompression**—increases storage capacity by reducing the space required to store data and programs.

To be a competent end user you need to be aware of the different types of secondary storage. You need to know their capabilities, limitations, and uses. There are four basic types: floppy disk, hard disk, optical disk, and magnetic tape.

OPTICAL DISKS

While floppy and hard disks use magnetic charges to represent data and programs, optical storage uses reflected light. Two types are **compact discs** and **digital versatile discs.**

Compact Disc
Compact discs or **CDs** are the most widely used optical disks today. With a typical capacity of 650 MB, there are three basic types:
- **CD-ROM**—used to distribute large databases, references, and application packages.
- **CD-R**—also known as **WORM,** used to create custom music CDs and to archive data.
- **CD-RW**—reusable; used to create and edit large multimedia presentations.

Digital Versatile Disc
Digital versatile discs or digital video discs or **DVDs** are similar to CDs with far greater capacity. Three basic types are:
- **DVD-ROM**—used to distribute full length feature films with theater-quality video and sound.
- **DVD-R**—expected to replace CD-R as prices decline.
- **DVD-RAM** and **DVD-RW**—two recently introduced standards for reusable DVDs; used to develop very large scale multimedia projects.

MAGNETIC TAPE

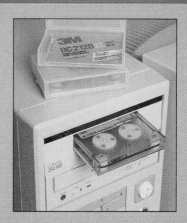

Magnetic tape is a sequential-access storage medium. It is used primarily to back up or duplicate data and programs. Two types are *tape streamers* and *tape reels.*

Magnetic Tape Streamers
Magnetic tape streamers are also known as **backup tape cartridge units.** Used almost exclusively with microcomputers, these units use tape cartridges to back up hard disks. **Digital audiotape (DAT)** is a high capacity, advanced technology for magnetic tape.

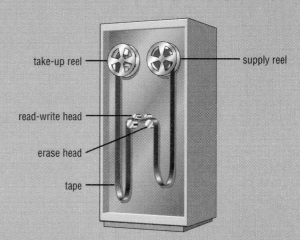

Magnetic Tape Reels
Magnetic tape reels are used to back up minicomputer and mainframe storage devices. The tape is run on **magnetic drives** or **magnetic tape units.**

Key Terms

access time (132)
backup tape cartridge unit (139)
CD (compact disc) (134)
CD-R (135)
CD-ROM (135)
CD-RW (135)
digital audiotape (DAT)
   drive (139)
disk (128)
disk caching (135)
diskette (128)
DVD (digital versatile, or video,
   disc) (136)
DVD-R (DVD-recordable) (138)
DVD-RAM (DVD–random-access
   memory) (138)
DVD-ROM (digital versatile
   disc–read only memory) (137)

DVD-RW (DVD-rewritable) (138)
erasable optical disk (135)
file compression (133)
file decompression (133)
fixed disk (130)
flexible disk (128)
floppies (128)
floppy disk (128)
floppy-disk cartridge (128)
hard disk (130)
hard-disk cartridge (131)
hard-disk pack (131)
head crash (130)
HiFD disk (128)
internal hard disk (130)
lands (134)
magnetic tape drive (140)
magnetic tape streamer (139)

magnetic tape unit (140)
nonvolatile storage (127)
1.44 MB 3½-inch disk (128)
pits (134)
redundant arrays of inexpensive
   disks (RAIDs) (133)
sector (129)
SuperDisk (128)
track (129)
2HD (128)
volatile storage (127)
WORM (135)
write-protect notch (128)
Zip disk (128)

Chapter Review

Reviewing Facts and Terms
Matching

Match each numbered item with the most closely related lettered item. Write your answers in the spaces provided.

1. Write once, read many. __P__

2. Optical disk that can store up to 4.7 gigabytes per side. __F__

3. Time between the computer's request for data from secondary storage and the completion of the data transfer. __A__

4. Helps increase storage capacity by reducing the amount of space required to store data and programs. __H__

5. Erasable optical disks. __C__

6. Happens when the surface of the read-write head or particles on its surface contact the magnetic disk surface. __K__

7. Invisible wedge-shaped sections on a floppy disk. __N__

8. Also known as diskettes, or simply as disks. __I__

9. Bumpy areas on the surface of an optical disk that represent portions of data. __M__

10. A device that allows you to duplicate or make a backup of the data on your hard disk on a tape cartridge. __L__

11. Magnetic tape drives that use 2-inch by 3-inch cassettes capable of storing 4 gigabytes or more. __D__

12. Improves hard-disk performance by anticipating data needs. __E__

13. Recordable DVDs. __G__

14. Data is recorded on a disk in rings. __O__

a. Access time
b. CD
c. CD-RW
d. DAT drive
e. Disk caching
f. DVD
g. DVD-R
h. File compression
i. Floppy disk
j. Hard disk
k. Head crash
l. Magnetic tape streamer
m. Pits
n. Sector
o. Track
p. WORM
q. Zip disk

15. Ridged metallic platter used to store and retrieve information. _J_

16. Floppy-disk cartridge that typically stores either 100 MB or 250 MB. _Q_

17. Optical disk that can store up to 650 megabytes per side. _B_

True/False

In the spaces provided, write T or F to indicate whether the statement is true or false.

1. Secondary storage holds information within the CPU. _F_

2. Floppy disks are also known as flexible disks and as floppies. _T_

3. Sectors are wedge-shaped sections on a disk. _T_

4. CD-Rs can be erased and used over and over again. _F_

5. Laser beams are used to record data on optical disks. _T_

6. An internal hard disk consists of one or more metallic platters sealed inside a container. _T_

Multiple Choice

Circle the letter of the correct answer.

1. Which of the following is exclusively a sequential-access storage media?
   a. floppy disk
   b. hard disk
   (c.) magnetic tape
   d. CD-ROM
   e. WORM

2. Data is stored on concentric circles called:
   a. platters
   (b.) tracks
   c. ovals
   d. wedges
   e. sectors

3. The disk with the largest capacity is:
   a. Zip
   (b.) 3½-inch two-sided, double-density
   c. SuperDisk
   d. CD-ROM
   e. DVD-ROM

4. The hard-disk type that has several platters aligned one above the other is the:
   a. Zip disk
   (b.) hard-disk pack
   c. floppy-disk array
   d. hard-disk cartridge
   e. disk cache

5. The method of improving hard-disk performance by anticipating data needs is:
   a. disk compression
   (b.) disk caching
   c. disk decompression
   d. RAIDs
   e. virtual processing

Completion

Complete each statement in the spaces provided.

1. A ___Compact___ disk is read by a laser projecting a beam of light.

2. Data is recorded on a disk in rings called ___Tracks___

3. Internal hard disks have two advantages over floppy disks: ___Capacity___ and speed.

4. ___Hard Disk Packs___ are removable storage devices with massive capacity.

5. The two forms of tape storage are magnetic tape streamers and magnetic tape ___Packs___.

6. Hard disks are typically used to store __programs__ and __large data files__.
7. The process of preparing a disk with tracks and sectors is called __formatting__.

Reviewing Concepts

LEVEL 2

Open-Ended

On a separate sheet of paper, respond to each question and statement.

1. Explain the difference between direct-access storage and sequential-access storage. Which is more apt to be identified with magnetic disk and which with magnetic tape?

2. What are three contenders to become the next floppy-disk standard? Define their storage capacities.

3. What are the three types of hard disks? What are their relative advantages and disadvantages?

4. What is so disastrous about a head crash?

5. What are the two types of optical disks? Discuss their differences and similarities.

6. Describe one important function that magnetic tape storage can provide.

Concept Mapping

On a separate sheet of paper, draw a concept map or a flowchart showing how the following terms are related. Show all relationships. Include any additional terms you can think of.

access time	DVD-R	head crash
backup tape cartridge unit	DVD-RAM	internal hard disk
CD	DVD-ROM	magnetic tape drive
CD-R	DVD-RW	pits
CD-ROM	file compression	RAIDs
CD-RW	file decompression	sector
DAT drive	floppy disk	track
disk	floppy disk cartridge	volatile storage
disk caching	hard disk	WORM
diskette	hard disk cartridge	write-protect notch
DVD	hard-disk pack	Zip disk

Critical Thinking Questions and Projects

LEVEL 3

Read each exercise and answer the related questions on a separate sheet of paper.

1. *Obtaining a DVD-ROM drive:* Suppose you have a microcomputer system that does not have a DVD-ROM drive, and you would like to have one. You have three options: (1) purchase an external DVD-ROM drive that connects to one of the ports in the back of your system unit; (2) purchase and install an internal DVD-ROM directly into your system unit; or (3) purchase a new microcomputer system that has a DVD-ROM drive.
   Research each of the three options and prepare a set of written guidelines that could be used to make the best choice.

2. *Privacy:* By the end of the year 2000, sales for DVD-ROM drives were over four times those for CD-ROM drives. Some experts believe the shift to DVD technology will have a dramatic and rapid impact. They point out that individuals will be able to own and access the contents of an entire library from a single disk. Of course, other types of information will be available as well.

2001
2002

Do you foresee any new businesses developing? What about the collection and distribution of information? Do you foresee any personal privacy concerns?

3. **CD-R Drives and Music from the Internet:** Creating custom music CDs from the Internet (See Making IT Work for You: CD-R Drives and Music from the Internet on pages 136–137.) and from original music CDs is relatively easy to do. The software industry has long faced the problem of software piracy, which is the illegal copying and distribution of programs. Now the music and motion picture industry face a very similar problem. With CD-R and DVD-R technology, music and full-length motion pictures can readily be copied and illegally sold.

   a. Do you think illegal copying of music and motion pictures is currently a serious problem? Why or why not?

   b. Do you think this problem will become more serious in the future? Why or why not?

   c. Do you think illegal copying of music and motion pictures can be controlled? Why or why not?

On the Web Exercises

1. DVD Entertainment Systems

The DVD home entertainment system is more than just a new entertainment standard. It is also a new standard for DVD-ROMs. To learn more about DVD home entertainment systems, visit our Web site at http://www.mhhe.com/it/oleary/exercise.mhtml to link to a site that specializes in DVD. Once at the site, explore and learn more about DVD. Be sure to check out the Frequently Asked Questions (FAQ) page. Print out the Web page that you find most informative. Write a paragraph addressing the following questions: What is DVD? How do DVDs compare to compact discs (CDs)?

2. JAZ Drives

The ever-increasing demands on hard-disk space has led to Iomega's innovations in high-capacity removable disk drives. For example, the Jaz drive can store 1 gigabyte on a single disk. To learn more about these drives, visit our Web site at http://www.mhhe.com/it/oleary/exercise.mhtml to link to Iomega's Web site. Once connected to that site, check out the available drives. Choose a drive that would be helpful to you in your career and print out its specifications. Write a paragraph discussing how this drive would help you and how it is superior to the other drives available.

3. Buying a Car?

Looking for a new or used car? Interested in the Kelly blue book trade-in price on your current car? Visit our Web site at http://www.mhhe.com/it/oleary/exercise.mhtml to a site that will answer many of your car-buying questions. Once at the site, find the best deal possible on a new or used car of your choice. Print out the information and write a brief paragraph on how you found it.

4. Internet Radio

You can locate, capture, and play music files using the Internet and your computer. (See Making IT Work for You: CD-R Drives and Music from the Internet on pages 136–137.) Did you know that you can also use the Internet to listen to radio broadcasts? Visit our Web site at http://www.mhhe.com/it/oleary/exercise.mhtml to link to a site that provides radio access. Once connected to a broadcast site, print out its opening Web page. Listen to a few minutes of one or two broadcasts and write a paragraph summarizing your listening experience. How did it compare to traditional radio? Write a second paragraph discussing the relative advantages of Internet Radio compared to traditional radio. How might you use this technology? Be specific.

CHAPTER 7

Communications and Connectivity

COMPETENCIES

After you have read this chapter, you should be able to:

1. Describe connectivity options: fax machines, e-mail, voice-messaging systems, videoconferencing systems, shared resources, and online services.

2. Describe conventional modems, T1, ISDN, DSL, cable modem, and satellite connections.

3. Describe the cable and air communications channels—telephone, coaxial, fiber-optic, microwave, and satellite.

4. Discuss bandwidth, serial versus parallel transmission, direction and modes of data transmission, and protocols.

5. Explain network architecture—configurations and strategies.

6. Describe local area, metropolitan area, and wide area networks.

A familiar instrument—the telephone—has extended our uses for the microcomputer enormously. With the telephone or other kinds of communications equipment, you can connect your microcomputer to the microcomputers of other people, to the Internet, and to other, larger computers located throughout the world. As we've mentioned earlier, this *connectivity* puts incredible power on your desk. The result is increased productivity—for you as an individual and for the groups and organizations of which you are a member. Connectivity has become

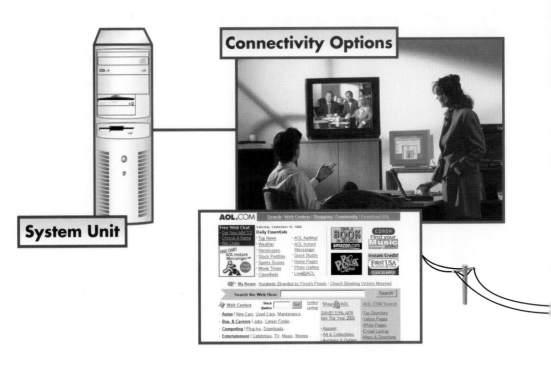

Connectivity Options

System Unit

148

particularly important in business, where individuals now find themselves connected in networks to other individuals and departments.

Data communications systems are the electronic systems that transmit data over communications lines from one location to another. You might use data communications through your microcomputer to send information to a friend using another computer. You might work for an organization whose computer system is spread throughout a building, or even throughout the country or world. Or you might use *telecommunications* lines—telephone lines—to tap into information located in an outside data bank. You could then transmit it to your microcomputer for your own reworking and analysis.

You can even set up a network in your home or apartment using existing telephone lines. Then you can share files, use one Internet connection, and play interactive games with others in your home.

Competent end users need to understand the concept of connectivity and the various connectivity options. Additionally, they need to know the essential parts of communications technology, including connections, channels, transmission, network architectures, and network types.

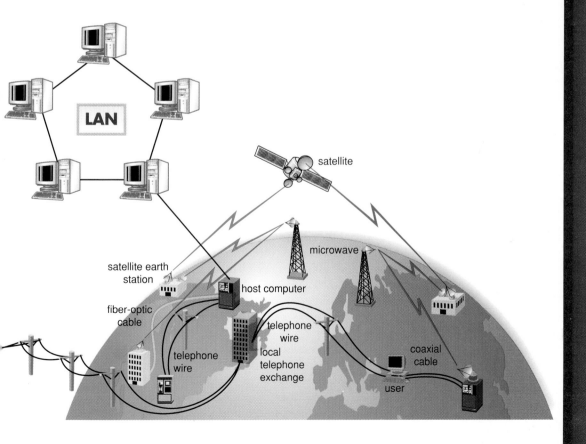

Communications and Connectivity

With communications capability, microcomputer users can transmit and receive data and gain access to electronic information resources.

You may have a desktop microcomputer next to a telephone. You may (or may someday) have a portable computer and a cellular phone in your car. Or you may have a microcomputer that is directly connected to other computers without telephone lines at all. (See Figure 7-1.) Whatever the case, communications systems present many opportunities for transmitting and receiving information, giving you access to many resources. This brings up the important revolution discussed in this chapter, that of connectivity.

Connectivity means you can connect your microcomputer by telephone or other telecommunications links to other computers and information sources almost anywhere. (See Figure 7-2.) With this connection, you are linked to the world of larger computers. This includes minicomputers and mainframes and their large storage devices, such as disk packs, and their enormous volumes of information. Thus, computer competence becomes a matter of knowing not only about microcomputers but also about larger computer systems and their information resources.

Let us consider the options that connectivity makes available to you. These include *fax machines, e-mail, voice-messaging systems, video-conferencing systems, shared resources,* and *online services.*

Fax Machines

Fax machines—facsimile transmission machines—have become essential machines in almost all offices. (See Figure 7-3.) As we mentioned in Chapter 6, these devices scan the image of a document. They convert the image to signals that can be sent over a telephone line to a receiving machine. This machine prints the image out on paper. Microcomputers, using fax/modem circuit boards, can be used to send and receive fax messages.

If you are in the business of meeting deadlines—as most people are—a fax machine is invaluable. Construction engineers can get cost estimates to major contractors. Advertising people can get prospective ad layouts to their clients. Just as important, because people often respond better to pictures than to text, fax can get a picture to them quickly. Rather than giving a friend detailed directions to your house, you could simply fax a map.

E-Mail

E-mail, also known as **electronic mail,** is a way of sending an electronic letter or message between individuals or computers. It is like an answering machine in that you can receive messages even when you are not home. Unlike an answering machine, e-mail can contain text, graphics, and images as well as sound. E-mail can also be used to communicate with more than one person at a time, to conveniently schedule meetings, to keep current on important events, and much more.

At one time, e-mail was used only by large organizations. Now, it is widely used by just about everyone. Students use it to communicate with their instructors and fellow students. (See Figure 7-4.) Parents use it to contact their children's teachers, and grandparents use e-mail to communicate with their grandchildren. Procedures for sending and receiving e-mail will be discussed in detail in Chapter 8.

FIGURE 7-1
Communications

FIGURE 7-2
Connectivity: electricity between computers

FIGURE 7-3
A fax machine

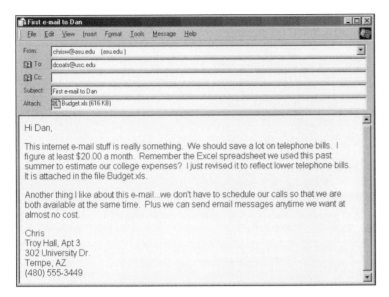

FIGURE 7-4
E-mail

Voice-Messaging Systems

Voice-messaging systems are computer systems linked to telephones that convert the human voice into digital bits. They resemble conventional answering machines and electronic-mail systems. However, they can receive large numbers of incoming calls and route them to the appropriate "voice mailboxes." They can deliver the same message to many people. These systems allow callers to leave "voice mail"—recorded voice messages. They can forward calls to your home or hotel, if you wish. When you check for your messages, you can speed through them or slow them down. You can dictate replies into the phone, and the system will send them out.

Videoconferencing Systems

Videoconferencing systems are computer systems that allow people located at various geographic locations to have in-person meetings. Many corporations have long used specially equipped videoconferencing rooms to hold large group meetings. Top-level executives scattered across the country and the world can meet periodically. Moreover, they can meet at a moment's notice to address emergency situations. Although very expensive, videoconferencing saves considerable travel time and costs.

Today, inexpensive desktop videoconferencing systems are widely used. These systems use microcomputers equipped with inexpensive video cameras and microphones that sit atop the computer monitor. Now, two or more people located in different cities or countries can conveniently and inexpensively meet face-to-face. (See Figure 7-5.)

FIGURE 7-5
Videoconferencing

Shared Resources

An important aspect of connectivity is that it lets microcomputer users share expensive hardware. We have mentioned many of these: laser printers, chain printers, disk packs, and magnetic tape storage. Only in rare instances would a single microcomputer user need the use of, say, a disk pack. However, several microcomputers linked in a network make this option not only feasible but in many cases even essential.

Another important aspect of connectivity is the ability to share data. You may have a personal database that only you use. However, it may also be data you share with others. The data might be stored on your microcomputer's hard disk. Or it might be located somewhere else. That is, you might use a shared database, like one a company might provide its employees so they can share information. This could be information stored on disk packs and accessible through the company's mainframe.

Online Services

Several businesses offer services specifically for microcomputer users. Four well-known service providers are America Online, AT&T WorldNet, CompuServe, and Microsoft Network. (See Figure 7-6.) Typical services provided are:

FIGURE 7-6
America Online

- **Teleshopping:** You dial into a database listing prices and descriptions of products such as appliances and clothes. You then order what you want and charge the purchase to a credit card. The merchandise is delivered later by a package delivery service.

- **Home banking:** If you arrange it with your bank, you may be able to use your microcomputer to pay some bills. Such bills include those owed to big department stores and utility companies. You can also make loan payments and transfer money between accounts.

- **Investing:** You can get access to current prices of stocks and bonds and enter buy and sell orders.

- **Travel reservations:** Just like a travel agent, you can get information on airline schedules and fares. You can also order tickets, charging the purchase to your credit card. (See Figure 7-7.)

- **Internet access:** The most-used service from any of the providers is access to the Internet. Chapter 8 is dedicated to the Internet and the Web.

See Figure 7-8 for a summary of the many connectivity options.

ON THE WEB EXPLORATIONS

America Online is one of the leading national online service providers. To learn more about the company, visit our Web site at

http://www.mhhe.com/it/oleary/ explore.mhtml

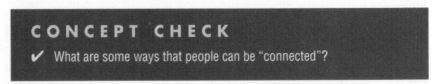

CONCEPT CHECK

✔ What are some ways that people can be "connected"?

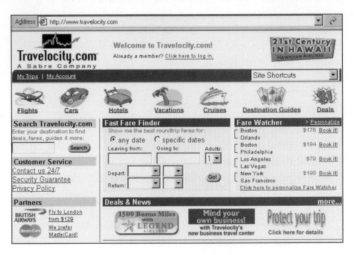

FIGURE 7-7
Travel-reservation Web site

Option	Function
Fax machines	Convert images to signals and send them over telephone lines to receiving fax machines
E-mail	Electronic letter or message sent between individuals or computers
Voice-mail messaging	Convert a voice message to digital bits and distribute to many locations over telephone lines
Videoconferencing systems	Systems allowing in-person electronic meeting between individuals or groups located in different places
Shared resources	Microcomputer users may share expensive hardware, data, and information
Online services	Provide wide range of activities including teleshopping, banking, investments, and travel reservations services

FIGURE 7-8
Summary of connectivity options

User Connection

Conventional modems convert analog and digital signals. Other user connections are T1, DSL, and cable modems.

A great deal of computer communications takes place over telephone lines. However, because the telephone was originally designed for voice transmission, telephones typically send and receive **analog signals.** (See Figure 7-9.) Computers, in contrast, send and receive **digital signals.** These represent the presence or absence of an electronic pulse—the on/off binary signals we mentioned in Chapter 4. To convert the digital signals of your microcomputer to analog and vice versa, you need a modem.

Modems

The word *modem* is short for "*mo*dulator-*dem*odulator." **Modulation** is the name of the process of converting from digital to analog. **Demodulation** is the process of converting from analog to digital. The modem enables digital microcomputers to communicate across analog telephone lines. Both voice communications and data communications can be carried over the same telephone line.

The speed with which modems transmit data varies. Communications speed is typically measured in *bits per second (bps)*. (See Figure 7-10.) The higher the speed, the faster you can send and receive information. For example, transferring an image like Figure 7-9, might take 75 seconds with a 33.6 kbps modem and only 45 seconds with a 56 kbps modem.

Unit	Speed
bps	bit per second
kbps	thousand bits per second
mbps	million bits per second
gbps	billion bits per second

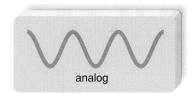
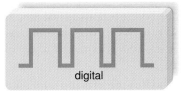

analog   digital

FIGURE 7-9
Analog versus digital signals

FIGURE 7-10
Communication speeds

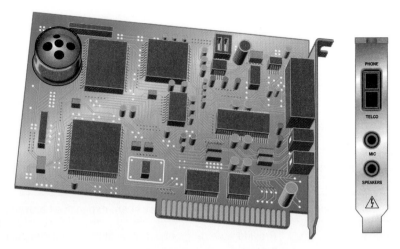

FIGURE 7-11
Internal modem

Types of Modems

There are three types of conventional modems: external, internal, and wireless.

- The **external modem** stands apart from the computer and is connected by a cable to the computer's serial port. Another cable connects the modem to the telephone wall jack.

- The **internal modem** consists of a plug-in circuit board inside the system unit. A telephone cable connects the modem to the telephone wall jack. (See Figure 7-11.)

- The **wireless modem** is very similar to the external modem. It connects to the computer's serial port. Unlike an external modem, it does not connect to telephone lines. Rather, wireless modems receive through the air. This is a new type of modem using new technology.

Types of Connections

Standard telephone lines and conventional modems provide what is called a **dial-up service.** Although still the most popular type of connection service, dial-up service is quite slow, and many users find it inadequate to meet their communication needs.

For years, large corporations have been leasing special high-speed lines from telephone companies. These lines—known as **T1, T2, T3,** and **T4 lines**—support all digital communications, do not require conventional modems, and provide very high capacity. Unfortunately, this type of connection is very expensive. For example, T1 lines provide a speed of 1.5 mbps (over 26 times as fast as a conventional modem) and cost several thousands of dollars. Some promising newer technologies include DSL, cable modems, and satellite/air connection.

- **Digital subscriber line (DSL)** uses existing telephone lines to provide high-speed connections as fast as a T1 connection at less cost. This is a new and rapidly evolving technology. Although presently limited to select cities, DSL is expected to become widely available in the next few years.

- **Cable modems** use existing television cables to provide high-speed connections as fast as a T1 or DSL connection, at a lower cost.

ON THE WEB
EXPLORATIONS

@Home is a leading cable modem service provider. To learn more about the company, visit its Web site at

http://www.mhhe.com/it/oleary/ explore.mhtml

Although cable connections reach 90 percent of the homes in America, all cable companies do not support cable modems. Industry observers, however, predict 100 percent availability within the next few years.

- **Satellite/air connection services** use satellites and the air to download or send data to users at a rate seven times faster than dial-up connections. Unfortunately, users cannot upload or send data to satellites and must rely on a dial-up connection. While slower than DSL and cable modems, satellite/air connections are available almost anywhere that a satellite-receiving disk can be aimed at the southern skies.

For a comparison of typical user connection costs and speeds, see Figure 7-12.

Type	Cost/Year	Speed	Seconds to Receive Image
Dial-up	$250	56 kbps	45.0 seconds
T1	17,150	1.5 mbps	1.7 seconds
DSL	750	1.5 mbps	1.7 seconds
Cable modem	750	1.5 mbps	1.7 seconds
Satellite/air	1,050	400 kbps	6.3 seconds

FIGURE 7-12
Typical user connection costs and speeds
(Cost/Year includes equipment and connection charges.)

CONCEPT CHECK

✔ Telephones typically send/receive _____ signals; computers send/receive _____ signals.

✔ What do modems do?

✔ Communications speed is typically measured in _____ ____ _____.

✔ Name three newer connection technologies.

Communications Channels

Data may flow through five kinds of communications channels: telephone lines, coaxial cable, fiber-optic cable, microwave, and satellite.

The two ways of connecting microcomputers with each other and with other equipment are through a cable and through the air. Specifically, five kinds of technology are used to transmit data. These are telephone lines (twisted pair), coaxial cable, fiber-optic cable, microwave, and satellite.

Telephone Lines

Most telephone lines you see strung on poles consist of **twisted pair** cable, which is made up of hundreds of copper wires. A single twisted pair culminates in a wall jack into which you can plug your phone. (See Figure 7-13.) Telephone lines have been the standard transmission medium for years for both voice and data. However, they are now being phased out by more technically advanced and reliable media.

FIGURE 7-13
Twisted pair cable

FIGURE 7-14
Coaxial cable

FIGURE 7-15
Fiber-optic cable

FIGURE 7-16
Microwave dish

Coaxial Cable

Coaxial cable, a high-frequency transmission cable, replaces the multiple wires of telephone lines with a single solid-copper core. (See Figure 7-14.) In terms of number of telephone connections, a coaxial cable has over 80 times the transmission capacity of twisted pair. Coaxial cable is often used to link parts of a computer system in one building.

Fiber-Optic Cable

In **fiber-optic cable,** data is transmitted as pulses of light through tubes of glass. (See Figure 7-15.) In terms of number of telephone connections, fiber-optic cable has over 26,000 times the transmission capacity of twisted pair.

However, it is significantly smaller. Indeed, a fiber-optic tube can be half the diameter of a human hair. Although limited in the distance they can carry information, fiber-optic cables have several advantages. Such cables are immune to electronic interference, which makes them more secure. They are also lighter and less expensive than coaxial cable and are more reliable at transmitting data. They transmit information using beams of light at light speeds instead of pulses of electricity, making them far faster than copper cable. Fiber-optic cable is rapidly replacing twisted pair telephone lines.

Microwave

In the case of the microwave communications channel, the medium is not a solid substance but rather the air itself. **Microwave communication** uses high-frequency radio waves that travel in straight lines through the air. Because the waves cannot bend with the curvature of the earth, they can be transmitted only over short distances. Thus, microwave is a good medium for sending data between buildings in a city or on a large college campus. For longer distances, the waves must be relayed by means of "dishes," or antennas. These can be installed on towers, high buildings, and mountaintops, for example. (See Figure 7-16.)

Satellite

Satellite communication uses satellites orbiting about 22,000 miles above the earth as microwave relay stations. (See Figure 7-17.) Many of these are offered by Intelsat, the *In*ternational *Tele*communications *Sat*ellite Consortium, which is owned by 114 governments and forms a worldwide

FIGURE 7-17
Satellite

Channel	Description
Twisted pair	Copper wire, standard voice telephone line
Coaxial cable	Solid copper core, more than 80 times capacity of twisted pair
Fiber-optic	Light carries data, more than 26,000 times capacity of twisted pair
Microwave	High-frequency radio waves carry data, travels in straight line through the air
Satellite	Microwave relay station in the sky, rotates at fixed point above the earth

FIGURE 7-18
Types of communications channels

communications system. Satellites rotate at a precise point and speed above the earth. This makes them appear stationary, so they can amplify and relay microwave signals from one transmitter on the ground to another. Thus, satellites can be used to send large volumes of data. Their major drawback is that bad weather can sometimes interrupt the flow of data.

For a summary of the five kinds of communication channels, see Figure 7-18.

CONCEPT CHECK

✔ List five kinds of communications channels.

✔ Which is the fastest form of cable communications channels?

✔ What is one limitation of microwave communications?

Data Transmission

Several technical matters affect data communications. They are bandwidth, serial versus parallel transmission, direction of flow, modes of transmission, and protocols.

Several factors affect how data is transmitted. They include bandwidth, serial or parallel transmission, direction of data flow, modes of transmitting data, and protocols.

Bandwidth

The different communications channels have different data transmission speeds. This bits-per-second transmission capability of a channel is called its **bandwidth.**

Bandwidth may be of three types:

- **Voiceband: Voiceband** is the bandwidth of a standard telephone line and used often for microcomputer transmission; typical speeds are 9,600 to 56 kbps, although with special equipment much higher speeds are possible.

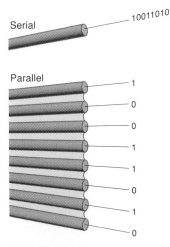

FIGURE 7-19
Serial versus parallel transmission

- **Medium band:** The **medium band** is the bandwidth of special leased lines used mainly with minicomputers and mainframe computers; the speed is 56 kbps to 264 mbps.

- **Broadband:** The **broadband** is the bandwidth that includes microwave, satellite, coaxial cable, and fiber-optic channels. It is used for very high speed computers whose processors communicate directly with each other. It is in the range of 264 mbps to 30 gbps.

Serial and Parallel Transmission

Data travels in two ways: serially and in parallel. (See Figure 7-19.)

- In **serial data transmission,** bits flow in a series or continuous stream, like cars crossing a one-lane bridge. Serial transmission is the way most data is sent over telephone lines. For this reason, external modems typically connect to a microcomputer through a serial port. More technical names for the serial port are **RS-232C connector** and **asynchronous communications port.**

- With **parallel data transmission,** bits flow through separate lines simultaneously. In other words, they resemble cars moving together at the same speed on a multilane freeway. Parallel transmission is typically limited to communications over short distances and typically is not used over telephone lines. It is, however, a standard method of sending data from a computer's CPU to a printer.

Direction of Data Transmission

There are three directions or modes of data flow in a data communications system. (See Figure 7-20.)

FIGURE 7-20
Simplex, half-duplex, and full-duplex communication

simplex

half-duplex

full-duplex

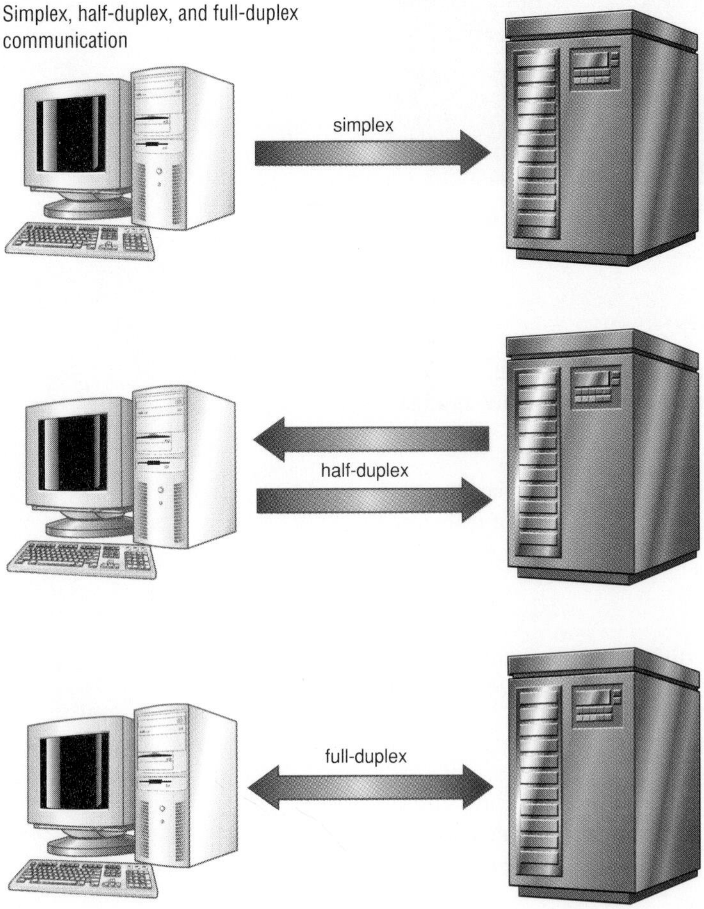

- **Simplex communication** resembles the movement of cars on a one-way street. Data travels in one direction only. The simplex mode is not frequently used in data communications systems today. One instance in which it is used may be in point-of-sale (POS) terminals in which data is being entered only.

- In **half-duplex communication,** data flows in both directions, but not simultaneously. That is, data flows in only one direction at any one time. This resembles traffic on a one-lane bridge. Half-duplex is very common and frequently used for linking micro-computers by telephone lines to other microcomputers.

- In **full-duplex communication,** data is transmitted back and forth at the same time, like traffic on a two-way street. It is clearly the fastest and most efficient form of two-way communication. Full-duplex has been widely used for mainframe communications for years. Now, it is becoming the standard mode for microcomputers as well.

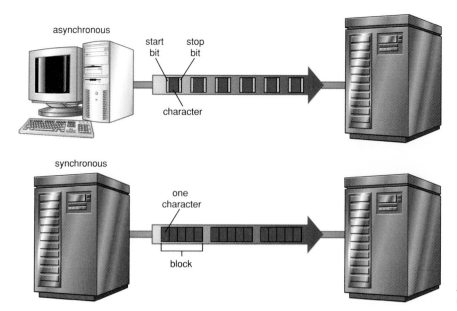

Modes of Data Transmission

Data may be sent by asynchronous or synchronous transmission. (See Figure 7-21.)

- In **asynchronous transmission,** the method frequently used with microcomputers, data is sent and received one byte at a time. Asynchronous transmission is often used for terminals with slow speeds. Its advantage is that the data can be transmitted whenever convenient for the sender. Its disadvantage is a relatively slow rate of data transfer.

- **Synchronous transmission** is used to transfer great quantities of information by sending several bytes or a block at a time. For the data transmission to occur, the sending and receiving of the blocks of bytes must occur at carefully timed intervals. Thus, the system requires a synchronized clock. Its advantage is that data can be sent very quickly. Its disadvantage is the cost of the required equipment.

Protocols

For data transmission to be successful, sender and receiver must follow a set of communication rules for the exchange of information. These rules for exchanging data between computers are known as the line **protocol.** A communications software package like Crosstalk helps define the protocol, such as speeds and modes, for connecting with another microcomputer.

When different types of microcomputers are connected in a network, the protocols can become very complex. Obviously, for the connections to work, these network protocols must adhere to certain standards. The first commercially available set of standards was IBM's Systems Network Architecture (SNA). This works for IBM's own equipment, but other machines won't necessarily communicate with them.

The International Standards Organization has defined a set of communications protocols called the Open Systems Interconnection (OSI). The purpose of the OSI model is to identify functions provided by any network. It separates each network's functions into seven "layers" of protocols, or communication rules. When two network systems communicate, their corresponding layers may exchange data. This assumes that the microcomputers and other equipment on each network have implemented the same functions and interfaces.

CONCEPT CHECK

✔ Bits-per-second transmission capability of a channel is called
_____.

✔ Data travels in two ways: _____ and _____.

✔ Data may be sent by _____ or _____ transmission.

✔ For data transmission to be successful, sender and receiver must
follow _____, or rules for exchanging data between computers.

Network Architecture

**Network architecture describes how a computer network is configured
and what strategies are used.**

Communications channels can be connected in different arrangements, or
networks, to suit different users' needs. A **computer network** is a communications system connecting two or more computers that work together to exchange information and share resources. **Network architecture**
describes how the network is arranged and how the resources are coordi-
nated and shared.

Terms

There are a number of specialized terms that describe computer networks.
Some terms often used with networks are:

- **Node:** A **node** is any device that is connected to a network. It could be
  a computer, printer, or data storage device.

- **Client:** A **client** is a node that requests and uses resources available
  from other nodes. Typically, a client is a user's microcomputer.

- **Server:** A **server** is a node that shares resources with other nodes.
  Depending on the resources shared, it may be called a file server,
  printer server, communication server, Web server, or database server.

- **Network Operating System (NOS):** Microcomputer operating systems
  interact with an application and a computer. On the other hand,
  network operating systems (NOS) control and coordinate the activities between computers on a network. These activities include electronic communication and the sharing of information and resources.

- **Distributed processing:** In a **distributed processing** system, computing
  power is located and shared at different locations. This type of system
  is common in decentralized organizations where divisional offices
  have their own computer systems. The computer systems in the
  divisional offices are networked to the organization's main or centralized computer.

- **Host computer:** A **host computer** is a large centralized computer,
  usually a minicomputer or a mainframe.

A network may consist only of microcomputers, or it may integrate microcomputers or other devices with larger computers. Networks can be controlled by all nodes working together equally or by specialized nodes
coordinating and supplying all resources. Networks may be simple or complex, self-contained or dispersed over a large geographical area.

Configurations

A network can be arranged or configured in several different ways. This arrangement is called the network's **topology.** The four principal network topologies are star, bus, ring, and hierarchical.

In a **star network,** a number of small computers or peripheral devices are linked to a central unit. (See Figure 7-22.) This central unit may be a *host computer* or a *file server.*

All communications pass through this central unit. Control is maintained by **polling.** That is, each connecting device is asked ("polled") whether it has a message to send. Each device is then in turn allowed to send its message.

One particular advantage of the star form of network is that it can be used to provide a **time-sharing system.** That is, several users can share resources ("time") on a central computer. The star is a common arrangement for linking several microcomputers to a mainframe that allows access to an organization's database.

In a **bus network,** each device in the network handles its own communications control. There is no host computer. All communications travel along a common connecting cable called a **bus.** (See Figure 7-23.) As the information passes along the bus, it is examined by each device to see if the information is intended for it.

The bus network is typically used when only a few microcomputers are to be linked together. This arrangement is common in systems for electronic mail or for sharing data stored on different microcomputers. The bus network is not as efficient as the star network for sharing common resources. (This is because the bus network is not a direct link to the resource.) However, a bus network is less expensive and is in very common use.

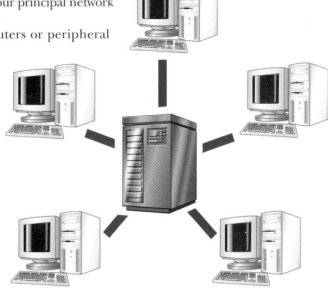

FIGURE 7-22
Star network

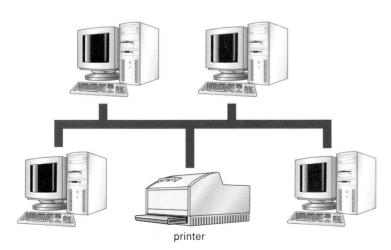

printer

FIGURE 7-23
Bus network

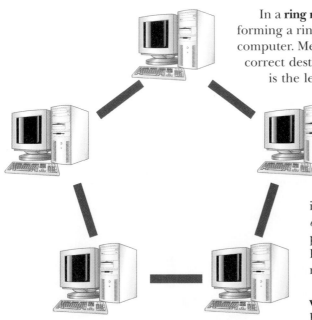

FIGURE 7-24
Ring network

In a **ring network,** each device is connected to two other devices, forming a ring. (See Figure 7-24.) There is no central file server or computer. Messages are passed around the ring until they reach the correct destination. With microcomputers, the ring arrangement is the least frequently used of the four networks. However, it often is used to link mainframes, especially over wide geographical areas. These mainframes tend to operate fairly autonomously. They perform most or all of their own processing and only occasionally share data and programs with other mainframes.

A ring network is useful in a decentralized organization because it makes possible a *distributed data processing system.* That is, computers can perform processing tasks at their own dispersed locations. However, they can also share programs, data, and other resources with each other.

The **hierarchical network**—also called a **hybrid network**—consists of several computers linked to a central host computer, just like a star network. However, these other computers are also hosts to other, smaller computers or to peripheral devices. (See Figure 7-25.)

Thus, the host at the top of the hierarchy could be a mainframe. The computers below the mainframe could be minicomputers, and those below, microcomputers. The hierarchical network allows various computers to share databases, processing power, and different output devices.

A hierarchical network is useful in centralized organizations. For example, different departments within an organization may have individual microcomputers connected to departmental minicomputers. The minicomputers in turn may be connected to the corporation's mainframe, which contains data and programs accessible to all.

For a summary of the network configurations, see Figure 7-26.

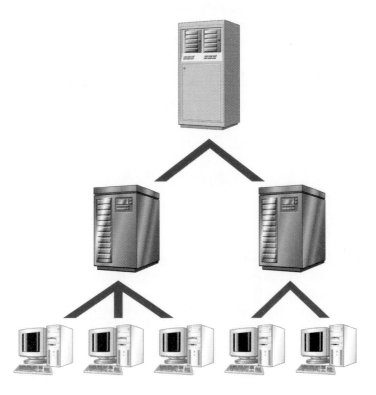

FIGURE 7-25
Hierarchical network

Type	Description
Star	Several computers connected to a central server or host; all communications travel through central server; good for sharing common resources
Bus	Computers connected by a common line; communication travels along this common line; less expensive than star
Ring	Each computer connected to two others forming a ring; communications travel around ring; often used to link mainframe computers in decentralized organizations
Hierarchical	One top-level host computer connected to next-level computers, which are connected to third-level computers; often used in centralized organizations

FIGURE 7-26
Principal network configurations

CONCEPT CHECK

✔ What is a network topology? List the four principal network topologies.

✔ What is one advantage of the star network topology?

✔ A ring network is useful because it makes possible a _____ _____ _____ _____.

✔ Where might you find a hierarchical network?

Strategies

Every network has a *strategy*, or way of coordinating the sharing of information and resources. The most common network strategies are terminal, peer-to-peer, and client/server systems.

In a **terminal network system,** processing power is centralized in one large computer, usually a mainframe. The nodes connected to this host computer are terminals with little or no processing capabilities. (See Figure 7-27.) The star and hierarchical networks are typical configurations with Unix as the operating system.

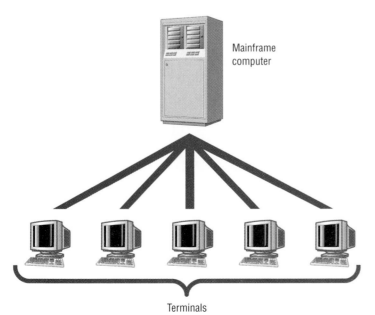

Mainframe computer

Terminals

FIGURE 7-27
Terminal network system

Many airline reservation systems are terminal systems. A large central computer maintains all the airline schedules, rates, seat availability, and so on. Travel agents use terminals to connect to the central computer and use it to schedule reservations. Although the tickets may be printed along with travel itineraries at the agent's desk, nearly all processing is done at the central computer.

One advantage of terminal network systems is the centralized location and control of technical personnel, software, and data. One disadvantage is the lack of control and flexibility for the end user. Another disadvantage is that terminal systems do not use the processing power available with microcomputers. Though the terminal strategy was once very popular, most new systems do not use it.

In a **peer-to-peer network system,** nodes can act as both servers and clients. For example, one microcomputer can *obtain* files located on another microcomputer and can also *provide* files to other microcomputers. (See Figure 7-28.) A typical configuration for a peer-to-peer system is the bus network. Commonly used network operating systems are Novell's NetWare Lite, Microsoft's Windows NT, and Apple's Macintosh Peer-to-Peer LANs.

There are several advantages to using this type of strategy. The networks are inexpensive and easy to install, and they usually work well for smaller systems with fewer than 10 nodes. As the number of nodes increases, however, the performance of the network declines. Another disadvantage is the lack of powerful management software to effectively monitor a large network's activities. For these reasons, peer-to-peer networks are typically used by small networks.

Client/server network systems use one powerful computer to coordinate and supply services to all other nodes on the network. The server provides access to centralized resources such as databases, application software, and hardware. (See Figure 7-29.) This strategy is based on specialization. Server nodes coordinate and supply specialized services, and client nodes request the services. Commonly used network operating systems are Novell's NetWare, Microsoft's Windows NT, IBM's LAN Server, and Banyan Vines.

One advantage of client/server network systems is their ability to handle very large networks efficiently. Another advantage is the powerful network management software that monitors and controls the network's activities. The major disadvantages are the cost of installation and maintenance.

For a summary of the network strategies, see Figure 7-30.

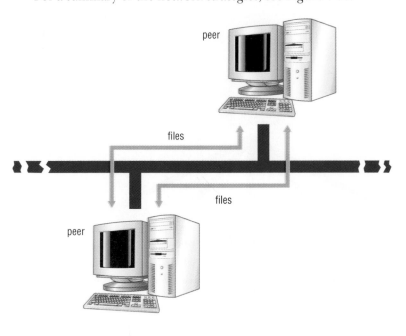

FIGURE 7-28
Peer-to-peer network system

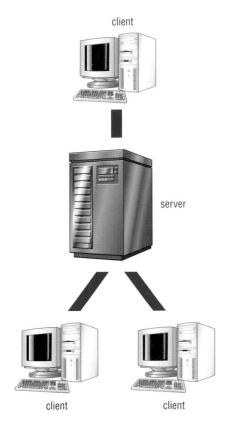

client

server

client   client

Type	Description
Terminal	One large computer provides all processing, strong central control, limited flexibility and control for users
Peer-to-peer	Computers act as both servers and clients, inexpensive and easy to install, works well in small networks
Client/server	Several clients or computers depend upon one central server or computer to coordinate and supply services

FIGURE 7-30
Common network strategies

FIGURE 7-29
Client/server network system

Computer networks in organizations have evolved over time. Most large organizations have a wide range of different network configurations, operating systems, and strategies. These organizations are moving toward integrating or connecting all of these networks together. That way, a user on one network can access resources available throughout the company. This is called **enterprise computing.**

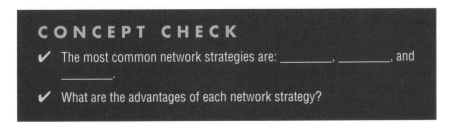

CONCEPT CHECK

✔ The most common network strategies are: _____, _____, and
_____.

✔ What are the advantages of each network strategy?

Network Types

Communications networks differ in geographical size. Three important types are LANs, MANs, and WANs.

Clearly, different types of channels—cable or air—allow different kinds of networks to be formed. Telephone lines, for instance, may connect communications equipment within the same building. You can even create your own network in your home or apartment. (See Making IT Work for You: Home Networking on pages 166–167.)

Networks may also be citywide and even international, using both cable and air connections. Here let us distinguish among three types: *local area networks, metropolitan area networks,* and *wide area networks.*

Local Area Networks

Networks with computers and peripheral devices in close physical proximity—within the same building, for instance—are called **local area networks (LANs).** Linked by cable—telephone, coaxial, or fiber-optic—LANs often use a bus form of organization.

Our illustration shows an example of a LAN. (See Figure 7-31.) This typical arrangement has two benefits. People can share different equipment, which lowers the cost of equipment. For instance, here the four microcomputers share the laser printer and the file server, which are expensive pieces

MAKING IT WORK FOR YOU

Home Networking

Computer networks are not just for corporations and schools anymore. If you have more than one computer, you can use a home network to share files and printers, to allow multiple users access to the Internet at the same time, and to play interactive computer games.

How It Works

Computers can be connected using a variety of ways including electrical wiring, special cables, and radio frequencies. One of the simplest ways, however, is to use your home's existing telephone wiring system. Using this approach, telephone network adapter cards are installed in each computer. The adapter cards are connected to the telephone wiring system using standard telephone connections. Without affecting normal telephone use, your computers are connected and can share resources.

Installing the Network

Various telephone home networking kits are available. These kits typically include adapter card(s), standard telephone extension lines, and installation software. Once the adapter cards have been installed and connected to the telephone wiring system, you need to run the installation software.

For example, using Intel's AnyPoint Home Network installation software, you could set up your home network as shown to the right.

1. Type a unique name for your computer. This is how others on the network will identify your computer.

2. Specify the resources you are willing to share on the network.

3. Select the available resources from other computers you would like to access.

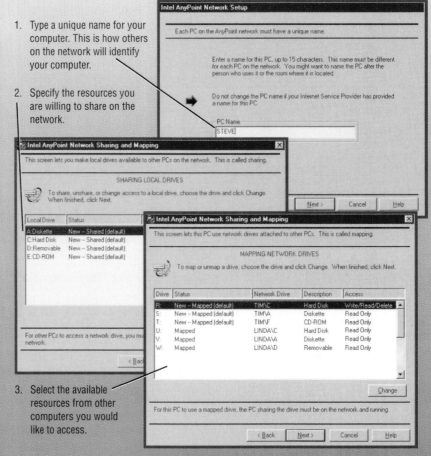

Using the Network

Now your computers are ready to share their resources. The four most common uses of a home network are to share files, printers, and Internet access, and to run multi-player computer games.

Sharing printers: Any computer can print files using any printer that is connected to a computer on the network. This is handy if you have only one printer, or would like to use separate printers for different tasks. For example, you might have one inexpensive black and white printer for written materials and a more expensive color printer for graphics.

Sharing files: You are able to access files stored on other computers on your network. For example, the computer identified as TIM can access files stored on LINDA's and STEVE's hard disk drives.

Sharing the Internet: Only one computer on the network connects directly to the Internet. It is known as the *server*. Whenever the server is connected to the Internet, each computer can surf the Internet independently at the same time. For example, the computer identified as STEVE is the server connected to the Internet. You could be researching a paper on the Web using TIM while someone else is using LINDA to send and receive e-mail.

Multi-player computer games: Several computer games have options to play against other people. Players interact using different computers on the network. For example, you could be using TIM while two friends are using LINDA and STEVE.

Home networks are continually changing and some of the specifics presented in this Making IT Work for You may have changed. See our Web site at http://www. mhhe.com/it/oleary/IT.mhtml for possible changes and to learn more about this application of technology.

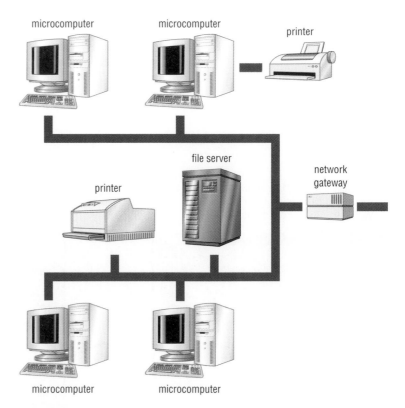

FIGURE 7-31
A local area network that includes a file server and network gateway

ON THE WEB EXPLORATIONS

What about LANs for the home? Intelogis and ShareWave are leaders in developing wireless home networks. To learn more about these companies, visit our Web site at

http://www.mhhe.com/it/oleary/ explore.mhtml

of hardware. (Individual microcomputers also often have their own less expensive printers, such as the ink-jet printer shown in our illustration.) Other equipment may also be added to the LAN—for instance, mini- or mainframe computers or optical-disk storage devices.

Note that the LAN shown in our illustration also features a **network gateway.** A LAN may be linked to other LANs or to larger networks in this manner. For example, the LAN of one office group may be connected to the LAN of another office group. It may also be connected to others in the wider world, even if their configurations are different. Alternatively, a **network bridge** would be used to connect networks with the same configurations.

Metropolitan Area Networks

The next step up from the LAN might be the **MAN**—the **metropolitan area network.** These networks are used as links between office buildings in a city. Cellular phone systems expand the flexibility of MANs by allowing links to car phones and portable phones.

Wide Area Networks

Wide area networks (WANs) are countrywide and worldwide networks. Among other kinds of channels, they use microwave relays and satellites to reach users over long distances—for example, from Los Angeles to Paris. (See Figure 7-32.) Of course, the widest of all WANs is the Internet, which spans the entire globe.

The primary difference between a LAN, MAN, and WAN is the geographical range. Each may have various combinations of hardware, such as microcomputers, minicomputers, mainframes, and various peripheral devices.

CONCEPT CHECK

✔ Distinguish the three types of networks described here.

✔ The primary difference between network types is the _____ _____.

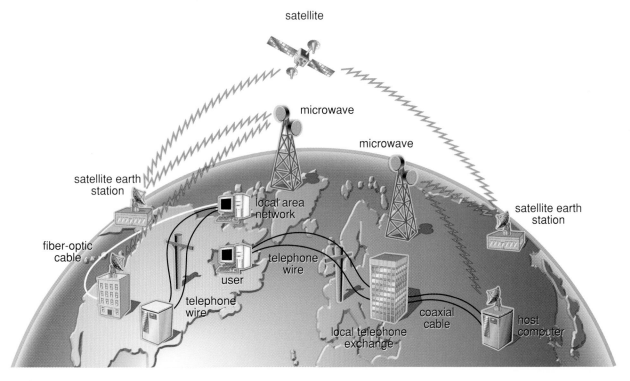

FIGURE 7-32
Wide area network

A LOOK TO THE FUTURE

Teledisc Corporation hopes to create an airborne Internet that will connect the world community.

Can you imagine what it would be like if all of the world's people could directly communicate with one another? If you wanted to know more about life in a remote African village, you could directly connect to members of that village to share experiences and explore the future. Medical researchers from every corner of the world could meet daily to coordinate and share their work. The world community could quickly react to and unite to face global crises.

Of course, that's what the Internet is all about. All the major cities in North America and Europe are wired for fast, efficient Internet access. Most of the rest of the world, however, is not. While satellites can be used to connect these other areas, most satellites orbit at such high altitudes that Internet connections are not effective. But that may be changing.

Bill Gates from Microsoft and Craig McCaw from Cellular Communications have formed a corporation called Teledisc. This corporation is investing over $9 billion to launch over 800 low-orbit satellites by the year 2003. These satellites will hover just 435 miles above the earth and provide direct effective Internet access. Once this airborne Internet is in place, all a user will need to link up to it from any location in the world is a microcomputer with a sender/receiver the size of a large pizza and a signal decoder.

Is there a need for this global communication system? Probably not right now. However, the Internet is growing fast and the demand for worldwide communication is growing even faster. Will Teledisc be a reality in the near future? This project requires a lot of groundbreaking technology and has had some serious financial challenges. We will have to wait and see.

VISUAL SUMMARY — Communications and Connectivity

CONNECTIVITY

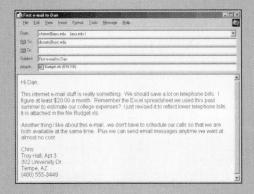

Connectivity is the act of linking computers to other electronic devices through communication systems to share access to information and data.

Available connection options include the following.

Fax Machines
Fax machines scan images and convert to signals to be transmitted over telephone lines.

E-Mail
E-mail or **electronic mail** is an electronic letter sent between individuals or computers.

Voice-Messaging Systems
Voice-messaging systems are computer systems linked to telephones that convert the human voice into digital bits.

Videoconferencing Systems
Videoconferencing systems allow people located at various geographic locations to have in-person meetings.

Shared Resources
Important aspects of connectivity are sharing expensive hardware and data.

Online Services
Online services provide a variety of services including **teleshopping, home banking, investing, travel reservations,** and **Internet access.**

USER CONNECTION

Communication Speeds	
Unit	Speed
bps	bit per second
kbps	thousand bits per second
mbps	million bits per second
gbps	billion bits per second

Users often connect to communication systems using standard telephone lines. The lines typically send and receive **analog** signals. Computers send and receive **digital** signals.

Modems
Modems convert digital signals to analog signals and vice versa. There are three types of conventional modems:
- **External**—outside the system cabinet connected by cable to serial port.
- **Internal**—card that plugs into a slot on the system board.
- **Wireless**—does not connect to telephone line; receives data through the air.

Other Connections
Dial-up connections connect standard telephone line to conventional modems.

Some connections are all digital. For example, **T1, T2, T3,** and **T4** support very high speed transmission without conventional modems. Some promising newer technologies include:
- **DSL**—**digital subscriber line.**
- **Cable modems**—use existing cable lines.
- **Satellite/air**—use satellites to download data.

To be a competent end user you need to understand the concept of connectivity and the various connectivity options. Additionally, you need to know the essential parts of communications technology including connections, channels, transmission, network architectures, and network types.

COMMUNICATIONS CHANNELS

Communications channels connect computers either through cable or through the air. The ways or channels for transmitting data include *telephone lines, coaxial cable, fiber-optic cable, microwave,* and *satellite.*

Telephone Lines
Telephone lines—older telephone lines consist of **twisted pairs** of copper wire. Although still very common, these lines are being replaced.

Coaxial Cable
Coaxial cable replaces the multiple wires of older telephone lines with a solid-copper core to provide higher transmission capacity.

Fiber-Optic Cable
Data is transmitted as pulses of light through a **fiber-optical cable** of glass. It is lighter and faster than coaxial cable and rapidly replacing twisted pair technology.

Microwave
Data is sent short distances through the air in **microwaves** or high-frequency radio waves.

Satellite
Satellites rotating above the earth send data over long distances through the air.

DATA TRANSMISSION

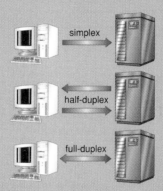

Several factors effect how data is actually transmitted from one point to another:

Bandwidth
Bandwidth is the bits-per-second transmission capability of a channel. Three bandwidths from lowest to highest are: **voiceband, medium band,** and **broadband.**

Serial and Parallel Transmission
With **serial transmission,** bits flow in a single continuous stream. With **parallel transmission,** bits flow through separate lines simultaneously.

Direction of Data Transmission
Three directions or modes of data flow are:
- **Simplex communication**—slower and older technology in which data travels in only one direction
- **Half-duplex communication**—allows data to flow in both directions but not simultaneously.
- **Full-duplex communication**—the fastest and becoming the standard for microcomputer communications. It allows data to travel in both directions simultaneously.

Modes of Data Transmission
Two modes are **asynchronous** (data is sent and received one byte at a time) and **synchronous** (several bytes, a block, are sent at a time).

Protocols
Protocols are rules for exchanging data between computers.

NETWORK ARCHITECTURE

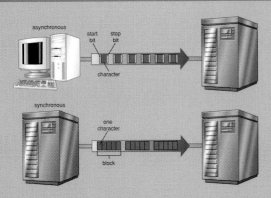

Network architecture describes how a **computer network** is configured and what strategies are used.

Terms
Terms often used with networks are: **node, client, server, network operating system, distributed processing,** and **host computer.**

Configurations
Networks can be configured in several ways:

- **Star network**—each device linked to a central unit; control maintained by **polling**.
- **Bus network**—each device handles its own communications along a common connecting cable called a **bus.**
- **Ring network**—each device connected to two other devices forming a ring.
- **Hierarchical network (hybrid network)**—connect smaller computers to a central host.

Strategies
Every network has a **strategy,** or way of sharing information and resources. Three widely used networking strategies are:

- **Terminal network system**—power is centralized in one large computer and distributed among several terminals.
- **Peer-to-peer network system**—each computer can act as both a server providing services and a client requesting services.
- **Client/server network system**—a computer is either a client requesting access to resources or a server providing access to centralized resources.

NETWORK TYPES

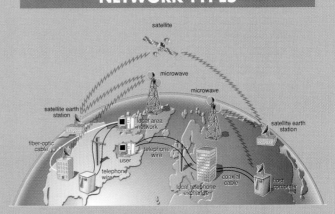

Networks can be citywide or even international using both cable and air connections. Three network types are: *local area networks, metropolitan area networks,* and *wide area networks.*

Local Area Networks
Local area networks or **LANs** connect devices that are located close to one another. Often these devices are located in the same office or floor. **Network gateways** and **bridges** link a LAN to other LANs or to larger networks.

Metropolitan Area Networks
Metropolitan area networks or **MANs** are used to link office building within a city. *Cellular phone systems* expand the flexibility of MANs by allowing links to car phones and portable phones.

Wide Area Networks
Wide Area Networks or **WANs** are the largest type. They span states and countries, or form worldwide networks. The Internet is the largest wide area network in the world.

Key Terms

analog signal (153)
asynchronous communications port (158)
asynchronous transmission (159)
bandwidth (157)
broadband (158)
bus (161)
bus network (161)
cable modem (155)
client (160)
client/server network system (164)
coaxial cable (162)
computer network (166)
data communications system (155)
demodulation (153)
dial-up service (154)
digital signal (153)
digital subscriber line (DSL) (154)
distributed processing (160)
electronic mail (150)
e-mail (150)

enterprise computing (165)
external modem (154)
fax machine (150)
fiber-optic cable (156)
full-duplex communication (158)
half-duplex communication (158)
hierarchical network (162)
host computer (160)
hybrid network (162)
internal modem (154)
local area network (LAN) (166)
medium band (158)
metropolitan area network (MAN) (168)
microwave communication (156)
modulation (15)
network (160)
network architecture (160)
network bridge (168)
network gateway (168)
network operating system (NOS) (160)
node (160)
parallel data transmission (158)

peer-to-peer network system (164)
polling (161)
protocol (159)
ring network (162)
RS-232C connector (158)
satellite communication (156)
satellite/air connection services (155)
serial data transmission (158)
server (160)
simplex communication (158)
star network (161)
synchronous transmission (159)
terminal network system (163)
time-sharing system (161)
T1, T2, T3, T4 lines (154)
topology (161)
twisted pair (155)
videoconferencing system (151)
voiceband (157)
voice-messaging system (151)
wide area network (WAN) (168)
wireless modem (154)

Chapter Review

LEVEL 1

Reviewing Facts and Terms
Matching

Match each numbered item with the most closely related lettered item. Write your answers in the spaces provided.

1. Rules for exchanging data between computers. _p_

2. In this channel, data is transmitted as pulses of light through tubes of glass. _i_

3. A node that shares resources with other nodes. _e_

4. In this method, data is sent and received one byte at a time. _a_

5. These represent the presence or absence of an electronic pulse. _g_

6. The arrangement or configuration of a network. _s_

7. Very similar to the external modem, but it receives through the air. _t_

8. Networks with computers and peripheral devices in very close proximity. _l_

9. A way of sending an electronic letter or message between individuals or computers. _h_

10. Network where a number of small computers are linked to a central unit. _R_

a. Asynchronous transmission
b. Bandwidth
c. Bus network
d. Cable modem
e. Client
f. Client/server network system
g. Digital signal
h. E-mail
i. Fiber-optic cable
j. Full-duplex communication
k. internal
l. Local area network
m. Network

2001 2002

11. Data is transmitted back and forth at the same time, like traffic on a two-way street. __J__

12. In this network system, nodes can act as both servers and clients. __O__

13. The bits-per-second transmission capability of a channel. __B__

14. Any device that is connected to a network. __N__

15. A node that requests and uses resources available from other nodes. ____

16. This network system uses one powerful computer to coordinate and supply services to all other nodes on the network. __Q__

17. A type of modem that consists of a plug-in circuit board inside the system unit. __K__

18. Communications system connecting two or more computers that work together to exchange information and share resources. __M__

19. Uses existing television cables to provide high-speed connections as fast as a T1 or DSL connection at a lower cost. __D__

20. In this network topology, each device handles its own communications control, there is no host, and communications travel along a common connecting cable. __C__

n. Node
o. Peer-to-peer network system
p. Protocol
q. Server
r. Star network
s. Topology
t. Wireless modem

True/False

In the spaces provided, write T or F to indicate whether the statement is true or false.

1. A local area network connects two or more computers within a limited area, such as within the same building. __T__

2. Modems are used to convert data on a CD-ROM to an internal hard disk. __F__

3. Frequently, computer communications over telephone lines require a modem. __F__

4. In half-duplex communication, data flows in both directions at the same time. __F__

5. In a client/server system, each node on the network has equal responsibility for coordinating the network's activities. __T__

6. Voice-messaging systems are computer systems linked to telephones that convert human voice into digital bits. __T__

Multiple Choice

Circle the letter of the correct answer.

1. Computer systems that allow people at different geographic locations to have in-person meetings:
   a. voice-messaging system
   b. online service
   c. fax
   d. videoconferencing system
   e. modem

2. This technology uses existing telephone lines to provide high-speed communication comparable to T1 lines:
   a. internal modems
   b. NET
   c. DDL
   d. cable modem
   e. MTV

3. What communications channel transfers data as pulses of light?

    a. telephone lines   **d.** microwave

    b. coaxial cable   **e.** satellite

    c. fiber-optic cable

4. Rules for exchanging data on a network:

    a. protocol   **d.** channel

    b. asynchronous transmission   **e.** serial transmission

    c. configuration

5. A system frequently used in decentralized organizations in which computing power is located and shared at different sites:

    a. client/server   **d.** mainframe

    b. ring   **e.** distributed

    c. centralized

6. A network strategy where processing power is centralized in one large computer, usually a mainframe, and the nodes connected to this host have little or no processing capability:

    a. terminal network system   **d.** bus network

    b. client/server network system   **e.** data communications system

    c. peer-to-peer network system

Completion

Complete each statement in the spaces provided.

1. _____ means the ability to connect your microcomputer by telecommunications links to other computers and information sources.

2. Internet access is a typical _____ service.

3. Because _____ travel in straight lines through the air and cannot bend with the curvature of the earth, they can transmit data only over short distances.

4. A _____ system controls and coordinates the activities on a network.

5. _____ are countrywide and worldwide networks that connect users over long distances.

6. _____ is the process of converting from analog signals to digital signals.

Reviewing Concepts

Open-Ended

On a separate sheet of paper, respond to each question and statement.

1. Describe typical online services. Have you used any of these services? If so, which ones and how?

2. What are the differences between a terminal, a peer-to-peer, and a client/server system?

3. Describe the differences between simplex, half-duplex, and full-duplex communication.

4. Discuss the four basic network configurations. When is each used?

5. Discuss the three most common network strategies.

6. List the factors that affect data transmission.

Concept Mapping

On a separate sheet of paper, draw a concept map or a flowchart showing how the following terms are related. Show all relationships. Include any additional terms you can think of.

analog signal	fiber-optic cable	polling
asynchronous transmission	full-duplex communication	protocol
bandwidth	half-duplex communication	ring network
broadband	hierarchical network	satellite
bus	host computer	serial data transmission
bus network	local area network	server
client	metropolitan area network	simplex communication
client/server network system	microwave	star network
coaxial cable	voice-messaging system	synchronous transmission
computer network	wide area network	terminal network system
data communications system	modulation	T1, T2, T3, T4 lines
demodulation	network	topology
digital signal	network architecture	voiceband
DSL	node	twisted pair
e-mail	parallel data transmission	videoconferencing system
	peer-to-peer network system	

Critical Thinking Questions and Projects

LEVEL 3

Read each exercise and answer the related questions on a separate sheet of paper.

1. *Connectivity options*: Which of the connectivity options have you used? Describe what you have used them for, when you used them, how you used them, and discuss their effectiveness. For those that you have not used, describe specific situations in which you might use them. Use personal and business examples.

2. *Ethics and privacy*: The FBI has long used court-approved wiretaps on telephone lines to monitor the communications of suspected criminals. Do you think the FBI also wiretaps electronic communications across computer networks? Do you think employers monitor e-mail sent across their internal networks? When you send e-mail or any other type of electronic information across the Internet, do you think other people could be listening in? Under what conditions do you think monitoring of electronic communications is acceptable and ethical? How might these conditions be enforced?

3. *Home Networking:* Most homes can be networked using existing telephone wiring systems (See Making IT Work for You: Home Networking on pages 166–167.). Many new homes today are designed for high-speed local area networks and home-based computer control systems. These control systems coordinate electronic devices in the home such as telephone, heating, air conditioning, and security systems. Prepare a short report describing control systems. Discuss what they are, what they can do, who might use them, and who manufactures them. (Hint: use the Web to check out http://www.smarthome.com and http://www.X10.com)

On the 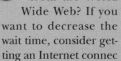 Exercises

1. Travel

You can find all sorts of travel information and advice on the Internet. Visit our Web site at http://www.mhhe.com/it/oleary/exercise.mhtml to link to one of the best travel el sites on the Web. Once connected to that site, you'll find a wide range of travel information from tickets to theatrical and sporting events to plane, hotel, and cruise schedules. Plan the vacation of your dreams. Print out the most interesting Web page, and write a paragraph describing your vacation plans, including estimated budget, destinations, airline (or cruise) departure times, and accommodations.

2. Satellite Link

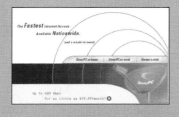

One way to speed up Internet access and delivery is by using a satellite link. All you need is a satellite service provider and satellite dish to download information from satellites and upload information via a phone line. To learn more about satellite links, visit our Web site at http://www.mhhe.com/it/oleary/exercise.mhtml to link to a site specializing in satellite links. Once connected to that site, explore and print out the Web page you find most informative. Write a paragraph listing the pros and cons of using a satellite link for your Internet connection.

3. Cable Connections

Are you tired of waiting for images and text to download from the World Wide Web? If you want to decrease the wait time, consider getting an Internet connec tion with a higher bandwidth. One way is to get a cable connection. To learn more about cable, visit our Web site at http://www.mhhe.com/it/oleary/exercise.mhtml to link to a site specializing in cable connections. Once connected to that site, explore and learn more about cable. Print out the Web page you find most informative. Write a paragraph on who should and who shouldn't get a cable modem and why. Consider individuals in different geographic locations, with different usage needs, and in large and small companies, and suggest alternatives to cable that might be more economical.

4. Home Networking

Home or consumer networking is rapidly growing in popularity. Users of these networks can readily share data, printers, and Internet connections. (See Making IT Work for You: Home Networking on pages 166–167.) These networks can linked by cable or through the air. Visit our Web site at http://www.mhhe.com/it/oleary/exercise.mhtml to link to a site that specializes in home and business networks. Once connected to that site, investigate various network alternatives. Print out the Web page that you found the most informative. Write a paragraph describing the different alternatives. Write a second paragraph discussing how you might use this technology and how a small business might use this technology.

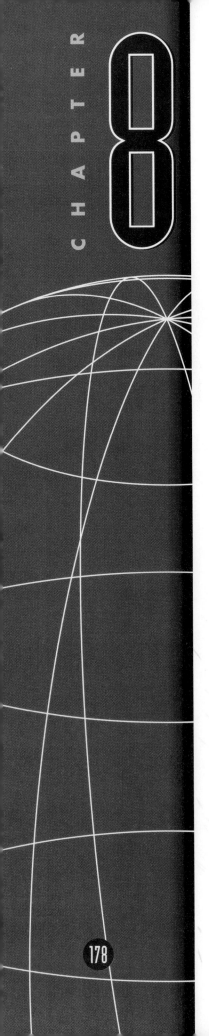

The Internet and the Web

COMPETENCIES

After you have read this chapter, you should be able to:

1. **Describe Internet providers, connections, and protocols.**

2. **Discuss e-mail, mailing lists, newsgroups, chat groups, and instant messaging.**

3. **Describe electronic commerce including Web storefronts, auctions, and electronic payment.**

4. **Describe Internet services: Telnet, FTP, Gopher, and the Web.**

5. **Discuss browsers, Web pages, and Web portals.**

6. **Compare the two types of search tools: indexes and search engines.**

7. **Discuss the two types of Web utilities: plug-ins and helper applications.**

8. **Describe intranets, extranets, and firewalls.**

W ant to communicate with a friend across town, in another state, or even in another country? Perhaps you would like to send a drawing, a photo, or just a letter. Looking for travel or entertainment information? Perhaps you're researching a term paper or

exploring different career paths. Where do you start? For these and other information-related activities, try the Internet and the Web. They are the 21st-century information resources designed for all of us to use.

The Internet is like a highway that connects you to millions of other people and organizations. Unlike typical highways that move people and things from one location to another, the Internet moves your *ideas* and *information*. Rather than moving through geographic space, you move through **cyberspace**—the space of electronic movement of ideas and information. The Web provides an easy-to-use, exciting, multimedia interface to

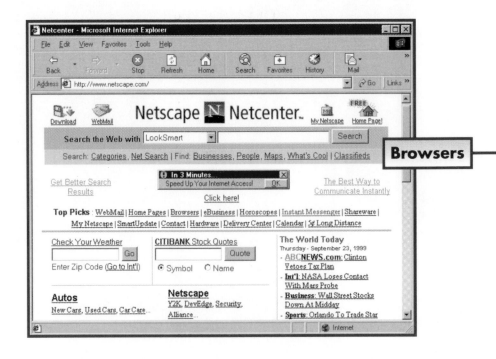

Browsers

connect to the Internet and to access the resources available in cyberspace.

The Internet was launched in 1969 when the United States funded a project that developed a national computer network called **Advanced Research Project Agency Network (ARPANET).** The **Web,** also known as **WWW** and the **World Wide Web,** was introduced in 1992 at the **Center for European Nuclear Research (CERN)** in Switzerland. Prior to the Web, the Internet was all text—no graphics, animations, sound, or video. The Web provided a multimedia interface to resources available on the Internet. From these research beginnings, the Internet and the Web have evolved as tools for all of us to use. For example, you can chat with friends and collaborate on group projects using the Internet and instant messaging.

Competent end users need to be aware of the resources available on the Internet and the Web. Additionally, they need to know how to access these resources, to effectively communicate electronically, to understand electronic commerce, and to efficiently use Web browsers, portals, pages, search tools, and utilities.

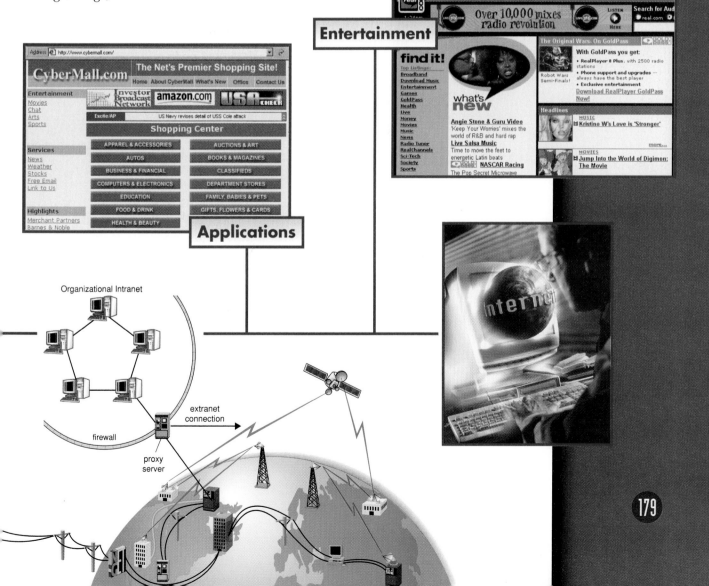

179

Internet Applications

The most common Internet applications are communicating, shopping, researching, and entertainment.

Every day over a billion users from every country in the world use the Internet and the Web. What are they doing? The most common uses are the following.

Communicating

Communicating via e-mail is the most popular Internet activity. You can exchange e-mail with your friends and family located almost anywhere in the world. You can also join and listen to discussions and debates on a wide variety of special-interest topics.

Shopping

One of the fastest-growing Internet applications is electronic commerce. You can visit a cybermall to windowshop at the best stores, look for the latest fashions, search for bargains, and make purchases. (See Figure 8-1.) You can purchase goods using checks, credit cards, or electronic cash.

Researching

How would you like to have one of the world's largest libraries available from home? Well, you can have several of them. By visiting virtual libraries, you can browse through the stacks, read selected items, and even check out books.

Entertainment

Do you like music, movies, books, magazines, or computer games? You'll find them all on the Internet waiting for you to locate and enjoy. You will find live concerts, movie previews, book clubs, and interactive live games. (See Figure 8-2.)

How can you get started using the Internet and the Web? The first step is to get connected, or to gain access to the Internet.

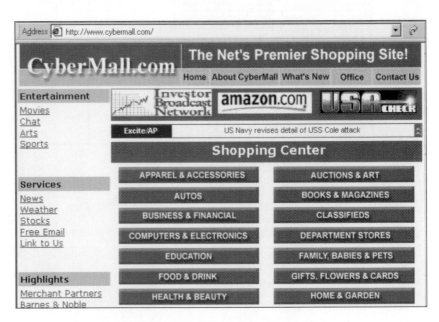

FIGURE 8-1
Cyber mall

FIGURE 8-2
Entertainment

CONCEPT CHECK

✔ What are the most common applications of the Internet?

Access

Providers give us access to the Internet. Internet connections are either direct, SLIP and PPP, or by terminal connection. TCP/IP is the standard Internet protocol.

The Internet and the telephone system are similar—you can connect a computer to the Internet much like you connect a phone to the telephone system. Once you are on the Internet, your computer becomes an extension of what seems like a giant computer—a computer that branches all over the world.

Providers

The most common way to access the Internet is through a **provider** or **host computer.** The providers are already connected to the Internet and provide a path or connection for individuals to access the Internet. There are three widely used providers.

- **Colleges and universities:** Most colleges and universities provide free access to the Internet through their local area networks. You may be able to access the Internet through your school or through other local colleges and universities.

- **Internet service providers:** An **Internet service provider (ISP)** offers access to the Internet for a fee. The best-known national Internet service providers are AT&T WorldNet, IBM Internet Connection, and MindSpring. Local providers are also available in many areas at a slightly lower cost.

TIPS Did you know that some Internet service providers are free? Users of free providers note numerous advertisements and sometimes slow service; however, free is free. If you're interested, check out these sites:

http://www.bluelight.com

http://www.netzero.com

Type	Provider	Site
ISP	AT&T WorldNet	att.net
	IBM Internet Connection	ibm.net
	MindSpring	mindspring.net
	Earthlink	earthlink.com
Online	America Online	aol.com
	CompuServe	compuserve.com
	MSN	msn.com
	Prodigy	prodigy.com

FIGURE 8-3
Service providers

- **Online service providers:** The most widely used source for access to the Internet is through **online service providers.** Like Internet service providers, they provide access to the Internet. Additionally, online service providers offer a variety of other services as well. The best known are America Online (AOL), CompuServe, MSN (Microsoft Network), and Prodigy.

To learn more about service providers, visit some of the sites presented in Figure 8-3.

Connections

To gain access to the Internet, you must have a connection. (See Figure 8-4.) This connection can be made either directly to the Internet or indirectly through a provider. There are three types of connections:

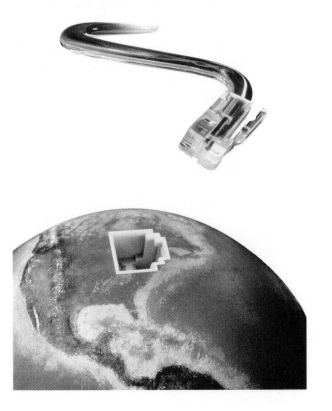

FIGURE 8-4
Connection

- **Direct or dedicated:** To have the most efficient access to all the functions on the Internet, you need a direct or dedicated link. Individuals rarely have direct connections because they are quite expensive. However, many organizations such as colleges, universities, service providers, and corporations do have direct links.

  The primary advantages of a direct link are complete access to Internet functions, ease of connection for individual users, and fast response and retrieval of information. The primary disadvantage is cost.

- **SLIP and PPP:** Using a high-speed modem and standard telephone lines, you can connect to a provider that has a direct connection to the Internet. This type of connection requires special software such as **serial line Internet protocol (SLIP)** or **point-to-point protocol (PPP).** Using this type of connection, your computer becomes part of a client/server network. The provider or host computer is the server providing access to the Internet. Your computer is the client. Using special client software, your computer is able to communicate with server software running on the provider's computer and on other Internet computers.

  This type of connection is widely used by end users to connect to the Internet. It provides a high level of service at a lower cost than a direct or dedicated connection. Of course, it is somewhat slower and may not be as convenient.

- **Terminal connection:** Another way to access the Internet using a high-speed modem and standard telephone lines is called a **terminal connection.** Using this type of connection, your computer becomes a part of a terminal network. Unlike with a SLIP or PPP connection, your computer's operations with a terminal connection are very limited. Your computer simply displays the communication that occurs between the provider and the other computers on the Internet. Compared to a SLIP or PPP connection, terminal connection is less expensive but not as fast or convenient.

For a summary of the typical costs and users of the three types of connections, see Figure 8-5.

TCP/IP

When information is sent over the Internet, it usually travels through numerous interconnected networks. Before a message is sent, it is broken down into small parts called **packets.** Each packet is then sent separately over the Internet, possibly traveling different routes to one common destination. At the receiving end, the packets are reassembled into the correct order. As we mentioned in Chapter 7, *protocols* control how messages are broken down,

Connection	User	Cost
Direct/Dedicated	Medium to large company	$4,000 to $15,000 per year
SLIP/PPP	Individual or small company	$100 to $300 per year, plus hourly charges in some cases
Terminal connection	Individual	$0 to $50 per year

FIGURE 8-5
Typical user and connection costs

sent, and reassembled. They govern how and when computers talk to one another. The standard protocol for the Internet is called **transmission control protocol/Internet protocol (TCP/IP).**

CONCEPT CHECK

✔ Name some common Internet providers.

✔ Before information is sent over the Internet, it is broken into _____. _____ control how messages are broken down, sent, and reassembled.

✔ What is the standard protocol for the Internet?

E-Mail

An e-mail message has three basic elements. Internet addresses use the domain name system.

As we noted earlier, communicating via e-mail is by far the most common Internet activity. You can communicate with anyone in the world who has an Internet address or e-mail account with a system connected to the Internet. All you need is access to the Internet and an e-mail program. Two of the most widely used e-mail programs are Microsoft's Outlook Express and Netscape's Navigator.

Suppose that you have a friend, Dan Coats, who is going to the University of Southern California. You and Dan have been calling back and forth at least once a week for the past month. Your telephone bill has skyrocketed. Fortunately, you both have Internet e-mail accounts through your schools. To save money, you and Dan agree to communicate via the Internet instead of the telephone. After exchanging e-mail addresses, you are ready to send your first Internet e-mail message to Dan.

Basic Elements

A typical e-mail message has three basic elements: header, message, and signature. (See Figure 8-6.) The **header** appears first and typically includes the following information:

- **Subject:** A one-line description, used to present the topic of the message. Subject lines typically are displayed when a person checks his or her mailbox.

- **Addresses:** Addresses of the persons sending, receiving, and, optionally, anyone else who is to receive copies.

- **Attachments:** Many e-mail programs allow you to attach files such as documents and worksheets. If a message has an attachment, the file name appears on the attachment line.

TIPS Did you know that a lot of e-mail never gets read? Busy people quickly scan their e-mail to select which messages to read. Here are a few simple tips to help ensure that your e-mail messages are read:

1. *Make the subject line precise.* Many people screen their e-mail by looking at the subject line. If the subject is unclear or appears to be unimportant, the e-mail may never be read.

2. *Write in short paragraphs.* It's also a good idea to skip a line between paragraphs. This will make your e-mail easier to read and more likely to be read.

3. *Be careful what you write.* You could be held criminally liable for messages that could be interpreted to be abusive, threatening, harassing, or bigoted.

4. *DO NOT TYPE IN ALL UPPERCASE CHARACTERS.* This is called shouting and is perceived as very aggressive. Also, do not type in all lowercase characters. This is perceived as very timid. Use a normal combination of uppercase and lowercase characters.

5. *Check spelling, punctuation, and grammar.* Also, think twice about the content of your message. Once it is sent, you can't get it back.

6. *Don't forward or copy unnecessary material.* This contributes to spam, the electronic equivalent to junk mail. Often used for commercial purposes, spam clutters in-boxes and slows down the Internet.

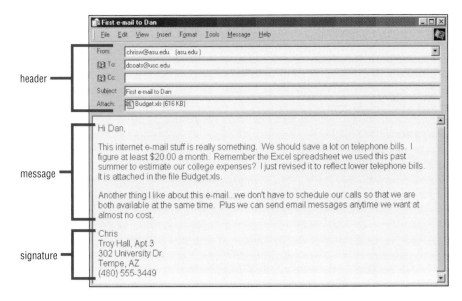

FIGURE 8-6
Basic elements of an e-mail message

The letter or **message** comes next. It is typically short and to the point. Finally, the **signature line** provides additional information about the sender. Typically, this information includes the sender's name, address, and telephone number.

Addresses

One of the most important elements of an e-mail message is the address of the person who is to receive the letter. The Internet uses an addressing method known as the **domain name system (DNS)** to assign names and numbers to people and computers. This system divides an address into three parts: user name, domain name, and domain code. (See Figure 8-7.)

Internet addresses typically are read backwards. The last part of the address, the **domain code,** identifies the geographical description or organizational identification. For example, *edu* in Figure 8-7 indicates an address at an educational and research institution. (For other codes, see Figure 8-8.)

Separated from the domain code by a dot (.) is the **domain name.** It is a reference to the particular organization. In this case, *usc* represents the

ON THE WEB EXPLORATIONS

Almost all ISPs and online service providers offer e-mail service to their customers. But you can get this service for free from Hotmail, Juno Online Services, USA Net Inc., and Yahoo. To learn more about these free services, visit our site at:

http://www.mhhe.com/it/oleary/ explore.mhtml

FIGURE 8-7
Parts of an Internet address

Domain	Identification
com	Commercial
edu	Educational and research
org	Other organizations
net	Major network centers
gov	Government

FIGURE 8-8
Common Internet domain codes

TIPS What if you don't know or have forgotten someone's e-mail address? You can go to e-mail address directories, also known as e-mail "white pages." These directories can be used much like you would use the telephone white pages. Here are three e-mail address directories you might try:

www.bigfoot.com

www.people.yahoo.com

www.infospace.com

University of Southern California. Separated from the domain name by an "at" (@) symbol, the **user name** identifies a unique person or computer at the listed domain. The address shown in Figure 8-7 is for Dan Coats (dcoats) at the University of Southern California (usc), which is an education and research (edu) organization.

CONCEPT CHECK

✔ What are the three basic elements of a typical e-mail message?

✔ The Internet uses an addressing method known as the _____ _____ _____ to assign names and numbers to people and computers.

✔ List the three parts of an address.

Discussion Groups

Mailing lists send e-mail to all members. Newsgroups use the UseNet. Chat groups support live conversations. Instant messaging provides greater control and flexibility. Lurking is good.

You can also use e-mail to communicate with people you do not know but with whom you wish to share ideas and interests. You can participate in discussions and debates that range from general topics like current events and movies to specialized forums like computer troubleshooting and Star Trek.

Mailing Lists

Mailing lists are one type of discussion group available on the Internet. Members of a mailing list communicate by sending messages to a **list address.** Each message is then copied and sent via e-mail to every member of the mailing list.

There are thousands of different mailing lists. To participate in one, you must first subscribe by sending an e-mail request to the mailing list **subscription address.** (See Figure 8-9.) Once you are a member of a list, you can expect to receive e-mail from others on the list. You may find the number of messages to be overwhelming. If you want to cancel a mailing list, send an e-mail request to "unsubscribe" to the subscription address.

Description	Subscription Address
Music and bands	dbird@netinfo.com.au
Movies	moviereview-request@cuenet.com
Jokes	dailyjoke@lists.ivillage.com
Travel	tourbus@listserv.aol.com

FIGURE 8-9
Popular mailing lists

Newsgroups

Newsgroups are a widely used type of discussion group. Unlike mailing lists, **newsgroups** use a special network of computers called the **UseNet.** Each of these computers maintains the newsgroup listing. There are over 10,000 different newsgroups organized into major topic areas that are further subdivided into subtopics.

This hierarchy system is similar to the domain name system. For example, the newsgroup specializing in motion picture discussions is categorized under the major topic *rec* (for "recreational"), then the subtopic *arts*, and then the further subdivision *cinema*. (See Figure 8-10.)

Contributions to a particular newsgroup are sent to one of the computers on the UseNet. This computer saves the messages on its system and periodically shares all its recent messages with the other computers on the UseNet. Unlike mailing lists, a copy of each message is not sent to each member of a list. Rather, interested individuals check contributions to a particular newsgroup, reading only those of interest.

There are thousands of newsgroups covering a wide variety of topic areas. (See Figure 8-11.)

Chat Groups

Chat groups are becoming a very popular type of discussion group. While mailing lists and newsgroups rely on e-mail, chat groups allow direct "live" communication. To participate, you join a chat group, select a **channel** or topic, and communicate live with others by typing words on your computer. Other members of your channel immediately see those words on their computers and can respond in the same manner.

By far the most popular chat service is called **Internet Relay Chat (IRC)**. To participate, you need access to a server or computer that supports IRC. This is done using special chat-client software. This software is available free from several locations on the Internet. Using the chat-client software, you log on to the server, select a channel or topic in which you are interested, and begin chatting.

FIGURE 8-10
Newsgroup hierarchy

Description	Newsgroups
Aerobics fitness	misc.fitness.aerobic
Cinema	rec.arts.movies
Mountain biking	rec.bicycles.off-road
Music	rec.music.hip-hop
Clip art	alt.binaries.clip-art

FIGURE 8-11
Popular newsgroups

Instant Messaging

Like chat groups, **instant messaging** allows one or more people to communicate via direct, "live" communication. Instant messaging, however, provides greater control and flexibility than chat groups. (See Making IT Work for You: Instant Messaging, shown below.)

To use instant messaging, you specify a list of friends, or "buddies," and register with an instant messaging server. Whenever you connect to the Internet, you use special software to tell your messaging server that you are online too. It notifies you if any of your buddies are online. At the same time, it notifies your buddies that you are online. You can then send messages back and forth to one another instantly.

MAKING IT WORK FOR YOU

Instant Messaging

Do you enjoy chatting with your friends? Are you working on a project and need to collaborate with others in your group? Perhaps instant messaging is just what you're looking for. It's easy and free with an Internet connection and the right software.

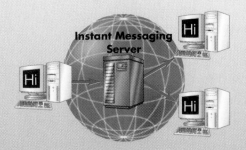

How It Works

Users register with an instant messaging server and identify friends and colleagues (buddies). Whenever a user is online, the instant messaging server notifies the user of all buddies who are also online and provides support for direct "live" communication.

Getting Started

The first step is to connect to one of the many Web sites that support instant messaging. Once at the site, register, download, and install instant messaging software, and create your buddy list.

For example, you can set up AOL Instant Messenger as shown below.

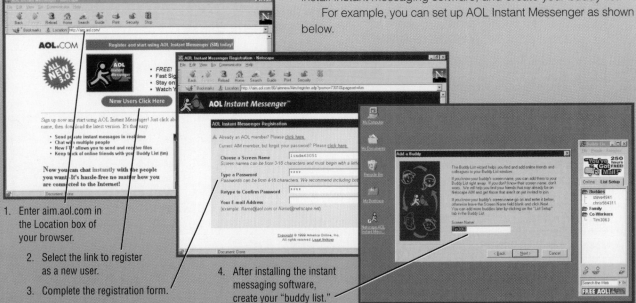

1. Enter aim.aol.com in the Location box of your browser.

2. Select the link to register as a new user.

3. Complete the registration form.

4. After installing the instant messaging software, create your "buddy list."

Communicating with a Friend

Once you have set up your instant messaging software, you can use it to communicate, live, with your online buddies. For example, you could use AOL Instant Messenger as follows:

1. Click the Instant Messenger icon on the desktop. Then enter your screen name and password.

3. Double click your friend's screen name.

4. Enter your message into the message window. Then click the Send button.

Your message is displayed along with your friend's reply.

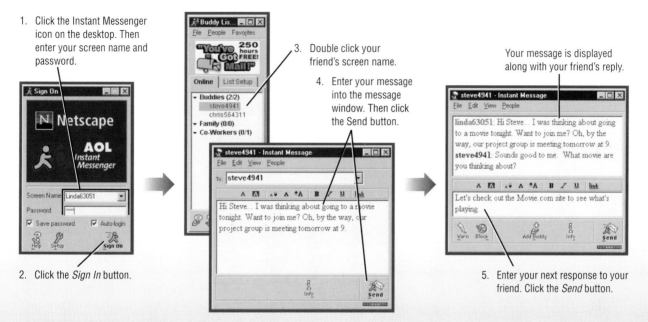

2. Click the *Sign In* button.

5. Enter your next response to your friend. Click the *Send* button.

Collaborating with a Group

You can just as easily communicate or collaborate with a group of people. To conduct a virtual group meeting, all group participants must be signed on and one participant acts as the coordinator.

For example, with AOL Instant Messenger, the coordinator begins as follows:

1. Select the screen name for each participant.

The invitation to join the meeting appears on each invited member's computer screen.

4. Each recipient clicks the *Go Chat* button to join the virtual meeting.

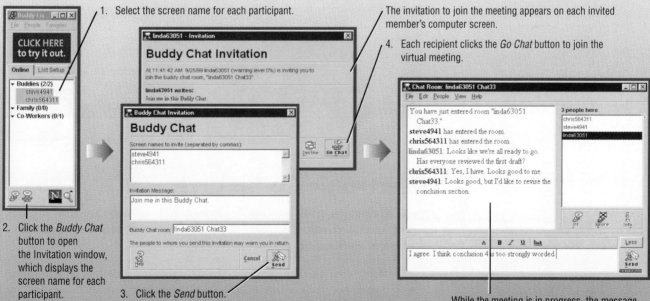

2. Click the *Buddy Chat* button to open the Invitation window, which displays the screen name for each participant.

3. Click the *Send* button.

While the meeting is in progress, the message window on each participant's screen displays the text of the conversation of all members.

Most instant messaging servers require that all participants use the same instant messaging software. However, as standards evolve this limitation will likely be overcome.

The Web is continually changing and some of the specifics presented in this Making IT Work for You may have changed. See our Web site at http://www.mhhe.com/it/oleary/IT.mhtml for possible changes and to learn more about this application of technology.

Term	Description
FAQ	Frequently asked question
Flaming	Insulting, putting down, or attacking
Lurking	Reading news but not joining in to contribute
RFD	Request for discussion
Saint	Someone who aids new users by answering questions
Thread	A sequence of ongoing messages on the same subject
Wizard	Someone who has comprehensive knowledge about a subject

FIGURE 8-12
Selected discussion-group terms

Terms

Before you submit a contribution to a discussion group, it is recommended that you observe or read the communications from others. This is called **lurking.**

By lurking, you can learn about the culture of a discussion group. For example, you can observe the level and style of the discussions. You may decide that a particular discussion group is not what you were looking for—in which case, unsubscribe. If the discussions are appropriate and you wish to participate, try to fit into the prevailing culture. Remember that your contributions will likely be read by hundreds of people.

For a list of some other commonly used discussion group terms, see Figure 8-12.

CONCEPT CHECK

✔ Give examples of the types of discussion groups you can find on the Internet.

✔ What is the most popular chat service on the Internet?

Electronic Commerce

Web storefronts offer goods and services. Web auctions are like traditional auctions. Electronic payment options include check, credit card, and electronic cash.

Electronic commerce, also known as **e-commerce,** is the buying and selling of goods over the Internet. Have you ever bought anything over the Internet? If you have not, there is a very good chance that you will within the next year or two. Shopping on the Internet is growing rapidly and there seems to be no end in sight. (See Figure 8-13.)

Web Storefronts

Web storefronts are virtual stores where shoppers can go to inspect merchandise and make purchases. (See Figure 8-14.) A new type of program

ON THE WEB EXPLORATIONS

Electronic commerce is one of the fastest growing Web applications. To learn more about it, visit our Web site at

http://www.mhhe.com/it/oleary/ explore.mhtml

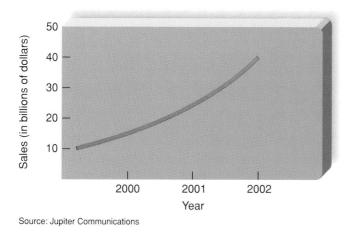

FIGURE 8-13
Online shopping

Source: Jupiter Communications

called **Web storefront creation packages** or **commerce servers** has recently evolved to help businesses create virtual stores. These packages create Web sites that allow visitors to register, browse, place products into virtual shopping carts, and purchase goods and services.

FIGURE 8-14
Web storefront

These programs do even more behind the scenes. They calculate taxes and shipping costs, handle a variety of payment options, update and replenish inventory, and ensure reliable and safe communications. Additionally, the storefront sites collect data about visitors and generate reports to evaluate the site's profitability. For a list of some of the most popular Web storefronts, see Figure 8-15.

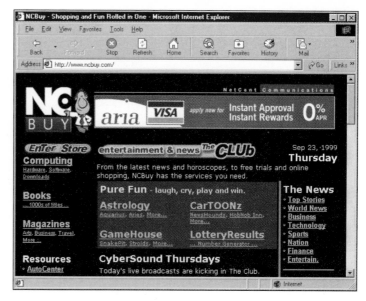

Web Auctions

A recent trend in electronic commerce is the growing popularity of Web auctions. **Web auctions** are similar to traditional auctions except that buyers and sellers seldom, if ever, meet face-to-face. Sellers post descriptions of products at a Web site and buyers submit bids electronically. Like traditional auctions, sometimes the bidding becomes highly competitive and enthusiastic. There are two basic types of Web auction sites:

- **Auction house sites** sell a wide range of merchandise directly to bidders. The auction house owner presents merchandise that is typically from a company's surplus stock. These sites operate like a traditional auction, and bargain prices are not uncommon. Auction house sites are generally considered safe places to shop.

- **Person-person auction sites** operate more like flea markets. The owner of the site provides a forum for numerous buyers and sellers to gather. While the owners of these sites typically facilitate the bidding process, they are not involved in completing transactions or in verifying the authenticity of the goods sold. As with purchases at a flea market, buyers and sellers need to be cautious.

Description	Site
Books	www.amazon.com
Music	www.cdnow.com
Computers and more	www.ncbuy.com
Gifts	www.buy.com
Alaskan products	www.alaskan.com

FIGURE 8-15
Web storefronts

FIGURE 8-16
Electronic payment

Electronic Payment

The single greatest challenge for electronic commerce is the development of fast, secure, and reliable payment methods for purchased goods. (See Figure 8-16.) The three basic payment options are check, credit card, and electronic cash.

- Checks are the most traditional and perhaps the safest. Unfortunately, check purchases require the longest time to complete. After selecting an item, the buyer sends a check through the mail. Upon receipt of the check, the seller verifies that the check is good. If it is good, then the purchased item is sent out.

- Credit card purchases are faster and more convenient than check purchases. Credit card fraud, however, is a major concern for both buyers and sellers. Criminals known as **carders** specialize in stealing, trading, and using stolen credit cards over the Internet.

- **Electronic cash,** or **e-cash,** is the Internet's equivalent to traditional cash. It is also known as **cybercash** and **digital cash**. Buyers purchase e-cash from a third party (a bank that specializes in electronic currency) by transferring funds from their banks. Buyers purchase goods using e-cash. Sellers convert the e-cash to traditional currency through the third party. Although not as convenient as credit card purchases, e-cash is more secure.

CONCEPT CHECK

✔ What is electronic commerce?

✔ What is the single greatest challenge for electronic commerce?

Internet Services

Telnet runs programs on remote computers. FTP transfers files. Gopher provides menus for available resources. The Web provides a multimedia interface to available resources.

There are numerous services available on the Internet. Four commonly used services are Telnet, FTP, Gopher, and the Web. (See Figure 8-17.)

Telnet

Many computers on the Internet will allow you to connect to them and to run selected programs on them. **Telnet** is the Internet service that helps you to connect to another computer (host) on the Internet and log on to that computer as if you were on a terminal in the next room. There are hundreds of computers on the Internet that you can connect to. Some allow limited free access, and others charge fees for their use.

FTP

File transfer protocol (FTP) is an Internet service for transferring files. Many computers on the Internet allow you to copy files to your computer. This is

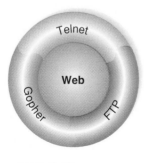

FIGURE 8-17
Available services

called **downloading.** You can also use FTP to copy files from your computer to another computer on the Internet. This is called **uploading.**

Gopher

Gopher is a software application that provides menu-based search and retrieval functions for a particular computer site. It was originally developed at the University of Minnesota in 1991. Internet **Gopher sites** are computers that provide menus describing their available resources and direct links to the resources. Essentially, these menus are a "table of contents" for organizing and locating information. In addition, these sites typically handle transferring of files (FTP) and connecting to other computers (Telnet).

The Web

The Internet service receiving the most attention today is the Web. It is easy to get the Internet and the Web confused, but they are not the same thing. The Internet is the actual physical network. It is made up of wires, cables, and satellites. It connects computers and resources throughout the world. The Web is a multimedia interface to resources available on the Internet.

See Figure 8-18 for a summary of Internet services.

Telnet	Runs programs on remote computers
FTP	Uploads and downloads files
Gopher	Provides menus for available resources for one computer site
Web	Uses a multimedia interface to link to resources located worldwide

FIGURE 8-18
Internet services

CONCEPT CHECK

✔ What are examples of common Internet services?

Browsers

Browsers connect to Web sites using URL addresses. Web portals offer services.

The Web is accessed through your computer using special software known as a **browser.** As we discussed in Chapter 3, this software connects you to remote computers, opens and transfers files, displays text and images, and provides in one tool an uncomplicated interface to the Internet and Web documents. Two well-known browsers are Netscape Navigator and Microsoft Internet Explorer. (See Figure 8-19.)

Uniform Resource Locators

For browsers to connect to other resources, the location or address of the resources must be specified. These addresses are called **uniform resource locators (URLs).** (See Figure 8-20.)

All URLs have at least two basic parts. (See Figure 8-21.) The first part presents the protocol used to connect to the resource. The protocol *http://,* is by far the most common. The second part presents the domain name or the name of the server where the resource is located. In Figure 8-21 the server is identified

FIGURE 8-19
Browser: Netscape Communicator

FIGURE 8-20
URL address

FIGURE 8-21
Two basic parts of a URL

as *www.netscape.com.* (Many URLs have additional parts specifying directory paths, file names, and pointers.)

The URL *http://www.netscape.com* connects your computer to a computer that provides information about Netscape. These informational locations on the Web are called **Web sites.** Moving from one Web site to another is called **surfing.**

Type	Site
Horizontal	www.excite.com
Horizontal portal	www.aol.com
Vertical for sports	www.cbs.sportsline.com
Vertical for news	www.usatoday.com
Vertical for computers	www.zdnet.com

FIGURE 8-22
Popular Web portals

Web Portals

Web portals are sites that offer a variety of services, typically including e-mail, sports updates, financial data, news, and links to selected Web sites. They are designed to encourage you to visit them each time you are on the Web, to act as your home base, and to use as a gateway to their resources. There are two types of portals. **Horizontal portals** are designed to appeal to mass audiences. They offer general-interest services and links. **Vertical portals** present focused content to appeal to special-interest groups. For a list of some popular portals, see Figure 8-22.

CONCEPT CHECK
- ✔ What is a browser?
- ✔ How do browsers locate/connect to other resources?
- ✔ What are Web portals?

Web Pages

Browsers interpret HTML documents to display Web pages.

Once the browser has connected to a Web site, a document file is sent to your computer. This document contains **Hypertext Markup Language (HTML)** commands. The browser interprets the HTML commands and displays the document as a **Web page.** Typically, the first page of a Web site is referred to as its home page. (See Figure 8-23.) The **home page** presents information about the site along with references and **hyperlinks,** or connections to other documents that contain related information—text files, graphic images, audio, and video clips.

These documents may be located on a nearby computer system or on one located halfway around the world. The references appear as underlined and

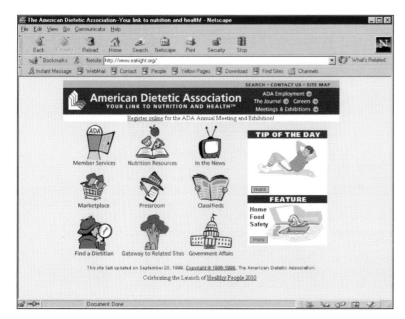

colored text and/or images on the Web page. To access the referenced material, all you do is click on the highlighted text or image. A link is automatically made to the computer containing the material, and the referenced material appears.

Applets and Java

Web pages can also contain links to special programs, called **applets,** written in a programming language called **Java.** These programs can be quickly downloaded and run by most browsers. Java applets are widely used to add interest and activity to a Web site by presenting animation, displaying graphics, providing interactive games, and much more.

CONCEPT CHECK

✔ Browsers interpret _____ documents to display Web pages.

✔ What are Java applets?

Search Tools

Indexes are organized by categories. Search engines are organized like a database.

The Web is a massive collection of interrelated Web pages. (See Figure 8-24.) With so much available information, locating the precise information you need can be difficult. Fortunately, a number of **search tools** have been developed and are available to you on certain Web sites. There are basically two types: indexes and search engines.

Indexes

Indexes, also known as **Web directories,** are organized by categories such as art, computers, entertainment, news, science, sports, and so on. Each category is further organized into subcategories. Using your browser, you select

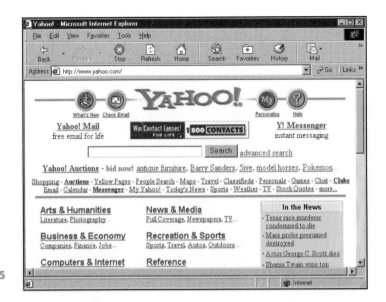

FIGURE 8-25
Index: Yahoo

a category and continue to select subcategories until your search has been nar-
rowed and a list of relevant documents appears. By selecting a document, you
cause the appropriate links to be made and the document appears. One of
the best known and most widely used indexes is Yahoo! (See Figure 8-25.)

Search Engines

Search engines are also known as **Web crawlers** and **Web spiders.** Information
is not organized by major categories. Rather, search engines are organized like
a database, and you search through them by entering key words and phrases. These databases are maintained by special programs called **agents, spiders,**
or **bots.** They automatically search for new information on the Web and
update the databases. Three widely used search engines are HotBot,
WebCrawler, and Alta Vista. (See Figure 8-26.)

CONCEPT CHECK

✔ What are some of the search tools available on the Web?

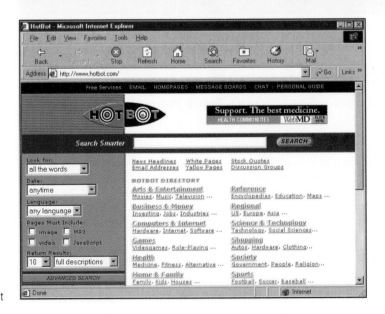

FIGURE 8-26
Search engine: HotBot

Web Utilities

Plug-ins are automatically loaded by your browser. Helper applications are independent programs executed from your browser.

Web utilities are programs that work with a browser to increase your speed, productivity, and capabilities. Many utilities are found in the latest versions of browsers, and others are available free or for a nominal charge. There are two categories of utilities: plug-ins and helper applications.

Plug-ins

Plug-ins are programs that are automatically loaded and operate as a part of your browser. Many Web sites require you to have one or more plug-ins to fully experience their content. Some widely used plug-ins include:

- Shockwave from Macromedia, used for a variety of Web-based games, live concerts, and dynamic animations.
- Quicktime from Apple, required by over 20,000 Web sites to display video and play audio.
- Cosmos from Silicon Graphics, displaying three-dimensional graphics and used in sites displaying virtual reality.

To learn more about plug-ins, visit some of the sites listed in Figure 8-27.

Source	Site
Shockwave	www.macromedia.com
Quicktime	www.quicktime.apple.com
Cosmos Player	www.cosmo.sgi.com

FIGURE 8-27
Plug-in sites

Helper Applications

Also known as **add-ons, helper applications** are independent programs that can be executed or launched from your browser. There are hundreds of helper applications, most of them designed to maximize your efficiency. Four of the most common types are off-line browsers, information pushers, off-line search utilities, and filters.

Off-Line Browsers

In order for a Web page to appear on your screen, its HTML document has to be downloaded from the Web site to your computer and executed. When the Internet is busy and/or the document is large, you spend a fair amount of time waiting. (See Figure 8-28.) Off-line browsers offer a solution.

FIGURE 8-28
Waiting for download

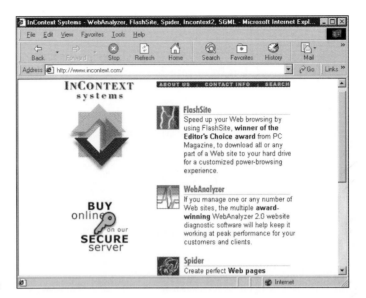

FIGURE 8-29
Off-line browser: InContext FlashSite

Off-line browsers, also known as **Web-downloading utilities** and **pull products,** are programs that automatically connect to selected Web sites, download HTML documents, and save them to your hard disk. You can view the Web pages later without being connected to the Internet and without waiting for documents to be downloaded. Two popular off-line browsers are InContext FlashSite and Teleport Pro. (See Figure 8-29.)

Information Pushers

Imagine a personalized newspaper containing only those articles that interested you the most. That is the basic idea behind information pushers.

You select topic areas known as *channels* that you are interested in. **Information pushers,** which are also known as **Web broadcasters** and **push products,** gather information on your topics and send it to your hard disk, where you can read it whenever you want. Two well-known information pushers are PointCast Network and BackWeb client.

Metasearch Utilities

One way to research a topic is to visit the Web site for several individual indexes and search engines. At each site, the search instructions are entered, and the search begins. After a few moments (or longer, depending on the particular search and the level of activity on the Internet), a list appears. This process can be quite time-consuming, and duplicate responses from the different search tools are inevitable. Metasearch utilities offer a solution.

Metasearch utilities are programs that automatically submit your search request to several indexes and search engines. They receive the results, sort them, eliminate duplicates, and create an index for your review. Three popular metasearch utilities are MetaCrawler, Dogpile, and SavvySearch. (See Figure 8-30.)

Filters

The Internet is an interesting and multifaceted arena. But one of those facets is a dark and seamy one. Parents, in particular, are concerned

---

TIPS   Do your children or little brother or sister use the Internet a lot? While you may want them to be able to take advantage of the educational and entertainment side of the Internet, you also want to protect them from the negative side. To help protect them from viewing inappropriate material, consider the following suggestions:

1. *Locate.* Place the computer in a common area where children can be easily supervised.

2. *Discuss.* Discuss browsing with children to explain what types of content you consider to be appropriate.

3. *Filter.* Use a filter program and/or investigate your browser's filtering capabilities.

Recent versions of Internet Explorer, for example, include a Content Advisor program.

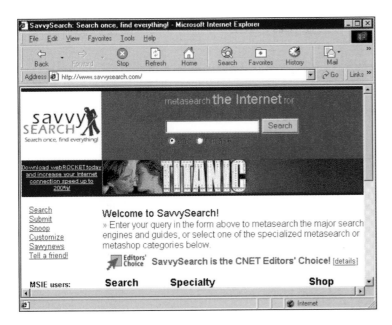

FIGURE 8-30
Metasearch utility: SavvySearch

about children roaming unrestricted across the Internet. **Filter** programs allow parents as well as organizations to block out selected sites and set time limits. Additionally, these programs can monitor use and generate reports detailing the total time spent on the Internet, and time spent at individual Web sites, chat groups, and newsgroups. Three well-known filters are Cyber Patrol, Cybersitter, and Net Nanny.

To learn more about helper applications, explore some of the sites presented in Figure 8-31.

CONCEPT CHECK

✔ What are Web utilities?

✔ What are the two categories of utilities?

Type	Source	Site
Off-line browsers	Incontext FlashSite	www.incontext.com
	Teleport Pro	www.tenmax.com
Information pushers	Entry Point	www.entrypoint.com
	BackWeb Client	www.backweb.com
Metasearch utilities	MetaCrawler	www.metacrawler.com
	Dogpile	www.dogpile.com
Filters	Cyber Patrol	www.cyberpatrol.com
	Cybersitter	www.cybersitter.com

FIGURE 8-31
Helper application sites

Organizational Internets: Intranets and Extranets

Intranets are private networks within an organization. Extranets are private networks connecting organizations. Firewalls use proxy servers to provide security.

Nearly every organization today uses the Internet to promote its products and to service its customers. These organizations have found the Internet and the Web to be powerful yet very easy to use tools for reaching the public. They have also found that they can apply Internet technologies within their organizations to connect their own employees and to connect to other organizations. (See Figure 8-32.) These networks are called intranets and extranets.

FIGURE 8-32
Connecting people and organizations

Intranets

An **intranet** is a *private* network within an organization that resembles the Internet. Like the *public* Internet, intranets use browsers, Web sites, and Web pages. Intranets typically provide e-mail, mailing lists, newsgroups, and FTP services accessible only to those within the organization.

Organizations use intranets to provide information to their employees. Typical applications include electronic telephone directories, e-mail addresses, employee benefit information, internal job openings, and much more. Employees find surfing their organizational intranets to be as easy and as intuitive as surfing the Internet.

Extranets

An **extranet** is a *private* network that connects *more than one* organization. Many organizations use Internet technologies to allow suppliers and others limited access to their networks. The purpose is to increase efficiency and reduce costs. For example, General Motors has thousands of suppliers for the parts that go into making an automobile. By having access to the production schedules, suppliers can schedule and deliver parts as they are needed at the General Motors assembly plants. In this way, General Motors can be assured of having adequate parts without maintaining large inventories.

Firewalls

Organizations have to be very careful to protect their information systems. A **firewall** is a security system designed to protect an organization's network against external threats. It consists of hardware and software that control access to a company's intranet or other internal networks.

Typically a firewall includes a special computer called a **proxy server.** This computer is a gatekeeper. All communications between the company's internal networks and the outside world must pass through it. By evaluating the source and the content of each communication, the proxy server decides whether it is safe to let a particular message or file pass into or out of the organization's network. (See Figure 8-33.)

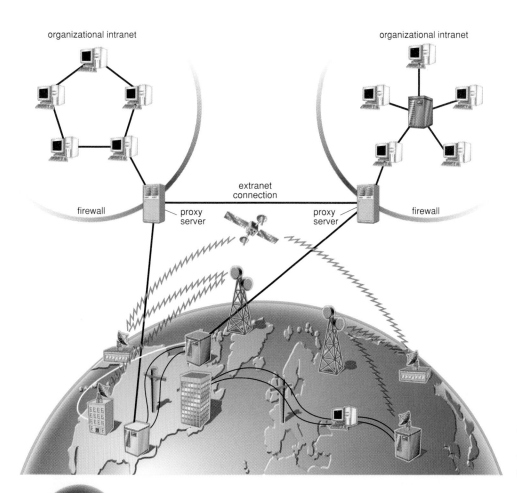

organizational intranet

organizational intranet

extranet
connection

firewall

proxy
server

proxy
server

firewall

FIGURE 8-33
Intranets, extranets,
firewalls, and proxy
servers

A LOOK TO THE FUTURE

Internet2 will be a private high-performance Internet.

Have you ever been unable to connect to the Internet? Have you ever had a long wait before a Web page or a graphic appeared on your screen? Almost all of us have experienced busy servers and slow access. Unfortunately, Internet service is expected to get worse. For organizations that depend on the Internet to reach customers and conduct other business activities, this trend is very concerning.

To address this concern, a separate, private Internet called Internet2 is being developed. It will be a high-speed network capable of dazzling feats that far exceed today's Internet

capabilities. Expected to be fully operational by the end of 2002, Internet2 will have limited access to those willing to pay more to get more. Access to today's Internet will remain public and available for a nominal fee.

The primary beneficiaries of Internet2 will be federal agencies and major corporations. Each will pay an annual fee of $500,000 for access to this network that combines high performance with tightly controlled security. One of the first to take advantage of Internet2 will be online publishers of books, photographs, and original artwork. Advanced virtual reality interfaces, called

nanomanipulators, are expected to be available. Researchers from different parts of the world will be able to share devices such as atomic microscopes and to jointly study, experience, and move within realistic virtual subatomic environments.

Will moving power users to Internet2 increase the performance of the public Internet? We will have to wait and see.

VISUAL SUMMARY The Internet and the Web

INTERNET APPLICATIONS

The most common Internet applications are **communicating, shopping, researching,** and **entertainment.**

ACCESS

Once connected to the Internet, your computer seemingly becomes an extension of a giant computer that branches all over the world.

Providers
The most common access is through a **provider** or **host computer.** Three widely used providers are:
- **Colleges and universities**—often offer free Internet access through their LANs.
- **Internet service providers (ISPs)**—offer access for a fee.
- **Online service providers**—offer access and a variety of other services for a fee.

Connections
To access the Internet, you need to connect to a provider. Three types of connections are **direct (dedicated), SLIP** and **PPP,** and **terminal.**

TCP/IP
TCP/IP is the standard **protocol** of the Internet.

E-MAIL

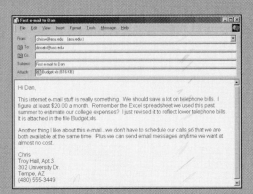

Sending and receiving **e-mail** is the most common Internet activity.

Basic Elements
E-mail messages have three basic elements:
- **Subject**—one line description.
- **Addresses**—sender, receiver, and anyone else receiving copies.
- **Attachments**—files.

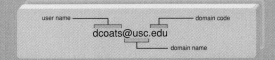

Addresses
The Internet uses the **domain name system (DNS)** for e-mail. The first part is the **user name,** followed by the **domain name,** and then the **domain code.**

DISCUSSION GROUPS

Discussion groups support electronic communication between individuals. Four types exist:
- **Mailing lists** use e-mail **subscription** and **list addresses.**
- **Newsgroups** are organized by major topic areas and use the **UseNet** network.
- **Chat groups** allow direct "live" communication.
- **Instant messaging** is for communicating and collaborating.

To be a competent end user you need to be aware of the basic Internet applications, to be able to access resources, to effectively communicate electronically, to understand electronic commerce, and to efficiently use Web browsers, pages, search tools, and utilities. Also, competent end users need to be aware of organizational intranets and extranets.

ELECTRONIC COMMERCE

Electronic commerce, or **e-commerce**, is the buying and selling of goods over the Internet.

Web Storefronts
Web storefronts are virtual stores for viewing and selecting goods. **Web storefront creation packages** or **commerce servers** are specialized packages for creating virtual stores.

Web Auctions
Web auctions are similar to traditional auction except that buyers and seller rarely meet face-to-face. Two basic types are **auction house sites** and **person-to-person sites.**

Electronic Payment
Three basic payment options are check, credit card, and **electronic cash (e-cash).**

SERVICES

There are four basic types of Internet services: **Telnet, FTP, Gopher,** and the **Web.**

Telnet	Runs programs on remote computers
FTP	Uploads and downloads files
Gopher	Provides menus for available resources for one computer site
Web	Uses a multimedia interface to link to resources located worldwide

BROWSERS

Browsers are programs used to access the Web and provide other Internet services.

Uniform Resource Locators
Uniform Resource Locators (URLs) are addresses of Web resources. The first part is the **protocol,** followed by the **domain name.** Moving from one site to another is called **surfing.**

Web Portals
Web portals are sites that provide a variety of services. **Horizontal portals** appeal to mass audiences. **Vertical portals** appeal to special interest groups.

WEB PAGES

Web pages are created in **HTML.** The first page at a Web site is its **home page. Hyperlinks** are connections to related sites and documents.

Applets and Java
Java is a programming language for creating special programs called **applets**.

SEARCH TOOLS

Search tools are used to locate information on the Web. There are two basic types:
- **Indexes (Web directories)** are organized by categories.
- **Search engines (Web crawlers** and **Web spiders)** are organized like a database.

WEB UTILITIES

Web utilities are programs that increase the performance of browsers. Two categories are: *plug-ins* and *helper applications.*

Plug-ins
Plug-ins are automatically loaded with the browser and are designed to better experience a Web site's contents.

Helper Applications
Helper applications (**add-ons**) are independent applications that can be launched from the browser.

Four types of helper applications exist: **off-line browsers, information pushers, metasearch programs,** and **filters.**

ORGANIZATIONAL INTERNETS

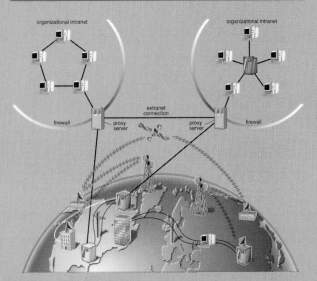

Many **organizations** use Internet technologies in their private networks. These include *Intranets, extranets,* and *firewalls.*

Intranets
Intranets are *private* networks within an organization that resemble the Internet. They use browsers, Web sites, and Web pages that are available only to those within the organization.

Extranets
Extranets are similar to intranets, except that extranets connect more than one organization. Extranets are often used to connect suppliers and producers.

Firewalls
A **firewall** is a security system to protect against external threats. It consists of both hardware and software. All communications into and out of an organization pass through a special security computer called a **proxy server**.

Key Terms

add-on (197)
Advanced Research Project Agency Network
  (ARPANET) (179)
agents (196)
applets (195)
auction house sites (191)
bots (196)
browser (193)
carders (192)
Center for European Nuclear Research
  (CERN) (179)
channel (187)
chat groups (187)
commerce servers (191)
cybercash (192)
cyberspace (178)
digital cash (192)
domain code (185)
domain name (185)
domain name system (185)
downloading (193)
e-cash (192)
e-commerce (190)
electronic cash (192)
electronic commerce (190)
extranet (200)
file transfer protocol (FTP) (192)
filter (199)
firewall (200)
Gopher (193)
Gopher site (193)
header (184)
helper application (197)
home page (194)
horizontal portal (194)
host computer (181)
hyperlinks (194)
Hypertext Markup Language (HTML) (194)
Internet Relay Chat (IRC) (187)
indexes (195)
information pusher (198)
instant messaging (188)
Internet service provider (ISP) (181)
intranet (200)
Java (195)
list address (186)

lurking (190)
mailing list (186)
message (185)
metasearch utility (198)
nanomanipulator (201)
newsgroup (187)
off-line browser (198)
online service provider (182)
packet (183)
person-to-person auction site (191)
plug-in (197)
point-to-point protocol (PPP) (183)
proxy server (200)
provider (181)
pull product (198)
push product (198)
search engines (196)
search tools (195)
serial line Internet protocol (SLIP) (183)
signature line (185)
spider (196)
subscription address (186)
surfing (194)
Telnet (192)
terminal connection (183)
transmission control protocol/Internet protocol
  (TCP/IP) (184)
uniform resource locator (URL) (193)
uploading (193)
UseNet (187)
user name (186)
vertical portal (194)
Web (179)
Web auction (191)
Web broadcaster (198)
Web crawler (196)
Web directory (195)
Web-downloading utility (198)
Web page (194)
Web portal (194)
Web site (194)
Web spiders (196)
Web storefront (190)
Web storefront creation package (191)
World Wide Web (179)
WWW (179)

Chapter Review

LEVEL 1

Reviewing Facts and Terms
Matching

Match each numbered item with the most closely related lettered item. Write your answers in the spaces provided.

1. The space of electronic movement of ideas and information. _C_

2. The standard protocol for the Internet. _P_

3. Security system designed to protect an organization's network against external threats. _F_

4. Special software used by your computer to access the Web. _B_

5. Identifies a unique person or computer at the listed domain. _S_

6. Programs that are automatically uploaded and operate as part of your browser to view certain Web content. _M_

7. Informational locations on the Web. _T_

8. Addressing method used by the Internet to assign names and numbers to people and computers. _D_

9. Copying files from your computer to another computer on the Internet. _R_

10. Widely used Internet programs that add interest and activity to a Web site by presenting animation, displaying graphics, providing interactive games, and more. _H_

11. The buying and selling of goods over the Internet. _e_

12. The type of commands interpreted by a browser to display Web pages. _G_

13. Programming language used to create applets. _j_

14. Internet service that helps you connect to another computer (host) on the Internet and log on to that computer. _Q_

15. A private network within an organization that resembles the Internet. _i_

16. One type of discussion group available on the Internet. _K_

17. Connections to other documents that contain related information. _A_

18. The small parts a message is broken into before it is sent over the Internet. _L_

19. Organized like a database, you search through them by entering key words and phrases. _O_

20. All communications between a company's internal networks and the outside world must pass through this special gatekeeping computer. _N_

a. Applets
b. Browser
c. Cyberspace
d. DNS
e. E-commerce
f. Firewall
g. HTML
h. Hyperlinks
i. Intranet
j. Java
k. Mailing list
l. Packets
m. Plug-in
n. Proxy server
o. Search engine
p. TCP/IP
q. Telnet
r. Uploading
s. User name
t. Web site

True/False

In the spaces provided, write T or F to indicate whether the statement is true or false.

1. The Internet is a huge network that connects computers worldwide. _T_
2. Newsgroups use a special network of computers called the NewsNet. _F_
3. URLs are used to route e-mail. _F_
4. Spiders are special programs that automatically search for new information on the Web. _T_
5. Intranet, extranet, and Internet all mean the same thing. _F_

Multiple Choice

Circle the letter of the correct answer.

1. If you wanted to window-shop on the Internet, you would visit a:
   - **a.** chat group
   - **b.** discussion group
   - **c.** Telnet
   - **(d.)** cybermall
   - **e.** FTP

2. In an e-mail message, the _____ provides a brief description of the message.
   - **(a.)** subject
   - **b.** closing
   - **c.** signature
   - **d.** message
   - **e.** attachment

3. The most popular chat service is:
   - **a.** Talk Today
   - **(b.)** Internet Relay Chat
   - **c.** browser
   - **d.** Yahoo!
   - **e.** Internet Explorer

4. The Web is accessed using a:
   - **(a.)** browser
   - **b.** newsgroup
   - **c.** WAIS
   - **d.** filter
   - **e.** search tool

5. Yahoo! is a(n):
   - **a.** information pusher
   - **b.** browser
   - **c.** chat group
   - **(d.)** home page
   - **e.** search tool

Completion

Complete each statement in the spaces provided.

1. _Internet_ is the electronic movement of ideas and information.
2. SLIP and _PPP_ connections are widely used to access the Internet.
3. The three parts of an Internet address are user name, domain name, and _Domain Code_
4. Off-line search utilities, information pushers, off-line browsers, and filters are _Helper Applications_
5. _Telnet_ is an Internet service that runs programs on remote computers.

Reviewing Concepts

LEVEL 2

Open-Ended

On a separate sheet of paper, respond to each question and statement.

1. Discuss four frequently used Internet services.

2. Describe how addresses are assigned under the domain name system.

3. What are the two types of Internet search tools? How do they differ?

4. Describe some typical Web utilities and how they can help you.

5. Discuss the similarities and differences between intranets and extranets.

Concept Mapping

On a separate sheet of paper, draw a concept map or a flowchart showing how the following terms are related. Show all relationships. Include any additional terms you can think of.

applets	FTP	plug-ins
browsers	host computer	TCP/IP
chat groups	HTML	Telnet
cyberspace	hyperlinks	uploading/downloading
e-mail	intranets	Web site
end user	ISP	World Wide Web
firewall	mailing lists	

Critical Thinking Questions and Projects

LEVEL 3

Read each exercise and answer the related questions on a separate sheet of paper.

1. *Going on an Internet scavenger hunt:* Use the Internet to find information about the following topics. (Record the URLs where you found the information, and write a short description of your findings.)

   a. Hotels in London

   b. Cast members for one of your favorite television programs

   c. MTV's news for this week

   d. Employment opportunities in a career of your choice

   e. Painting known as the *Mona Lisa*

   f. The weather conditions for your city (or the nearest large city)

2. *Ethics and privacy:* While visiting a Web site, it is not uncommon for the site to be collecting information about you. This information is saved in a file that is typically stored on your hard disk to be retrieved and used again the next time you visit that Web site. What type of information do you think is being collected? Why do you suppose it is being collected? Do you think the site has an ethical obligation to inform visitors that it is collecting information? Is there an ethical obligation to inform visitors before using their computer resources? Do you see this activity as an invasion of your privacy? Why or why not?

3. *Privacy and security:* Some who cite the right to free speech and personal privacy argue that the Internet should be totally unencumbered and uncensored by government restrictions and legislation. Do you agree with this position? Do you think religious groups should be allowed to distribute information over the Internet? What about doomsday cult groups? Do you think vendors of pornographic materials should be allowed to use the Internet? What about child pornography? If you think some regulation is required, who should determine what and how restrictions are to be imposed and enforced?

4. *Instant Messaging:* Instant messaging is a popular way to have "live" communication over the Internet. (See Making IT Work for You: Instant Messaging on pages 188–89.)

   a. Have you used instant messaging? If so, describe how and why you used it and whether you found it useful.

   b. Discuss the differences between instant messaging and chat groups. What are the relative advantages and disadvantages?

   c. Can you foresee any situations in which you would not want to be on another person's buddy list? If so, describe the situation.

   d. Describe three situations in which you might use instant messaging. Be specific.

   e. Do you think that the popularity of instant messaging will continue in the future? Why or why not?

On the Web Exercises

1. Off-Line Browsing

Off-line browsing offers a great promise of increased productivity. What could be better than to have your computer searching the Web for you while you sleep? To learn more about off-line browsing, visit our site at http://www.mhhe. com/it/oleary/exercise.mhtml to link a site that specializes in off-line browsing. Once connected, explore and learn more about off-line browsing. Print out the most informative Web page you find and write a brief paragraph summarizing the pros and cons of off-line browsing.

2. Adventure Tours

Travel to exotic places around the world without ever leaving home. To do this, visit our site at http://www.mhhe.com/ it/oleary/exercise.mhtml to link to a site specializing in exotic virtual travel. Once connected to the site, explore and learn more about adventure tours. Print out the Web page you find the most informative, and write a brief paragraph describing a virtual tour and discuss what you think the future is for this type of activity.

3. E-Cash

Several companies, including Cyber-Cash and Digi-Cash, are working on different variations of e-cash. Visit our site at http://www. mhhe.com/it/oleary/ exercise.mhtml to link to a site that specializes in e-cash. Once connected, explore and learn more about e-cash and its alternatives. Print out the most informative Web page you find and write a brief paragraph describing how e-cash works.

4. Instant Messaging

Instant messaging is widely used to directly communicate with friends, to conduct meetings, and to share files on the Internet. (See Making IT Work for You: Instant Messaging on pages 188–89.) Most servers that support instant messaging require participants to use the same instant messaging software. To learn more about different instant messaging software, visit the Yahoo site at http://www.yahoo.com and look at the subject area "Computers and Internet: Software: Internet: Instant Messaging." Explore and print out the Web page you found the most informative. Write a paragraph comparing the different instant messaging programs that are available.

Multimedia, Web Authoring, and More

COMPETENCIES

After you have read this chapter, you should be able to:

1. **Describe multimedia, including story boards and authoring programs.**
2. **Explain Web authoring, Web site design, and Web authoring programs.**
3. **Describe desktop publishing, image editors, and illustration programs.**
4. **Define virtual reality, VRML, and V-R authoring programs.**
5. **Discuss artificial intelligence: robotics and knowledge-based systems.**
6. **Define project management, Gantt charts, and PERT charts.**

Expect surprises—exciting ones, positive ones. This is the view to take in achieving computer competency. If at first the surprises worry you, that's normal. Most people wonder how well they can handle something new. But the latest technological developments also offer you new opportunities to vastly extend your range. As we show in this chapter, software that for years was available only for mainframes has recently become available for microcomputers. Here's a chance to join the competent computer users of tomorrow.

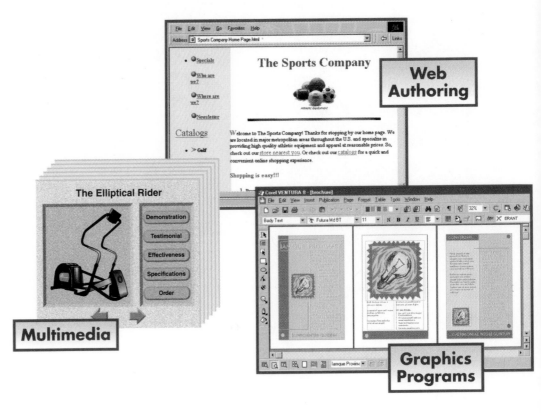

Specialized applications: This is the characterization we have given to a whole new generation of software only recently available for microcomputers. For example, you can create your own personalized Web site for free. Is there really a need to know anything about these new developments? There is if you want to be like those professionals in every area who are at the forefront of their disciplines. They are there because they have found more efficient ways to use their time and talents. You owe it to yourself, therefore, to at least be aware of what this software and hardware can do.

Competent end users need to be aware of specialized applications. They need to know who uses them, what they are used for, and how they are used. These advanced applications include multimedia, Web authoring, graphics programs, virtual reality, artificial intelligence, and project managers.

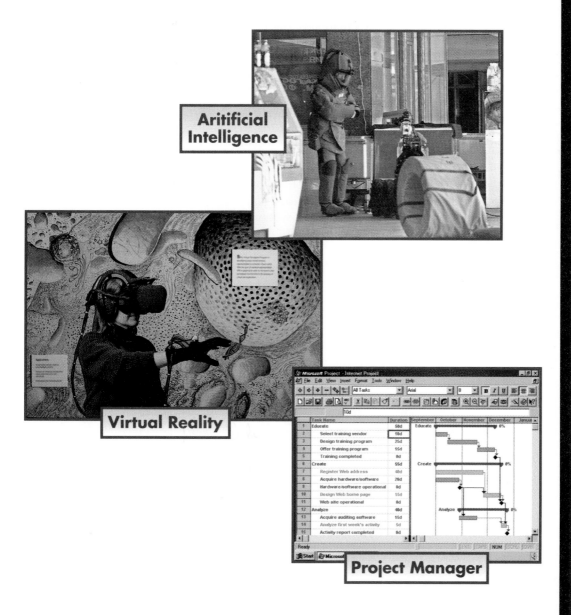

Aritificial Intelligence

Virtual Reality

Project Manager

Multimedia

Multimedia integrates all kinds of information. Pages are linked by buttons. Story boards show logic, flow, and structure. Authoring programs create presentations.

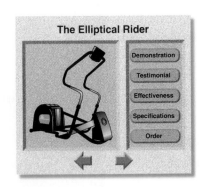

FIGURE 9-1
Opening page of a multimedia presentation

Multimedia, also called **hypermedia,** is the integration of all sorts of media into one presentation. These media may include video, music, voice, graphics, and text. An essential and unique feature of multimedia is user participation, or **interactivity.** When experiencing a multimedia presentation, users typically can control the flow and content by selecting options that customize the presentation to their needs. (See Figure 9-1.)

Once used almost exclusively for computer games, multimedia is now widely used in business, education, and the home. Business uses include high-quality interactive presentations, product demonstrations, and Web page design. In education, multimedia is used for in-class presentation as well as individual study. In the home, multimedia is primarily used for entertainment. In the very near future, however, higher-end multimedia applications such as interactive home shopping are expected to be commonplace.

Links and Buttons

A multimedia presentation is typically organized as a series of related pages. Each page presents information and provides **links** or connections to related information. These links can be to video, sound, graphics, and text files, to other pages, and to other resources.

By clicking special areas called **buttons** on a page, you can make appropriate links and "navigate" through a presentation to locate and discover information. Typically, there are several buttons on a page. You can select one, some, or none of them. You are in control. You direct the flow and content of the presentation.

Story Boards

Story boards are used in the early planning phase of a multimedia project. They are a design tool used to record the intended overall logic, flow, and structure of a multimedia presentation. (See Figure 9-2.) Individual story

ON THE WEB
EXPLORATIONS

Macromedia Inc. is one of the leaders in the development of multimedia authoring programs. To learn more about the company, visit our web site at

http://www.mhhe.com/it/oleary/ explore.mhtml

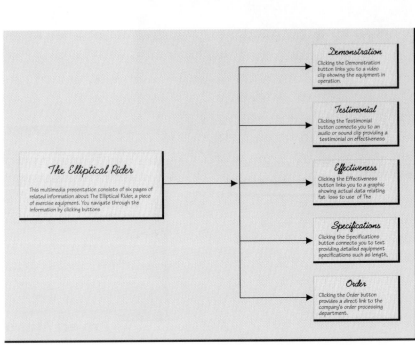

FIGURE 9-2
Partial story board for multimedia project

boards specify the content, style, and design of each display along with the links to video, audio, graphics, text, or any other media.

Multimedia Authoring Programs

Multimedia authoring programs are special programs used to create multimedia presentations. They bring together all the video, audio, graphics, and text elements into an interactive framework. Widely used authoring programs include Macromedia Director, Authorware, and Toolbook.

To learn more about multimedia applications, study Figure 9-3.

TIPS Do you think you might ever be involved with creating a multimedia presentation? Here are a few suggestions:

1. *Design.* Use story boards to graphically represent flow, style, and content. Large projects use design teams consisting of a content expert, a multimedia specialist, and a multimedia programmer.

2. *Produce.* Create and integrate media using a multimedia authoring program. Commands or scripts link and control the various types of media.

3. *Distribute.* Use CDs or DVD-ROMs to distribute completed presentation.

CONCEPT CHECK

✔ What is meant by the term *multimedia?*

✔ Name an essential and unique feature of multimedia.

FIGURE 9-3
How multimedia presentations work

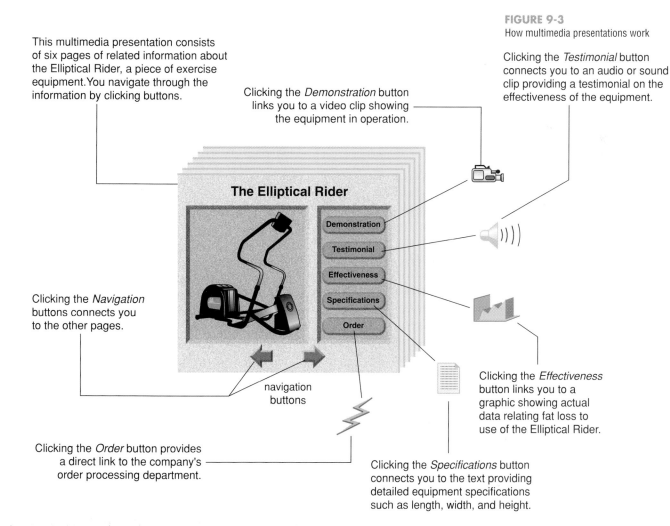

This multimedia presentation consists of six pages of related information about the Elliptical Rider, a piece of exercise equipment. You navigate through the information by clicking buttons.

Clicking the *Demonstration* button links you to a video clip showing the equipment in operation.

Clicking the *Testimonial* button connects you to an audio or sound clip providing a testimonial on the effectiveness of the equipment.

Clicking the *Navigation* buttons connects you to the other pages.

The Elliptical Rider

Demonstration
Testimonial
Effectiveness
Specifications
Order

navigation buttons

Clicking the *Effectiveness* button links you to a graphic showing actual data relating fat loss to use of the Elliptical Rider.

Clicking the *Order* button provides a direct link to the company's order processing department.

Clicking the *Specifications* button connects you to the text providing detailed equipment specifications such as length, width, and height.

Web Authoring

Creating Web sites is called Web authoring. Graphical maps present overall site design. Web authoring programs support design and HTML coding.

There are over half a million commercial Web sites on the Internet, and hundreds more being added every day. Corporations use the Web to reach new customers and to promote their products. Individuals can even create their own personal sites on the Web. (See Making IT Work for You: Personal Web Sites on pages 216–17.) Have you ever wondered how these sites are created or ever thought of creating your own Web site? Creating a site is called **Web authoring.** It begins with Web site design followed by creation of a document file that displays the Web site's content. (See Figure 9-4.)

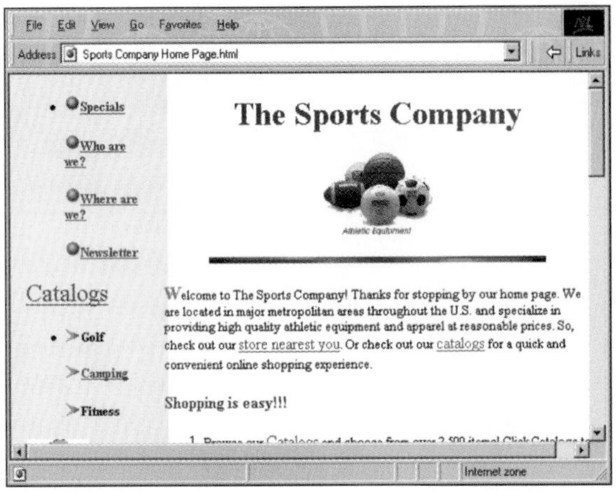

FIGURE 9-4
Web page

Web Site Design

A Web site is an interactive multimedia form of communication. Designing a Web site begins with determining the site's overall content. The content is then broken down into a series of related pieces of information. The overall site design is commonly represented in a **graphical map.** Blocks in the map represent the pieces of information. Lines joining the blocks represent the relationships between the blocks or pieces of information.

The Web site's home page or opening page serves as a table of contents. The following pages present the specific pieces or blocks of information. Links connect related blocks or pages of information. Multimedia elements including sound, video, and animation are added to individual pages to enhance interest and interactivity.

Web Authoring Programs

As we mentioned earlier, Web pages are displayed using HTML documents. With knowledge of HTML and a simple text editor, you can create Web pages. Even without knowledge of HTML you can create Web pages using one of many word processing packages.

More specialized and powerful programs, called **Web authoring programs,** are typically used to create sophisticated commercial sites. Also known as **Web page editors** and **HTML editors,** these programs provide support for Web site design and HTML coding. Widely used Web authoring programs include Adobe PageMill, Corel WebSite Builder, and Microsoft FrontPage.

To learn more about FrontPage and Web authoring, study Figure 9-5.

TIPS Are you thinking about creating your own Web page? Perhaps you already have one and would like to spruce it up a bit? Here are a few suggestions that might help.

1. *Use a common design and theme.* Consistency in the use of colors, fonts, background designs, and navigation features gives a Web site a unified feeling and makes it easier to use.

2. *Use graphics and animations to add interest.* Graphics and animations add interest and focus the user's attention. However, they take time to download and the wait time frustrates users. Be selective and limit the size of graphics. Also reuse graphics from one page to another.

3. *Make navigating your Web site easy.* Create a simple method of navigating that allows users to get to their desired information as quickly as possible. None of your content should be more than three clicks from the home page.

4. *Design your page to fit on a standard 800 × 600 display.* To maximize the impact of your site for the largest number of users, design it to be viewed by a monitor with a standard 800 × 600 resolution.

CONCEPT CHECK

✔ What steps are involved in Web authoring?

✔ What is a Web site?

HTML coding

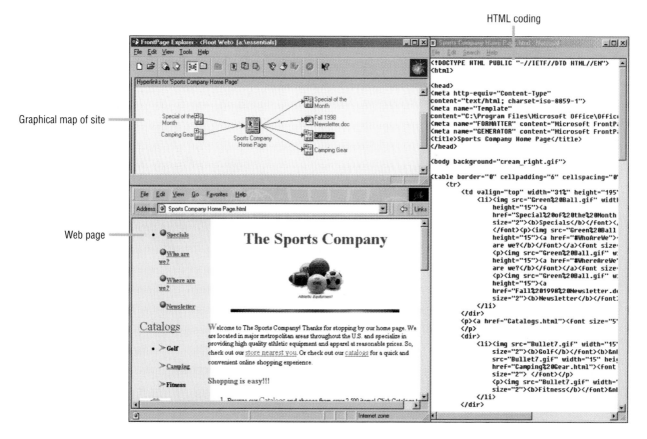

Graphical map of site

Web page

FIGURE 9-5
Web authoring program: Microsoft FrontPage

Graphics Programs

Desktop publishers mix text and graphics. Image editors modify bitmap image files. Illustrators modify vector files. Graphics suites bundle separate programs.

In Chapter 2, we discussed analytical and presentation graphics; which are widely used to analyze data and to create professional-looking presentations. Here we focus on more specialized graphics programs used by professionals in the graphic arts profession. These graphics programs are desktop publishing, image editors, and illustration programs. While desktop publishing is used to produce professional publications, image editors and illustration programs are used to modify graphic images.

Desktop Publishing

Desktop publishing programs allow you to mix text and graphics to create publications of professional quality. While *word processors* focus on creating text and have the ability to combine text and graphics, *desktop publishers* focus on page design and layout and provide greater flexibility.

Desktop publishing programs are widely used by graphic artists to create brochures, newsletters, newspapers, and textbooks. Popular desktop publishing programs include Adobe FrameMaker, Adobe PageMaker, Corel Ventura, and QuarkXPress. (See Figure 9-6.)

FIGURE 9-6
Desktop publisher: Corel Ventura

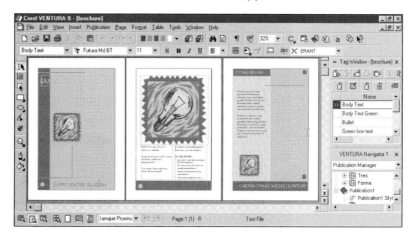

Personal Web Sites

Do you have anything to share with the world? Would you like a personal Web site, but don't want to deal with learning HTML and paying for server time? Many services are available to get you started—for FREE!

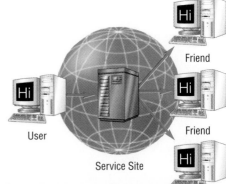

How It Works

A service site on the Web provides access to tools to create personal Web pages. After registering with the site, you create your Web pages. Once completed, the service site acts as a host for your personal Web site and others are free to visit it from anywhere on the Web.

Getting Started

The first step to creating your own Web site is to register with a service site. One of the most popular Personal Web site services is Homestead. To connect to and register for your Web site, follow the instructions below.

1. Connect to www.homestead.com.

2. Click *SIGN UP*, complete the registration procedure, and LOG in.

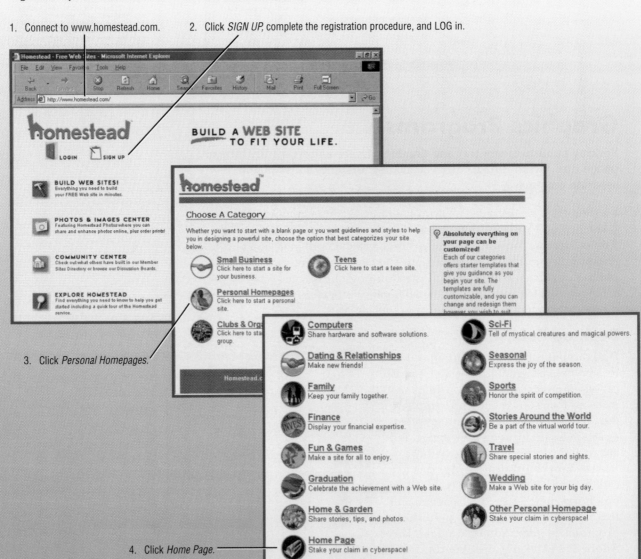

3. Click *Personal Homepages.*

4. Click *Home Page.*

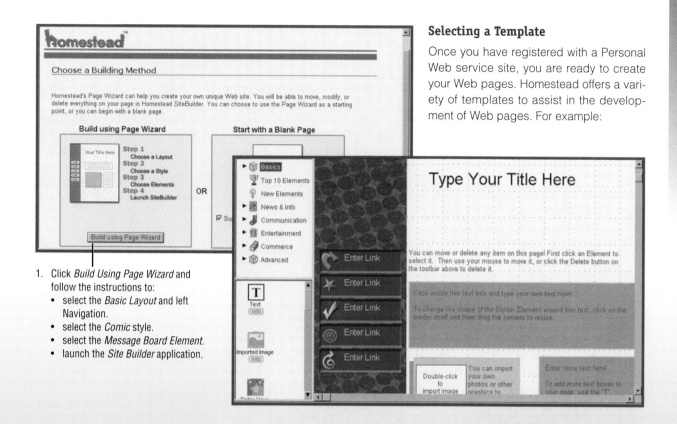

Selecting a Template

Once you have registered with a Personal Web service site, you are ready to create your Web pages. Homestead offers a variety of templates to assist in the development of Web pages. For example:

1. Click *Build Using Page Wizard* and follow the instructions to:
   - select the *Basic Layout* and left Navigation.
   - select the *Comic* style.
   - select the *Message Board Element*.
   - launch the *Site Builder* application.

Creating Your Web Pages

Once you have selected the template, you are ready to customize it by adding elements, photos, text and/or links to your Web page. After completing the page(s), save it to make it available to anyone over the Internet.

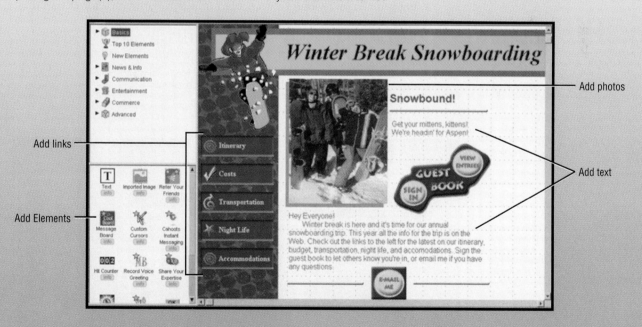

The Web is continually changing and some of the specifics presented in this Making IT Work for You may have changed. See our Web site at http://www.mhhe.com/it/oleary/IT.mhtml for possible changes and to learn more about this application of technology.

Image Editors

Image editors, also known as **paint programs,** are used to create and to modify bitmap image files. In a **bitmap file,** the image is made up of thousands of dots or pixels—much like how a computer screen displays images. Bitmapped graphic files are very common and are ideally suited to represent realistic images such as photographs.

Graphic artists use image editing programs to correct or change colors and to create special effects. One type of special effect called **morphing** allows you to smoothly blend two images so that one image seems to melt into the next, often producing amusing results.

Paintbrush for Windows is a low-end image editor. Popular professional image editor programs include Adobe Photoshop, Corel Photo, and Macromedia xRes. (See Figure 9-7.)

Illustration Programs

Illustration programs, also known as **draw programs,** are used to modify **vector images.** In a vector file, the image is composed of a collection of objects such as lines, rectangles, and ovals. A vector file contains all the shapes, colors, and starting and ending points necessary to recreate the image. Graphic artists use illustration programs to create line art for magazines, books, and special publications. Engineers and programmers use them to create 3-D models. One exciting application is to create virtual worlds or virtual reality, as we describe later in this chapter.

Popular professional illustration programs include Adobe Illustrator, CorelDraw, Macromedia FreeHand, and Micrografx Designer. (See Figure 9-8.)

Graphics Suites

Some companies are combining or bundling their separate graphics programs as a group called **graphics suites.** The advantage of the graphics suites is that you can buy a larger variety of graphics programs at a lower cost than if purchased separately.

The most popular suite is from the Corel Corporation. The suite, called CorelDraw, includes five individual Corel graphics programs plus a large library of clip art, media clips, and fonts. Two other popular suites are Corel's Graphics Pack and Micrografx's ABC Graphics Suite.

FIGURE 9-7
Image editor: Corel Photo

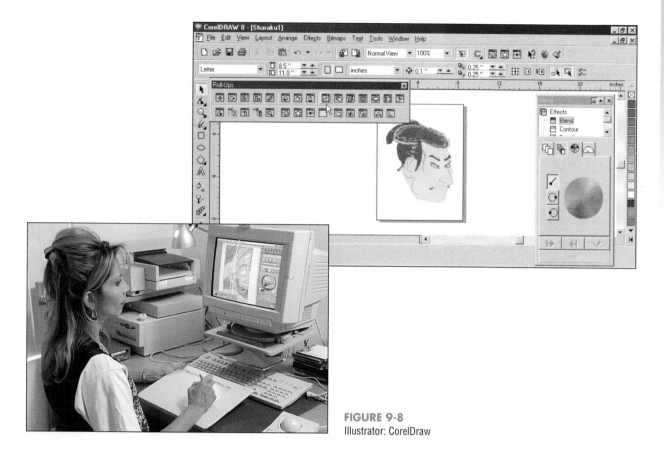

FIGURE 9-8
Illustrator: CorelDraw

CONCEPT CHECK

✔ List different types of specialized graphics programs.

✔ What are graphics suites?

Virtual Reality

Virtual reality creates simulated environments. VRML is a programming language to produce virtual reality applications. Virtual reality authoring programs are used to create applications.

Suppose you could create and virtually experience any new form of reality you wished. You could see the world through the eyes of a child, a robot—or even a lobster. You could explore faraway resorts, the moon, or inside a nuclear waste dump, without leaving your chair. This simulated experience is possible with *virtual reality.*

Virtual reality is also known as **VR, artificial reality,** or **virtual environments.** Virtual reality hardware includes headgear and gloves. The headgear (one type is called Eyephones) has earphones and three-dimensional stereoscopic screens. The gloves (DataGlove) have sensors that collect data about your hand movements. Coupled with software (such as a program called Body Electric), this interactive sensory equipment lets you immerse yourself in a computer-generated world.

ON THE WEB EXPLORATIONS

Cosmo Software is one of the leaders in the development of virtual reality on the Web. To learn more about the company, visit our Web site at

http://www.mhhe.com/it/oleary/ explore.mhtml

VRML

Virtual reality modeling language (VRML), is used to create real-time animated 3-D scenes. Hundreds of sites exist on the Web with virtual reality applications. Users are able to experience these applications with browsers that support VRML.

Virtual Reality Authoring Programs

Creating virtual reality programs once required very high end software costing several thousands of dollars. Recently, several lower-cost yet powerful authoring programs have been introduced. These programs utilize VRML and are widely used to create Web-based virtual reality applications. Three of the best known are Cosmo Worlds from Cosmo Software, Platinum VRCreator from Platinum Technology, and V-Realm Builder from Ligos Corporation.

Applications

An example of virtual reality is shown in Figure 9-9. In the photo is a person wearing an interactive sensory headset and glove. The rest of the photo shows what the person is actually seeing—in this instance a molecular view of a human cell. In Figure 9-10, two colleagues share this dynamic world.

FIGURE 9-9
Virtual reality: looking inside a molecule

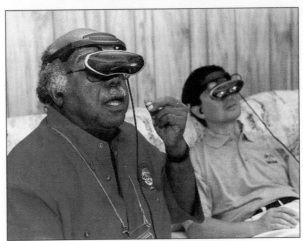

FIGURE 9-10
Virtual reality: sharing the view

There are any number of possible applications for virtual reality. The ultimate recreational use might be something resembling a giant virtual amusement park. More serious applications can simulate important experiences or training environments such as aviation, surgical operations, spaceship repair, or nuclear disaster cleanup.

CONCEPT CHECK

✔ What is virtual reality?

✔ The language used to create real-time 3-D scenes is _____.

Artificial Intelligence

Artificial intelligence attempts to simulate human thought processes and actions. Two areas are knowledge-based (expert) systems and robotics.

Does human intelligence really need the presence of "artificial intelligence," whatever that is? Indeed, you might worry, do we need the competition? Actually, the goal of artificial intelligence is not to replace human intelligence, which is probably not replaceable. Rather, it is to help people be more productive.

In the past, computers used calculating power to solve *structured* problems, which can be broken down into a series of well-defined steps. People—using intuition, reasoning, and memory—were better at solving *unstructured* problems, whether building a product or approving a loan. Organizations have long been able to computerize the tasks once performed by clerks. Now, knowledge-intensive work, such as that performed by many managers, is being automated.

The field of computer science known as **artificial intelligence (AI)** is moving into the mainstream. AI attempts to develop computer systems that can mimic or simulate human thought processes and actions. These include reasoning, learning from past actions, and using senses such as vision and touch. True artificial intelligence that corresponds to human intelligence is still a long way off. However, several tools that emulate human problem solving and information processing have been developed. Many of these tools have practical applications for business, medicine, law, and so on.

Let us now consider two areas in which human talents and abilities have been enhanced with "computerized intelligence": knowledge-based systems and robotics.

Knowledge-Based (Expert) Systems

People who are expert in a particular area—certain kinds of law, medicine, accounting, engineering, and so on—are generally well paid for their specialized knowledge. Unfortunately for their clients and customers, they are expensive, not always available, and hard to replace when they move on.

What if you were to somehow *capture* the knowledge of a human expert and make it accessible to everyone through a computer program? This is exactly what is being done with so-called *knowledge-based* or *expert systems*. **Expert systems** are computer programs that provide advice to decision makers who would otherwise rely on human experts. These expert systems use **knowledge bases** that contain specific facts, rules to relate these facts, and user

222

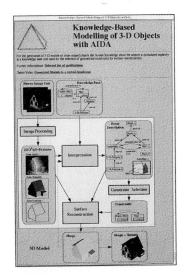

FIGURE 9-11
Knowledge-based system: creating 3-D images

input to formulate recommendations and decisions. The rules are used only when needed. The sequence of processing is determined by the interaction of the user and the knowledge base.

Many expert systems use so-called **fuzzy logic,** which allows users to respond to questions in a very humanlike way. For example, if an expert system asked how your classes were going, you could respond, "great," "OK," "terrible," and so on.

Over the past decade, expert systems have been developed in areas such as medicine, geology, chemistry, photography, and architecture. (See Figure 9-11.) There are expert systems with such names as Oil Spill Advisor, Bird Species Identification, and even Midwives Assistant. A system called Grain Marketing Advisor helps farmers select the best way to market their grain. Another, called Senex, shows how to treat breast cancer based on advanced treatment techniques.

Robotics

Robotics is the field of study concerned with developing and using robots. **Robots** are computer-controlled machines that mimic the motor activities of humans. Some toylike household robots (such as the Androbots) have been made for entertainment purposes. (See Figure 9-12.) Most, however, are used in factories and elsewhere. They differ from other assembly-line machines in that they can be reprogrammed to do more than one task. Robots are often used to handle dangerous, repetitive tasks. There are three types of robots:

- **Perception systems:** Perception system robots imitate some of the human senses. For example, robots with television-camera vision systems are particularly useful. They can be used for guiding machine tools, for inspecting products, for identifying and sorting parts, and for welding. (See Figure 9-12.) Other kinds of perception systems rely on a sense of touch, such as those used on microcomputer assembly lines to put parts into place.

FIGURE 9-12
Perception system: vision-system robot, used for welding

FIGURE 9-13
Industrial robot: handling dangerous materials

- **Industrial robots:** Industrial robots are used in factories to perform a variety of tasks. Examples are machines used in automobile plants to do painting and polishing. In the garment industry, robot pattern cutters create pieces of fabric for clothing. Some types of robots have claws for picking up objects.

- **Mobile robots:** Mobile robots act as transporters, such as "mailmobiles." They carry mail through an office, following a preprogrammed route. Specialized mobile robots with perception capabilites are used for military and police applications such as locating and disarming explosive devices. (See Figure 9-13.)

CONCEPT CHECK

✔ What does artificial intelligence attempt to simulate?

✔ Name two areas in which human talents and abilities have been enhanced with artificial intelligence.

Project Management

Project management software allows you to plan projects, schedule people, and control resources.

There are many occasions in business where managers need to watch projects to avoid delays and cost overruns. A **project** may be defined as a one-time operation composed of several tasks that must be completed during a stated period of time. Examples of large projects are found in construction, aerospace, and political campaigns. Examples of smaller jobs might occur in advertising agencies, corporate marketing departments, and management information systems departments. As a student, you may have projects like term papers and lab experiments.

Project Managers

Project managers are programs designed to assist in planning, scheduling, and controlling the people, resources, and costs needed to complete a project on time. For instance, a contractor building a housing development might use it to keep track of the materials, dollars, and people required for success.

bars represent individual tasks; length of bar indicates time to complete tasks

project timeline September through January

table lists task names and duration (length of time to complete)

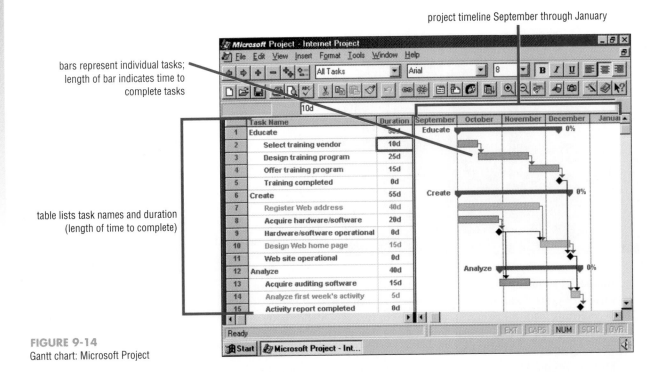

FIGURE 9-14
Gantt chart: Microsoft Project

Examples of project management software are Harvard Project Manager, Microsoft Project, Project Scheduler, SuperProject, and Time Line.

A typical use of project management software is to show the scheduled beginning and ending dates for each task in order to complete a particular project on time. It then shows the dates when each task was actually completed. Two important tools found in project management software are Gantt charts and PERT charts.

Gantt Charts

A **Gantt chart** uses bars and lines to represent a project. It has two parts. The first part is a table that lists information about each task. The second part is a bar chart that displays task durations, start dates, and finish dates on a timeline. The relative positions of the task bars reflect the order in which each task is to be performed. (See Figure 9-14.)

PERT Charts

A **Program Evaluation Review Technique (PERT) chart** uses boxes and lines to represent a project. A PERT chart shows not only the timing of a project but also the relationships among its tasks. The chart identifies which tasks must be completed before others can begin. The relationships are represented by lines that connect boxes stating the tasks, completion times, and dates. The **critical path,** the sequence of tasks that takes the longest to complete, is also identified. (See Figure 9-15.)

CONCEPT CHECK

✔ What are the functions of project management software?

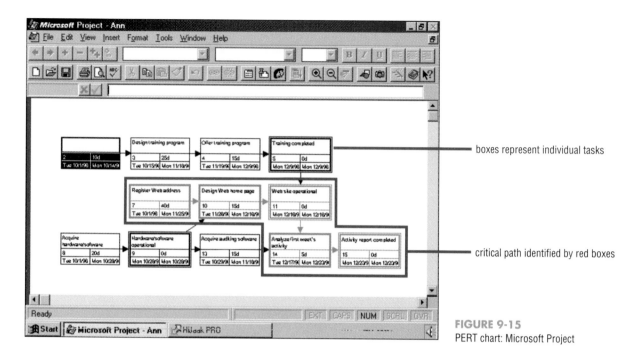

boxes represent individual tasks

critical path identified by red boxes

FIGURE 9-15
PERT chart: Microsoft Project

A LOOK TO THE FUTURE

IPIX, a new immersive photographic technique, may allow you to take virtual tours and exotic vacations.

There are few applications with more exciting potential than virtual reality. Although some virtual reality applications do exist, few present lifelike images and fluid motion. This is due in part to the expense and difficulty in capturing 3-D data.

Some experts predict that we will see an explosion of high-quality, photo-realistic virtual reality applications in the near future. One key is a recently released immersive photographic technique called IPIX.

Using a special fish-eye (180-degree) camera, two opposite facing images are recorded digitally and stored in memory. IPIX software takes the two images, corrects for distortion, and joins them into a single image.

Using special software, you can view this image on a computer screen, changing the angle, looking up, down, left, and right. You see almost exactly what you would see if you were standing in one location

and looking in every direction. Data for a virtual tour of any environment can be readily captured by taking successive photographs while walking through that environment.

Will we be able to take realistic tours of famous museums, explore classical architecture, or even take exotic vacations without leaving our home? Who knows, but some observers say we will within the next two years.

VISUAL SUMMARY

Multimedia, Web Authoring, and More

MULTIMEDIA

Multimedia or **hypermedia** integrates all sorts of media into an interactive presentation. This media may include video, music, voice, graphics, and text.

An essential feature is user participation or **interactivity.**

Links and Buttons
A multimedia presentation is organized as a series of related pages. Pages are **linked,** or connected, by **buttons.**

Story Boards
Story boards are design tools to record the intended overall logic, flow, and structure of a multimedia presentation. Individual storyboards specify links, content, style, and design of each display.

Multimedia Authoring Programs
Multimedia authoring programs are special programs to create multimedia presentations. They bring together all the video, audio, graphics, and text elements in an interactive framework.

WEB AUTHORING

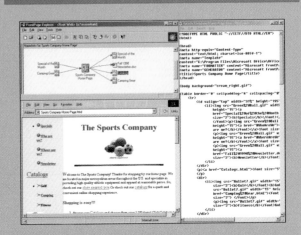

Creating Web sites is called **Web authoring.** It begins with Web site design followed by creating a document file that displays the Web site content.

Web Site Design
Web site design begins with determining overall content.

Graphical maps present overall Web site design. The home page presents overall content much like a table of contents. Links connect related blocks or pages of information.

Web Authoring Programs
Web sites can be created using a simple text editor or word processor. **Web authoring programs,** also known as **Web page editors** or **HTML editors,** are more powerful and specialized. They provide support for Web site design and HTML coding.

To be a competent end user you need to be aware of advanced applications. You need to know who uses them, what they are used for, and how they are used. Advanced applications include multimedia, Web authoring, graphics programs, virtual reality, artificial intelligence, and project managers.

GRAPHICS PROGRAMS

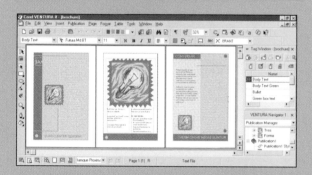

Professionals in graphic arts use specialized **graphics programs.** These programs are desktop publishers, image editors, and illustration programs.

Desktop Publishers
Desktop publishers mix text and graphics to create professional publications. They are widely used by graphic artists to create brochures, newsletters, newspapers, and textbooks.

Image Editors
Image editors, also known as **paint programs**, create and modify **bitmap image files**. In bitmap files, images are recorded as thousands of dots or pixels. **Morphing** is a special effect that blends two images into one.

Illustration Programs
Illustration programs, also known as **draw programs**, modify vector images. In a vector file, images are recorded as a collection of objects such as lines, rectangles, and ovals.

Graphics Suites
A **graphics suite** is a collection of individual graphics programs sold as a unit.

VIRTUAL REALITY

Virtual reality is also known as **VR, artificial reality,** or **virtual environments.** It consists of interactive sensory equipment to simulate alternative realities to the physical world. VR creates computer-generated simulated environments.

VRML
VRML (virtual reality modeling language) is used to create real-time virtual reality applications on the Web.

VR Authoring Programs
Virtual reality authoring programs use VRML and are widely used to create Web-based virtual reality applications.

Applications
Applications include virtual amusement parks, aviation, surgical operations, spaceship repairs, and nuclear disaster cleanup.

ARTIFICIAL INTELLIGENCE

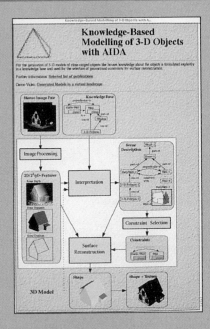

Artificial intelligence is a research field to develop computer systems that simulate human thought processes and actions. Two areas are knowledge-based systems and robotics.

Knowledge-Based (Expert) Systems
Knowledge-based (expert) systems are programs that duplicate the knowledge humans have for performing specialized tasks.

Expert systems provide advice to users who would normally rely on human experts. **Knowledge bases** contain specific facts and rules to relate to these facts. **Fuzzy logic** is used to allow humanlike input and output.

Robotics
Robots are computer-controlled machines that mimic the motor activities of humans. Three types are:
- Industrial robots—widely used in factories.
- Perception systems—simulate human senses.
- Mobile robots—act as transporters.

PROJECT MANAGEMENT

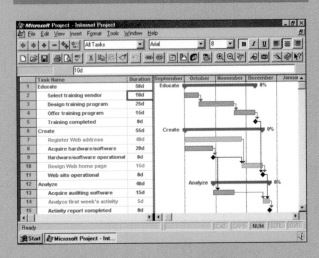

A **project** can be defined as a one-time operation composed of several tasks that must be completed during a stated period of time. Project management is a common business activity.

Project Managers
Project managers are programs designed to assist in the planning, scheduling, and controlling of people, resources, and costs of a project. Two important tools are Gantt and PERT charts.

Gantt Charts
A **Gantt chart** uses bars and lines to represent a project. Bars represent the length of time to complete each task and their starting and ending dates. The lines connecting the bars indicate the order that the tasks must be completed.

PERT Charts
A **Program Evaluation Review Technique (PERT) chart** uses boxes and lines to represent a project. Each box includes a description of the task, projected completion times, and dates. Lines between boxes represent the order of completion. The **critical path** is the sequence of tasks that takes the longest to complete.

Key Terms

artificial intelligence (AI) (221)
artificial reality (219)
bitmap file (218)
button (212)
critical path (224)
desktop publishing (215)
draw program (218)
expert system (221)
fuzzy logic (222)
Gantt chart (224)
graphical map (214)
graphics suite (218)
HTML editor (214)

hypermedia (212)
illustration program (218)
image editor (218)
interactivity (212)
knowledge base (221)
link (212)
morphing (218)
multimedia (212)
multimedia authoring
    program (213)
paint program (218)
Program Evaluation Review Technique (PERT) chart (224)

project (223)
project managers (223)
robot (222)
robotics (222)
story board (212)
vector images (218)
virtual environment (219)
virtual reality (VR) (219)
Virtual reality programming
    language (VRML) (220)
Web authoring (214)
Web authoring program (214)
Web page editor (214)

2001
2002

Chapter Review

LEVEL 1

Reviewing Facts and Terms
Matching

Match each numbered item with the most closely related lettered item. Write your answers in the spaces provided.

1. Creates simulated environments. _R_

2. Images composed of a collection of objects such as lines, rectangles, and ovals. _Q_

3. Programs designed to assist in planning, scheduling, and controlling the people, resources, and costs needed to complete a project on time. _N_

4. Used in the early planning phase of a multimedia - project. _P_

5. Uses bars and lines to represent a project. _G_

6. An essential and unique feature of multimedia. _K_

7. Field of study concerned with developing and using robots. _A_

8. Program used to modify vector images. _I_

9. Creating a Web site. _S_

10. Uses boxes and lines to represent a project. _M_

11. Programs that allow you to mix text and graphics to create publications of professional quality. _D_

12. Computer programs that provide advice to decision makers who would otherwise rely on human experts. _E_

13. In this file type, the image consists of thousands of pixels. _B_

14. Used to create sophisticated Web sites while providing support for Web site design and HTML coding. _T_

15. Attempts to simulate human thought processes and actions. _O_

a. Artificial intelligence
b. Bitmap file
c. Buttons
d. Desktop publishing
e. Expert systems
f. Fuzzy logic
g. Gantt chart
h. Graphics suite
i. Illustration program
j. Image editor
k. Interactivity
l. Multimedia
m. PERT chart
n. Project managers
o. Robotics
p. Story board
q. Vector images
r. Virtual reality
s. Web authoring
t. Web page editors

16. Bundle of graphics programs—you can buy a larger variety at a lower cost this way. _H_

17. The integration of different media into one presentation. _L_

18. Special areas on a Web page that can link you to related Web sites or to additional areas within a Web page. _U_

19. Used to create and modify bitmap image files. _J_

20. Allows computer users to respond to questions in a very humanlike way. _f_

True/False

In the spaces provided, write T or F to indicate whether the statement is true or false.

1. Desktop publishing programs combine text and graphics to create professional-quality publications. _T_

2. Illustrators are used to create HTML code. _____

3. Image editors are used to create Web pages. _F_

4. Virtual reality employs various sensory equipment to allow you the experience of a computer-generated world. _T_

5. Expert systems are programs that give advice to individuals who would otherwise rely on human experts. _T_

Multiple Choice

Circle the letter of the correct answer.

1. The application that can link all sorts of media into one form of presentation:
   a. image editor
   b. word processor
   c. multimedia
   d. spreadsheet
   e. virtual reality

2. _____ are a multimedia design tool.
   a. story boards
   b. bitmaps
   c. VRML
   d. video clips
   e. animations

3. _____ is a research field that develops computer systems to simulate human thought processes.
   a. cyberspace
   b. illustrators
   c. AI
   d. cybernetics
   e. cryonics

4. A one-time operation composed of several tasks that must be completed during a stated period:
   a. plan
   b. image editor
   c. manager
   d. schedule
   e. project

5. The _____ chart uses bars and lines to represent a project.
   a. Gantt
   b. analytical
   c. directional
   d. presentational
   e. pie

6. An area of artificial intelligence that simulates certain experiences using special headgear, gloves, and software that translates data into images:
   - **a.** virtual reality
   - **b.** expert system
   - **c.** natural language processing
   - **d.** hypermedia
   - **e.** knowledge-based systems

Completion

Complete each statement in the spaces provided.

1. _____ _____ are used to modify bitmap images.

2. _____ is a programming language used to create Web pages.

3. _____ is the integration of text, graphics, animation, video, music, and voice into one presentation.

4. Unlike most assembly-line machines, _____ can be reprogrammed to do more than one task.

5. _____ systems are programs that duplicate the knowledge of highly trained people.

Reviewing Concepts

Open-Ended

On a separate sheet of paper, respond to each question and statement.

1. What is multimedia? What is it used for?

2. What are Web authoring programs? What are they used for?

3. Describe desktop publishers, image editors, and illustration programs. How do they differ?

4. What are the two areas of artificial intelligence? Describe each.

5. Explain what project management software does. Give an example of how one might be used.

Concept Mapping

On a separate sheet of paper, draw a concept map or a flowchart showing how the following terms are related. Show all relationships. Include any additional terms you can think of.

artificial intelligence	HTML editor	paint program
bitmap file	hypermedia	PERT chart
button	illustration program	project managers
desktop publishing	image editor	robotics
draw program	interactivity	story board
expert system	knowledge base	vector images
fuzzy logic	link	virtual reality
Gantt chart	multimedia	VRML
graphical map	multimedia authoring	Web authoring programs
graphics suite	program	Web page editor

Critical Thinking Questions and Projects

2001 2002

LEVEL 3

Read each exercise and answer the related questions on a separate sheet of paper.

1. *New areas for expert systems:* There are numerous expert systems designed to pick winning stocks in the stock market. However, not everyone using these systems has become rich. Why? List and discuss three other areas in which you think it would be difficult to devise an expert system.

2. *Tools for entrepreneurs:* Have you ever thought of starting a new business? More and more people are doing so. Advanced applications like multimedia, Web authoring, graphics programs, virtual reality, artificial intelligence, and project management provide tools for entrepreneurs to get a business off the ground and moving toward success. Think of a new business. It can be almost anything, from an auto parts store to a catering service to a consulting company. Briefly describe the business and then discuss how you might use any or all of the specialized applications discussed in this chapter to start and to run the business.

3. *Ethics:* One way to learn how to create interesting and dynamic Web sites is to study how professionals create their sites. Once connected to a Web site, you can examine the HTML code used to create that site. With Netscape, for example, select Document Source from the View submenu. Do you see any ethical issues that relate to examining the HTML code? What about copying part of parts of that code? What about copying all of the code? Defend your position(s).

4. *Personal Web Sites:* Not all Web sites are for commercial purposes. Many individuals maintain Web sites to communicate and share with friends and family. (See Making IT Work for You: Personal Web Sites on pages 216–17.) Others use Web sites to promote personal and humanitarian causes. For example, visit the Hunger site at http://www.thehungersite.com, which began as a personal project for one individual.

 a. Have you created or visited a personal Web site? If so, describe the site and it's objectives. Did you find the site effective? Be specific.

 b. Describe three situations in which you might create a personal Web site to communicate with friends, family, etc. Specify the objectives of the site and how the site helps to achieve those objectives. Be specific.

 c. Describe three situations in which you might create a personal Web site for humanitarian reasons. Specify the objectives of the site and how the site helps to achieve those objectives. Be specific.

 d. Do you think more individuals in the future will have their own Web sites? Why or why not?

On the Web Exercises

1. Multimedia

Macromedia Corporation is a leader in multimedia software. Visit our Web site at http://www.mhhe.com/it/oleary/exercise.mhtml to link to their site.

Once connected to that site, preview the latest software and choose a program that would be useful for creating multimedia presentations. Print out a page from the site that describes the software and write a paragraph discussing how you could use this software to create or enhance a multimedia presentation.

2. Shareware and Freeware

Shareware and freeware are application programs available to the public at little or no cost. Visit our site at http://www.mhhe.com/it/oleary/exercise.mhtml for a link to a site

providing an extensive list of shareware and freeware that you can download to your computer system. Once connected to this site, find a list of the most popular downloads for your operating system. Print out the list and write a paragraph detailing the three most interesting programs you found at this site.

3. Graphics Programs

Adobe Systems is a leader in high-end graphics software. Visit our site at http://www.mhhe.com/it/oleary/exercise.mhtml to link to their site. Once connected to that site,

explore and select one software program that you find interesting. Print out the Web page that best describes the product. Write a paragraph describing the program's basic features and capabilities. Write a second paragraph discussing two or three specific applications where you might use the product.

4. Creating Web Pages

Nearly every organization today has a Web site, and it is very likely that you will be involved with either creating Web sites or modifying them. (See Making IT

Work for You: Personal Web Sites on pages 216–17.) Perhaps you will work with a specialist or work on your own. In either case, it is worthwhile to know what makes a site interesting and attractive. Visit our Web site at http://www.mhhe.com/it/oleary/exercise.mhtml to link to a site that provides guidance for Web page design, style, and content. Print out the first page of this site, explore the site, and write a paragraph describing some of the programs for creating Web pages.

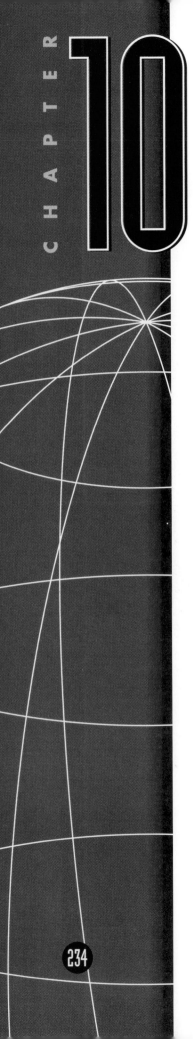

CHAPTER 10

Privacy, Security, Ergonomics, and the Environment

COMPETENCIES

After you have read this chapter, you should be able to:

1. Discuss the privacy issues raised by the presence of large databases, electronic networks, the Internet, and the Web.

2. List the major laws on privacy.

3. Explain the effects of computer crimes, including the spreading of computer viruses.

4. Describe other hazards to the computer.

5. Discuss security measures that may be taken.

6. Describe ergonomics and how it helps avoid physical and mental risks.

7. Describe the four ethical issues: privacy, accuracy, property, and access.

8. Discuss what the computer industry is doing to help protect the environment.

9. Discuss what you can do to help protect the environment.

The tools and products of the information age do not exist in a world by themselves. As we said in Chapter 1, a computer system consists not only of software, hardware, data, and procedures but also of *people*. Because of people, computer systems may be used for both good and bad purposes.

There are more than 300 million microcomputers in use today. What are the consequences of the widespread presence of this technology? Does technology make it easy for others to invade our personal privacy? When we apply for a loan or for a driver's license or when we

Privacy

Security

```
PROGRAM MAIN
  REAL X, A, B, WIDTH, *AREA, F
INTEGER I, N
F(X) = SQRT (ABS (4 – X **2)
PRINT *, 'VIRUS'
    *, A, B
    *, 'YOUR COMPUTER IS DEAD'
    *, N
WIDTH = (B – A)/N
X = A
AREA = 0
DO 20 I = 1, N
  AREA = AREA + WIDTH * F (X + WIDTH/2)
  X = X + WIDTH
CONYINUE
PRINT *, 'SUPER HACKER'
END
```

Data

check out at the supermarket, is that information about us being distributed and used without our permission? When we use the Web, is information about us being collected and shared with others?

Does technology make it easy for others to invade the security of business organizations like our banks or our employers? What about health risks to people who use computers? What about the environment? Do computers pose a threat to our ecology?

Lots of questions—very important questions. Perhaps some of the most important questions for the 21st century. Competent end users need to be aware of the potential impact of technology on people and how to protect themselves on the Web. They need to be sensitive to and knowledgeable about personal privacy, organizational security, ergonomics, and the environmental impact of technology.

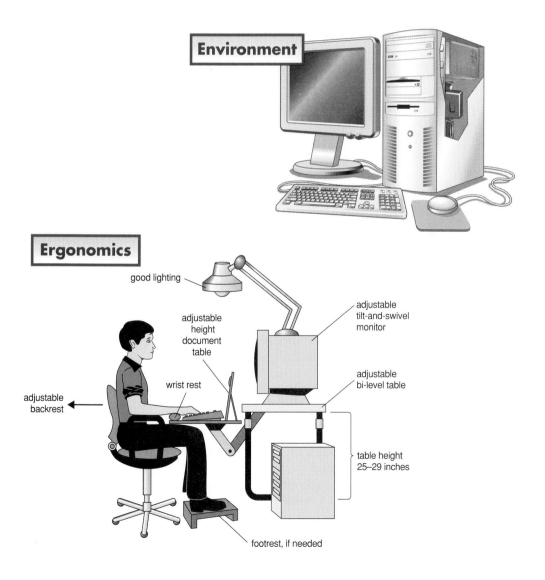

Environment

Ergonomics

good lighting

adjustable height document table

adjustable tilt-and-swivel monitor

wrist rest

adjustable bi-level table

adjustable backrest

table height 25–29 inches

footrest, if needed

Privacy

Every computer user should be aware of ethical matters, including how databases and networks are used and the major privacy laws.

What do you suppose controls how computers can be used? You probably think first of laws. Of course that is right, but technology is moving so fast that it is very difficult for our legal system to keep up. The essential element that controls how computers are used today is ethics.

Ethics, as you may know, are standards of moral conduct. *Computer ethics* are guidelines for the morally acceptable use of computers in our society. There are four primary computer ethics issues:

- **Privacy** concerns the collection and use of data about individuals.
- **Accuracy** relates to the responsibility of those who collect data to ensure that the data is correct.
- **Property** relates to who owns data and rights to software.
- **Access** relates to the responsibility of those who have data to control who is able to use that data.

We are all entitled to ethical treatment. This includes the right to keep personal information, such as credit ratings and medical histories, from getting into the wrong hands. Many people worry that this right is severely threatened. Let us see what some of the concerns are.

Large Databases

Large organizations are constantly compiling information about us. The federal government alone has over 2,000 databases. Our social security numbers are now used routinely as key fields in databases for organizing our employment, credit, and tax records. Indeed, even children are now required to have social security numbers. Shouldn't we be concerned that cross-referenced information might be used for the wrong purposes?

Every day, data is gathered about us and stored in large databases. For example, for billing purposes, telephone companies compile lists of the calls we make, the numbers called, and so on. A special telephone directory (called a *reverse directory*) lists telephone numbers followed by subscriber names. Using it, government authorities and others could easily get the names, addresses, and other details about the persons we call. Credit card companies keep similar records. Supermarket scanners in grocery checkout counters record what we buy, when we buy it, how much we buy, and the price. (See Figure 10-1.) Publishers of magazines, newspapers, and mail-order catalogs have our names, addresses, phone numbers, and what we order.

A vast industry of data gatherers or "information resellers" now exists that collects such personal data. They then sell it to direct marketers, fund-raisers, and others. Did you know that for a nominal fee someone could obtain a copy of your monthly bank and credit card statements? If you own stocks, bonds, and/or mutual funds, they could also obtain a list of your financial holdings along with specific account numbers.

Your personal information including preferences, habits, and financial data has become a marketable commodity. This raises many issues, including:

FIGURE 10-1
Large organizations are constantly compiling information about us, such as the kinds of products we buy

- **Spreading information without personal consent:** How would you feel if your name and your taste in movies were made available nationwide? For a while, Blockbuster, a large video rental company, considered doing just this.

What if a great deal of information about your shopping habits—collected about you without your consent—was made available to any microcomputer user who wanted it? Before dropping the project, Lotus Development Corporation and Equifax Inc. planned to market disks containing information on 120 million American consumers. (Lotus claimed it was providing small businesses with the same information currently available to larger organizations.)

How would you feel if an employer were using your *medical* records to make decisions about hiring, placement, promotion, and firing? A University of Illinois survey found that half the Fortune 500 companies were using employee medical records for that purpose.

- **Spreading inaccurate information:** How *accurate* is the information being circulated? Mistakes that creep into one computer file may find their way into other computer files. For example, credit records may be in error. Moreover, even if you correct an error in one file, the correction may not be made in other files. Indeed, erroneous information may stay in computer files for years. It's important to know, therefore, that you have some recourse.

The law allows you to gain access to those records about you that are held by credit bureaus. Under the Freedom of Information Act (described shortly), you are also entitled to look at your records held by government agencies. (Portions may be deleted for national security reasons.)

Private Networks

Suppose you use your company's electronic mail system to send a co-worker an unflattering message about your supervisor. Later you find the boss has been spying on your exchange. Or suppose you are a subscriber to an online discussion group. You discover that the company that supports the discussion group screens all your messages and rejects those it deems inappropriate. Both these situations have actually happened.

The first instance, of firms eavesdropping on employees, has inspired attempts at federal legislation. One survey revealed that over 20 percent of businesses search employees' electronic mail and computer files using so-called **snoopware.** These programs can record virtually everything you do on your computer. Currently this is legal. One proposed law would not prohibit electronic monitoring but would require employers to provide prior written notice. They would also have to alert employees during the monitoring with some sort of audible or visual signal.

The second instance, in which online information services restrict libelous, obscene, or otherwise offensive material, exists with most commercial services. In one case, the Prodigy Information Service terminated the accounts of eight members who had been using the electronic mail system to protest Prodigy's rate hikes. Prodigy executives argued that the U.S. Constitution does not give members of someone's private network the right to express their views without restrictions. Opponents say that the United States is becoming a nation linked by electronic mail. Therefore, the government has to provide protection for users against other people reading or censoring their messages. (See Figure 10-2.)

The Internet and the Web

When you send e-mail on the Internet or browse the Web, do you have any concerns about privacy? Most people do not. They think as long as they are selective about disclosing their name or other personal information,

FIGURE 10-2
Computer security

then little can be done to invade their personal privacy. Experts call this the *illusion of anonymity* that the Internet brings.

As discussed earlier, it is a common practice in many organizations to monitor e-mail content on messages sent within their private electronic networks. Likewise, for some unscrupulous individuals, it is also a common practice to eavesdrop or snoop into the content of e-mail sent across the Internet.

Furthermore, when you browse the Web, your activity is monitored. Whenever you visit a Web site, your browser stores critical information onto your hard disk, typically without your permission or knowledge. For example, your browser creates a **history file** that includes the locations of sites visited by your computer system. Additionally, many Web sites have specialized programs called **cookies** that record how often you visit a site, what you do there, and any other information that you provide, such as credit card numbers. Although these programs are intended to provide better service to you when you revisit a Web site, you may consider some of this information private, and you may not like others saving this information on your hard disk without your approval.

Recently, companies have been created that specialize in monitoring Internet and Web site activities. Many of these firms sell e-mail mailing lists and individualized personal profiles without obtaining permission. How can you protect yourself?

Major Laws on Privacy

Some federal laws governing privacy matters (summarized in Figure 10-3) are as follows:

- **Fair Credit Reporting Act:** The **Fair Credit Reporting Act of 1970** is intended to keep inaccuracies out of credit bureau files. Credit agencies are barred from sharing credit information with anyone but

ON THE WEB EXPLORATIONS

The Center for Democracy and Technology monitors privacy issues and legislation. To learn more about this organization, visit our Web site at

http://www.mhhe.com/it/oleary/
explore.mhtml

FIGURE 10-3
Summary of privacy laws

Law	Protection
Fair Credit Reporting Act	Gives right to review and correct personal credit records; restricts sharing of personal credit histories
Freedom of Information Act	Gives right to see personal files collected by federal agencies
Privacy Act	Prohibits use of federal information for purposes other than original intent
Right to Financial Privacy Act	Limits federal authority to examine personal bank records
Computer Fraud and Abuse Act	Allows prosecution of unauthorized access to computers and databases
Electronic Communications Privacy Act	Protects privacy on public electronic-mail systems
Video Privacy Protection Act	Prevents sale of video-rental records
Computer Matching and Privacy Protection Act	Limits government's authority to match individual's data
Computer Abuse Amendments Act	Outlaws transmission of viruses
National Information Infrastructure Protection Act	Protects computer systems, networks, and information
No Electronic Theft (NET) Act	Prevents unauthorized distribution of copyrighted software on the Internet

238

authorized customers. Consumers have the right to review and correct their records and to be notified of credit investigations for insurance and employment.

Drawback: Credit agencies may share information with anyone they reasonably believe has a "legitimate business need." *Legitimate* is not defined.

- **Freedom of Information Act:** The **Freedom of Information Act of 1970** gives you the right to look at data concerning you that is stored by the federal government.

  Drawback: Sometimes a lawsuit is necessary to pry data loose.

- **Privacy Act:** The **Privacy Act of 1974** is designed to restrict federal agencies in how they share information about American citizens. It prohibits federal information collected for one purpose from being used for a different purpose.

  Drawback: Exceptions written into the law permit federal agencies to share information anyway.

- **Right to Financial Privacy Act:** The **Right to Financial Privacy Act of 1979** sets strict procedures that federal agencies must follow when seeking to examine customer records in banks.

  Drawback: The law does not cover state and local governments.

- **Computer Fraud and Abuse Act:** The **Computer Fraud and Abuse Act of 1986** allows prosecution of unauthorized access to computers and databases.

  Drawback: The act is limited in scope. People with legitimate access can still get into computer systems and create mischief without penalty.

- **Electronic Communications Privacy Act:** The **Electronic Communications Privacy Act of 1986** protects the privacy of users on public electronic-mail systems.

  Drawback: The act is limited to public electronic communications mail systems. It does not cover communication within an organization's internal electronic communications.

- **Video Privacy Protection Act:** The **Video Privacy Protection Act of 1988** prevents retailers from selling or disclosing video-rental records without customer consent or a court order.

  Drawback: The same restrictions do not apply to even more important files, such as medical and insurance records.

- **Computer Matching and Privacy Protection Act:** The **Computer Matching and Privacy Protection Act of 1988** sets procedures for computer matching or searching of federal data. Such matching can be for verifying a person's eligibility for federal benefits or for recovering delinquent debts. Individuals are given a chance to respond before the government takes any adverse action against them.

  Drawback: Many possible computer matches are not affected, including those done for law-enforcement or tax reasons.

- **Computer Abuse Amendments Act:** The **Computer Abuse Amendments Act of 1994** amends the 1984 act to outlaw transmission of viruses and other harmful computer code.

  Drawback: Locating the origins and individuals responsible for creating computer viruses is very difficult.

- **National Information Infrastructure Protection Act:** The **National Information Infrastructure Protection Act of 1996** provides penalties for trespassing on computer systems, making threats against computer networks, and stealing information.

  Drawback: It is difficult to gather sufficient information to prosecute.

ON THE WEB EXPLORATIONS

The Center for Computers & Democracy monitors current and pending federal privacy laws. To connect to the Center and to learn more about privacy laws, visit our Web site at

http://www.mhhe.com/it/oleary/ explore.mhtml

TIPS What can you do to protect your privacy while on the Web? Here are a few suggestions.

1. *Encrypt sensitive e-mail.* Encrypt or code sensitive e-mail using special encryption programs.

2. *Shield your identity.* Use an anonymous remailer or special Web site that forwards your e-mail without disclosing your identity.

3. *Block cookies.* Use the newer browsers that allow you to block Web sites from depositing cookies on your hard disk.

4. *Notify providers.* Instruct your service provider or whomever you use to link to the Internet not to sell your name or any other personal information.

5. *Be careful.* Never disclose your telephone number, passwords, or other private information to strangers.

Principle	Description
No secret databases	There must be no record-keeping systems containing personal data whose very existence is kept secret.
Right of individual access	Individuals must be able to find out what information about them is in a record and how it is used.
Right of consent	Information about individuals obtained for one purpose cannot be used for other purposes without their consent.
Right to correct	Individuals must be able to correct or amend records of identifiable information about them.
Assurance of reliability and proper use	Organizations creating, maintaining, using, or disseminating records of identifiable personal data must make sure the data is reliable for its intended use. They must take precautions to prevent such data from being misused.

FIGURE 10-4
Principles of the Code of Fair
Information Practice

- **No Electronic Theft (NET) Act:** The **No Electronic Theft (NET) Act of 1997** provides penalties for unauthorized distribution of copyrighted materials (software) on the Internet.

  Drawback: It is often difficult to pinpoint the source illegally distributing the copyrighted materials.

Currently, privacy is primarily an *ethical* issue, for many records stored by nongovernment organizations are not covered by existing laws. Yet individuals have shown that they are concerned about controlling who has the right to personal information and how it is used. A Code of Fair Information Practice is summarized in Figure 10-4. The code was recommended in 1977 by a committee established by former Secretary of Health, Education, and Welfare Elliott Richardson. It has been adopted by many information-collecting businesses, but privacy advocates would like to see it written into law.

CONCEPT CHECK

✔ What are four primary computer ethics issues?

✔ Name some privacy issues associated with large databases, private networks, and the Internet.

✔ Which privacy laws discussed in this chapter most apply to you?

Security

Threats to computer security are computer crimes, including viruses, electronic break-ins, and natural and other hazards. Security measures consist of encryption, restricting access, anticipating disasters, and making backup copies.

We are all concerned with having a safe and secure environment to live in. We are careful to lock our car doors and our homes. We are careful about where we walk at night and who we talk to. This is physical security. What about computer security? Does it matter if someone gains access to personal information about you? What if someone learns your credit card number or your checking account number? What if a mistake is made and your credit history shows a number of large unpaid loans? What if all your school records are lost? These are just a few of the reasons to be concerned about computer security. (See Figure 10-5.)

Threats to Computer Security

Keeping information private depends on keeping computer systems safe from criminals, natural hazards, and other threats.

Computer Criminals

A **computer crime** is an illegal action in which the perpetrator uses special knowledge of computer technology. Computer criminals are of four types:

FIGURE 10-5
There are numerous threats to computer security

- **Employees:** The largest category of computer criminals consists of those with the easiest access to computers—namely, employees. Sometimes the employee is simply trying to steal something from the employer—equipment, software, electronic funds, proprietary information, or computer time. Sometimes the employee may be acting out of resentment and is trying to get back at the company.

- **Outside users:** Not only employees but also some suppliers or clients may have access to a company's computer system. Examples are bank customers who use an automated teller machine. Like employees, these authorized users may obtain confidential passwords or find other ways of committing computer crimes.

- **Hackers and crackers:** Some people think of these two groups as being the same, but they are not. **Hackers** are people who gain unauthorized access to a computer system for the fun and challenge of it. **Crackers** do the same thing but for malicious purposes. They may intend to steal technical information or to introduce what they call a bomb—a destructive computer program—into the system.

- **Organized crime:** Members of organized crime groups have discovered that they can use computers just as people in legitimate businesses do, but for illegal purposes. For example, computers are useful for keeping track of stolen goods or illegal gambling debts. In addition, counterfeiters and forgers use microcomputers and printers to produce sophisticated-looking documents such as checks and driver's licenses.

Computer Crime

Computer crime can take various forms, as follows:

- **Damage:** Disgruntled employees sometimes attempt to destroy computers, programs, or files. Hackers and crackers are notorious for creating and distributing malicious programs known as viruses.

  Viruses are programs that *migrate* through networks and operating systems and attach themselves to different programs and databases. There are four basic types of viruses: boot sector, file, Trojan horse, and macro. Creating and knowingly spreading a virus is a very serious crime and a federal offense punishable under the Computer Abuse Amendments Act of 1994.

  A variant on the virus is the **worm.** This destructive program fills a computer system with self-replicating information, clogging the system so that its operations are slowed or stopped. The most infamous to date is known as the Internet Worm. In 1988, it traveled across North America, stopping thousands of computers along its way.

Name	Description
Concept	Makes Word's normal.dot template display a pop-up box containing the number 1; infects all documents saved using the Save As command
Cap	Replaces all Word macros with its own macros
Wazzu	Damages every Word document you open by moving around words and randomly inserting the word Wazzu
NPAD	Displays the message "DOEUNPAD94, v2.21" every 23rd time you open a Word document
MDMA	On systems using Windows 95, it changes key, display, and network log-on settings, and deletes Windows Help files and certain system files; triggers the first day of the month

FIGURE 10-7
Norton AntiVirus

FIGURE 10-6
Commonly encountered viruses

Viruses and worms typically find their way into microcomputers through copied floppy disks or programs downloaded from the Internet. Because viruses can be so serious—certain "disk-killer" viruses can destroy all the information on one's system—computer users are advised to exercise care in accepting new programs and data from other sources. See Figure 10-6 for a list of common viruses.

Detection programs called *virus checkers* are available to alert users when certain kinds of viruses enter the system. (See Making IT Work for You: Virus Protection on pages 244–45.) Four of the most widely used virus checkers are Dr. Solomon's Anti-Virus, McAfee VirusScan, eSafe, and Norton AntiVirus. (See Figure 10-7.) Unfortunately, new viruses are being developed all the time, and not all viruses can be detected.

To learn more about viruses and how you could get one, study Figure 10-8.

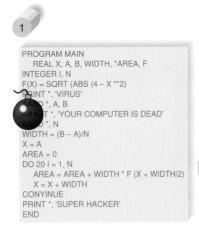

```
PROGRAM MAIN
   REAL X, A, B, WIDTH, *AREA, F
INTEGER I, N
F(X) = SQRT (ABS (4 – X **2)
PRINT *, 'VIRUS'
    *, A, B
    *, 'YOUR COMPUTER IS DEAD'
    *, N
WIDTH = (B – A)/N
X = A
AREA = 0
DO 20 I = 1, N
   AREA = AREA + WIDTH * F (X + WIDTH/2)
   X = X + WIDTH
CONYINUE
PRINT *, 'SUPER HACKER'
END
```

① A virus begins when a cracker or programmer writes a program that attaches itself to an operating system, another program, or piece of data.

② The virus travels via floppy disk or downloading from networks or bulletin boards anywhere that the operating system, program, or data travels.

③ The virus is set off. A nondestructive virus may simply print a message ("Surprise!"). A destructive virus may erase data, destroy programs, and even (through repeated reading and writing to one location) wear out a hard disk. The virus may be set off either by a time limit or by a sequence of operations by the user.

FIGURE 10-8
How a computer virus can spread

- **Theft:** Theft can take many forms—of hardware, of software, of data, of computer time. Thieves steal equipment, of course, but there are also white-collar crimes. Thieves steal data in the form of confidential information such as preferred-client lists. They also use (steal) their company's computer time to run another business.

  Unauthorized copying of programs for personal gain is a form of theft called **software piracy.** According to the **Software Copyright Act of 1980,** it is legal for a program owner to make only his or her own backup copies of that program. *It's important to note that none of these copies may be legally resold or given away. This may come as a surprise to those who copy software from a friend, but that's the law.*

  Pirated software accounts for over 40 percent of software used in the United States. The incidence of pirated software is even higher overseas in such countries as Italy (82 percent) and Thailand (92 percent). Penalties for violating this law are up to $250,000 in fines and five years in prison.

- **Manipulation:** Finding entry into someone's computer network and leaving a prankster's message may seem like fun, which is why hackers do it. It is still against the law. Moreover, even if the manipulation seems harmless, it may cause a great deal of anxiety and wasted time among network users.

  The Computer Fraud and Abuse Act of 1986 makes it a crime for unauthorized persons even to *view*—let alone copy or damage—data using any computer across state lines. It also prohibits unauthorized use of any government computer or computer used by any federally insured financial institution. Offenders can be sentenced to up to 20 years in prison and fined up to $100,000.

  Of course, using a computer in the course of performing some other crime, such as selling fraudulent products, is also illegal.

TIPS Are you concerned about catching a virus? Here are a few suggestions that might help:

1. *Purchase antivirus programs.* Install antivirus programs on all computer systems you use.
2. *Check disks.* Before using any floppy or CD, check for viruses.
3. *Enable write protection.* Protect data and programs on floppy disks by enabling write protection.
4. *Check all downloads.* Check all files downloaded from the Internet.
5. *Use an antivirus program.* Run your antivirus program frequently.
6. *Update your antivirus program.* New viruses are being developed daily, and the virus programs are continually being revised. Update your antivirus program frequently.

Other Hazards

There are plenty of other hazards to computer systems and data besides criminals. They include the following:

- **Natural hazards:** Natural forces include fires, floods, wind, hurricanes, tornadoes, and earthquakes. Even home computer users should store backup disks of programs and data in safe locations in case of fire or storm damage.

- **Civil strife and terrorism:** Wars, riots, and other forms of political unrest are real risks in some parts of the world. Even people in developed countries, however, must be mindful that acts of sabotage are possible.

- **Technological failures:** Hardware and software don't always do what they are supposed to do. (See Figure 10-9.) For instance, too little electricity, caused by a brownout or blackout, may cause the loss of data in primary storage. Too much electricity, as when lightning or other electrical disturbance affects a power line, may cause a **voltage surge,** or **spike.** This excess of

FIGURE 10-9
Hardware and software don't always do what they are supposed to do

Virus Protection

Worried about computer viruses? Did you know that others could be intercepting your private e-mail? It is even possible for them to gain access and control over your computer system. Fortunately, Internet security suites are available to help ensure your safety while you are on the Internet.

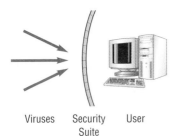

Viruses   Security   User
          Suite

How It Works

Internet security suites are collections of programs that create a protective barrier around your computer system. These suites, such as Norton AntiVirus and eSafe, are designed to protect against computer viruses and to ensure security and privacy of computer system resources.

Getting Started

The first step is to install an Internet security suite. Once installed, the software will continually work to ensure security and privacy. One such suite, eSafe, has a version available free from the Internet. To install this suite, follow the instructions below.

1. Connect to ealaddin.com/esafe and click *Downloads*.

2. Click *Home Users*.

3. Complete the information form and follow instructions to download.

4. Once you are returned to the desktop, double-click the eSafe icon and follow the instructions to complete the installation process.

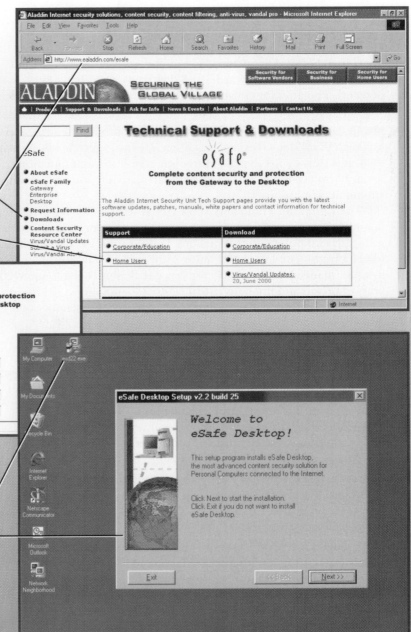

eSafe

Numerous security files have been installed. One of these, Desktop Watch, runs continually to search for privacy and security violations to the computer system. Another program, Desktop Configuration, provides a menu to access some of eSafe's most powerful applications including Sandbox, Personal Firewall, and Anti-Virus.

SANDBOX

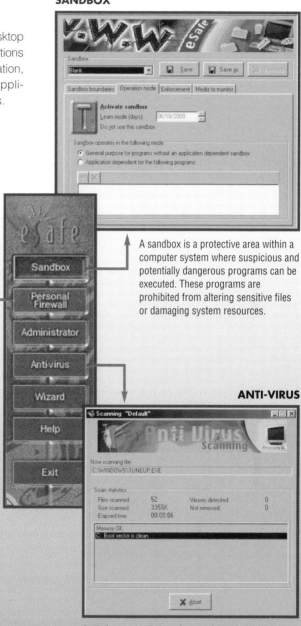

A sandbox is a protective area within a computer system where suspicious and potentially dangerous programs can be executed. These programs are prohibited from altering sensitive files or damaging system resources.

PERSONAL FIREWALL

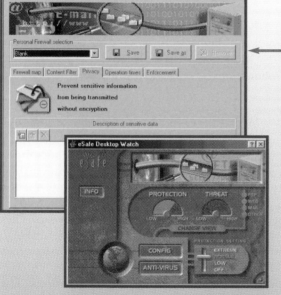

Personal Firewalls are programs that monitor all inbound and outbound traffic to a computer system. They limit access to only authorized users, automatically check files for viruses, and filter out unwanted content.

ANTI-VIRUS

Anti-virus controls how frequently the computer system is searched for computer viruses. When a file is checked, it is compared to the profile of over 6,000 known viruses. Once a virus is detected, it is typically either eliminated from the file or the entire file is deleted.

The Web is continually changing and some of the specifics presented in this Making IT Work for You may have changed. See our Web site at http://www.mhhe.com/it/oleary/IT.mhtml for possible changes and to learn more about this application of technology.

electricity may destroy chips or other electronic components of a computer.

Microcomputer users should use a **surge protector,** a device that separates the computer from the power source of the wall outlet. When a voltage surge occurs, it activates a circuit breaker in the surge protector, protecting the computer system.

Another technological catastrophe is when a hard-disk drive suddenly "crashes," or fails, perhaps because it has been bumped inadvertently. If the user has forgotten to make backup copies of data on the hard disk, data may be lost.

- **Human errors:** Human mistakes are inevitable. Data-entry errors are probably the most commonplace. Programmer errors also occur frequently. Some mistakes may result from faulty design, as when a software manufacturer makes a deletion command closely resembling another command. Some errors may be the result of sloppy procedures. One such example occurs when office workers keep important correspondence under file names that no one else in the office knows.

Measures to Protect Computer Security

Security is concerned with protecting information, hardware, and software. They must be protected from unauthorized use as well as from damage from intrusions, sabotage, and natural disasters. (See Figure 10-10.) Considering the numerous ways in which computer systems and data can be compromised, we can see why security is a growing field. Some of the principal measures to protect computer security are the following.

Encrypting Messages

Whenever information is sent over a network, the possibility of unauthorized access exists. The longer the distance the message has to travel, the higher the security risk is. For example, an e-mail message on a LAN meets a limited number of users operating in controlled environments such as offices. An e-mail message traveling across the country on the information superhighway affords greater opportunities for the message to be intercepted.

Businesses have been **encrypting,** or coding, messages for years. They have become so good at it that some law enforcement agencies are unable to wiretap messages from suspected criminals. Some federal agencies have suggested that a standard encryption procedure be used so that law enforcement agencies can monitor suspected criminal communications. The government is encouraging businesses that utilize the Internet to use a special encryption program. This program is available on a processor chip called the Clipper chip and is also known as the Key Escrow chip.

FIGURE 10-10
Disasters—both natural and man-made—can play havoc with computers

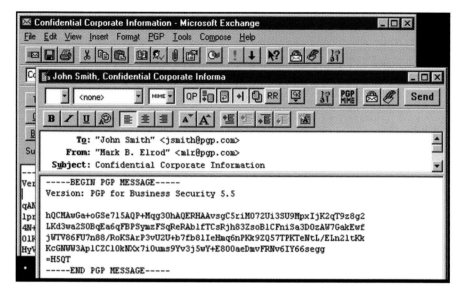

FIGURE 10-11
Encrypted e-mail

Individuals are also using encryption programs to safeguard their private communications. One of the most widely used personal encryption programs is Pretty Good Privacy. (See Figure 10-11.)

Restricting Access

Security experts are constantly devising ways to protect computer systems from access by unauthorized persons. Sometimes security is a matter of putting guards on company computer rooms and checking the identification of everyone admitted. Oftentimes it is a matter of being careful about assigning passwords to people and of changing them when people leave a company. **Passwords** are secret words or numbers that must be keyed into a computer system to gain access. In some dial-back computer systems, the user telephones the computer, punches in the correct password, and hangs up. The computer then calls back at a certain preauthorized number.

As mentioned in Chapter 8, most major corporations today use special hardware and software called **firewalls** to control access to their internal computer networks. Firewalls act as a security buffer between the corporation's private network and all external networks, including the Internet. All electronic communications coming into and leaving the corporation must be evaluated by the firewall. Security is maintained by denying access to unauthorized communications. (See Figure 10-12.)

Anticipating Disasters

Companies (and even individuals) that do not make preparations for disasters are not acting wisely. **Physical security** is concerned with protecting hardware from possible human and natural disasters. **Data security** is concerned with protecting software and data from unauthorized tampering or damage. Most large organizations have a **disaster recovery plan** describing ways to continue operating until normal computer operations can be restored.

Hardware can be kept behind locked doors, but often employees find this restriction a hindrance to their work, so security is lax. Fire and water (including the water from ceiling sprinkler systems) can do great damage to equipment. Many companies therefore will form a cooperative arrangement to share equipment with other companies in the event of catastrophe. Special emergency facilities called **hot sites** may be created that are fully equipped back up computer centers. They are called **cold sites** if they are empty shells in which hardware must be installed.

ON THE WEB EXPLORATIONS

Pretty Good Privacy (PGP) is one of the leaders in the development of encryption programs. To learn more about the company, visit our Web site at

http://www.mhhe.com/it/oleary/ explore.mhtml

FIGURE 10-12
Restricting access is one way to protect computer systems

Backing Up Data

Equipment can always be replaced. A company's *data*, however, may be irreplaceable. Most companies have ways of trying to keep software and data from being tampered with in the first place. They include careful screening of job applicants, guarding of passwords, and auditing of data and programs from time to time. The safest procedure, however, is to make frequent backups of data and to store them in remote locations.

Security for Microcomputers

If you own a microcomputer system, there are several procedures to follow to keep it safe:

- **Avoid extreme conditions:** Don't expose the computer to extreme conditions. Direct sun, rain from an open window, extreme temperatures, cigarette smoke, and spilled drinks or food are harmful to microcomputers. Clean your equipment regularly. Use a surge protector to protect against voltage surges.

- **Guard the computer:** Put a cable lock on the computer. If you subscribe or belong to an online information service, do not leave passwords nearby in a place accessible to others. Etch your driver's license number or social security number into your equipment. That way it can be identified in the event it is recovered after theft. (See Figure 10-13.)

- **Guard programs and data:** Store disks properly, preferably in a locked container. Make backup copies of all your important files and programs. Store copies of your files in a different—and safe—location from the site of your computer.

See Figure 10-14 for a summary of the different measures to protect computer security.

CONCEPT CHECK

✔ Give examples of threats to computer security that involve computer crimes.

✔ Give examples of other hazards that can threaten computer security.

✔ What are some of the measures taken to ensure computer security?

FIGURE 10-13
Label your computer equipment to make it easy to identify in case of theft

Measure	Description
Encrypting	Coding all messages sent over a network
Restricting	Limiting access to authorized persons using such measures as passwords, dial-back systems, and firewalls
Anticipating	Preparing for disasters by ensuring physical security and data security through a disaster recovery plan
Backing up	Routinely copying data and storing it at a remote location
Securing	Protecting a microcomputer by avoiding extreme conditions, guarding the computer, programs, and data

FIGURE 10-14
Measures to protect computer security

Ergonomics

Ergonomics helps computer users take steps to avoid physical and mental health risks and to increase productivity.

Even though computers have decreased significantly in cost, they are still expensive. Why have them, then, unless they can make workers more effective? Ironically, there are certain ways in which computers may actually make people *less* productive. Many of these problems will most likely affect workers in positions that involve intensive data entry, such as clerks and word processor operators. However, they may also happen to anyone whose job involves heavy use of the computer. As a result, there has been great interest in a field known as ergonomics.

Ergonomics (pronounced "er-guh-*nom*-ix") is defined as the study of human factors related to things people use. It is concerned with fitting the job to the worker rather than forcing the worker to contort to fit the job. As computer use has increased, so has interest in ergonomics. People are devising ways that computers can be designed and used to increase productivity and avoid health risks.

ON THE WEB EXPLORATIONS

Workplace Designs is one of the leading distributors of ergonomic equipment. To learn more about the company, visit our Web site at

http://www.mhhe.com/it/oleary/ explore.mhtml

Physical Health

Sitting in front of a screen in awkward positions for long periods may lead to physical problems. These can include eyestrain, headaches, and back pain. Users can alleviate these problems by taking frequent rest breaks and by using well-designed computer furniture. Some recommendations by ergonomics experts for the ideal setup for a microcomputer are illustrated in Figure 10-15.

The physical health matters related to computers that have received the most attention recently are the following:

- **Avoiding eyestrain and headache:** Our eyes were made for most efficient seeing at a distance. However, monitors require using the eyes at closer range for a long time, which can create eyestrain, headaches, and double vision.

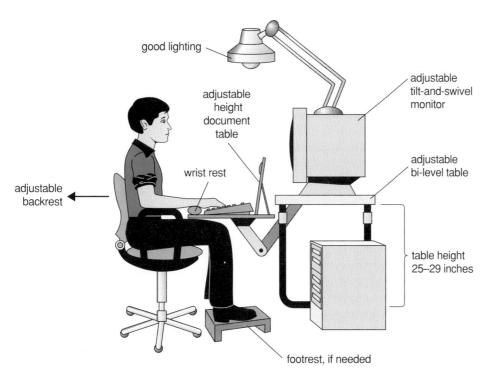

good lighting

adjustable height document table

adjustable tilt-and-swivel monitor

wrist rest

adjustable backrest

adjustable bi-level table

table height 25–29 inches

footrest, if needed

FIGURE 10-15
Recommendations for the ideal microcomputer work environment

ON THE WEB EXPLORATIONS

The Mayo Clinic and the American Optometric Association provide free electronic newsletters that describe computer-related eyestrain. To learn more about these organizations, visit our Web site at

http://www.mhhe.com/it/oleary/ explore.mhtml

To make the computer easier on the eyes, take a 15-minute break every hour or two. Avoid computer screens that flicker. Keep computer screens away from windows and other sources of bright light to minimize reflected glare on the screen. Special antiglare screen coatings and glare shields are also available. Make sure the screen is three to four times brighter than room light. Keep everything you're focusing on at about the same distance. For example, the computer screen, keyboard, and a document holder containing your work might be positioned about 20 inches away. Clean the screen of dust from time to time.

- **Avoiding back and neck pain:** Many people work at monitors and keyboards that are in improper positions. The result can be pains in the back and neck.

    To avoid such problems, make sure equipment is adjustable. You should be able to adjust your chair for height and angle, and the chair should have good back support. The table on which the monitor stands should also be adjustable, and the monitor itself should be of the tilt-and-swivel kind. The monitor should be at eye level or slightly below eye level. Keyboards should be detachable. Document holders should be adjustable.

- **Avoiding effects of electromagnetic fields:** Like many household appliances, monitors generate invisible electromagnetic field (EMF) emissions, which can pass through the human body. Some observers feel that there could be a connection between these EMF emissions and miscarriages (and even some cancers). A study by the government's National Institute of Occupational Safety and Health found no statistical relationship between monitors and miscarriages. Even so, several companies have introduced low-emission monitors. They state that no health or safety problems exist with older monitors; rather, they are merely responding to market demands.

    One recommendation is that computer users should follow a policy of "prudent avoidance" in reducing their exposure to EMF emissions. (See Figure 10-16.) They should try to sit about 2 feet or more from the computer screen and 3 feet from neighboring terminals. The strongest fields are emitted from the sides and backs of terminals. Pregnant women should be particularly cautious and are encouraged to consult with their physician.

- **Avoiding repetitive strain injury:** Data-entry operators may make as many as 23,000 keystrokes a day. Some of these workers and other heavy keyboard users have fallen victim to a disorder known as repetitive strain injury.

    Repetitive strain injury (RSI)—also called **repetitive motion injury** and **cumulative trauma disorder**—is the name given to a number of injuries. These result from fast, repetitive work that can cause neck, wrist, hand, and arm pain. RSI is by far the greatest cause of workplace illnesses in private industry. It accounts for *billions* of dollars in compensation claims and lost productivity every year. Some RSI sufferers are slaughterhouse, textile, and automobile workers, who have long been susceptible to the disorder. One particular type of RSI, **carpal tunnel syndrome,** found among heavy computer users, consists of damage to nerves and tendons in the hands. Some victims report the pain is so intense that they cannot open doors or shake hands and that they require corrective surgery.

    Before the computer, typists would stop to change paper or make corrections, thus giving themselves short but frequent rest periods.

FIGURE 10-16
Computer users should avoid exposure to EMF emissions

Ergonomically correct keyboards have recently been developed to help prevent injury from heavy computer use. (See Figure 10-17.) But, in addition, because RSI is caused by repetition and a fast work pace, you should remember to take frequent short rest breaks. Experts also advise getting plenty of sleep and exercise, watching your weight, sitting up straight, and learning stress-management techniques.

Mental Health

Computer technology offers not only ways of improving productivity but also some irritants that may be counterproductive.

FIGURE 10-17
Ergonomic keyboard

- **Avoiding noise:** Computing can be quite noisy. Voice input and output can be distracting for co-workers. Working next to a printer for several hours can leave one with ringing ears. Also, users may develop headaches and tension from continual exposure to the high-frequency, barely audible squeal produced by cooling fans and vibrating parts inside the system unit. This is particularly true for women, who hear high-frequency sounds better than men do. They may be affected by the noise even when they are not conscious of hearing it.

  Head-mounted microphones and earphones greatly reduce the effect of voice input and output. Acoustical tile and sound-muffling covers are available for reducing the noise from co-workers and impact printers. Tightening loose system unit components will reduce high-frequency noise.

- **Avoiding stress from excessive monitoring:** Research shows that workers whose performance is monitored electronically suffer more health problems than do those watched by human supervisors. For instance, a computer may monitor the number of keystrokes a data-entry clerk completes in a day. It might tally the time a customer-service person takes to handle a call. The company might then decide to shorten the time allowed and to continue monitoring employees electronically. By so doing, it may force a pace leading to physical problems, such as RSI, and mental health difficulties. One study found that electronically monitored employees reported more boredom, tension, extreme anxiety, depression, anger, and severe fatigue than those who were not electronically monitored.

  Recently it has been shown that electronic monitoring actually is not necessary. For instance, both Federal Express and Bell Canada replaced electronic monitoring with occasional monitoring by human managers. They found that employee productivity stayed up and even increased.

  A new word—*technostress*—has been proposed to describe the stress associated with computer use that is harmful to people. **Technostress** is the tension that arises when we have to unnaturally adapt to computers rather than having computers adapt to us.

Design

Electronic products from microwave ovens to VCRs to microcomputers offer the promise of more efficiency and speed. Often, however, the products are so overloaded with features that users cannot figure them out. Because a microprocessor chip handles not just one operation but several, manufacturers feel obliged to pile on the bells and whistles. Thus, many home and office products, while being fancy technology platforms, are difficult for humans to use.

A recent trend among manufacturers is to deliberately strip down the features offered, rather than to constantly do all that is possible. In appliances,

Problem	Remedy
Eyestrain and headache	Take frequent breaks, avoid screen glare, place object at fixed focal distance
Back and neck pain	Use adjustable equipment
Electromagnetic field emissions	Sit 2 feet or more from screen
Repetitive strain injury	Use ergonomically correct keyboards, take frequent breaks
Noise	Use head-mounted microphones and earphones, install acoustical tile and sound-muffling covers, tighten system unit components
Stress from excessive monitoring	Remove electronic monitoring

FIGURE 10-18
Summary of ergonomic concerns

this restraint is shown among certain types of high-end audio equipment, which come with fewer buttons and lights. In computers, there are similar trends. Surveys show that consumers want plug-and-play equipment—machines that they can simply turn on and quickly start working. Thus, computers are being made easier to use, with more menus, windows, icons, and pictures.

For a summary of ergonomic concerns, see Figure 10-18.

CONCEPT CHECK

✔ What is ergonomics and why is it important?

✔ What are some of the most significant physical and mental health concerns?

The Environment

Computer industry has responded to the Energy Star program with the Green PC. You can help by conserving, recycling, and educating.

What do you suppose uses the greatest amount of electricity in the workplace? Microcomputers do. They account for 5 percent of the electricity used. Increased power production translates to increased air pollution, depletion of nonrenewable resources, and other environmental hazards.

The Environmental Protection Agency (EPA) has created the **Energy Star** program to discourage waste in the microcomputer industry. Along with over 50 manufacturers, the EPA has established a goal of reducing power requirements for system units, monitors, and printers. The industry has responded with the concept of the **Green PC.** (See Figure 10-19.)

FIGURE 10-19
The Green PC

The Green PC

The basic elements of the Green PC are:

- **System Unit:** Using existing technology from portable computers, the system unit (1) uses an energy-saving microprocessor that requires a minimal amount of power, (2) employs microprocessor and hard-disk drives that shift to an energy-saving or sleep mode when not in operation, (3) replaces the conventional supply unit with an adapter that requires less electricity, and (4) eliminates the cooling fan.

- **Display:** Displays have been made more energy efficient by using (1) flat panels that require much less energy than the traditional monitors, (2) special power-down monitors that automatically reduce power consumption when not in use, and (3) screen-saver software that clears the display whenever it is not in use.

- **Manufacturing:** Computer manufacturers such as Intel, Apple, Compaq, and others are using fewer harmful chemicals in production. Particular attention is given to **chlorofluorocarbons (CFCs)** in solvents and cleaning agents. (CFCs can travel into the atmosphere and are suspected by some in the scientific community to deplete the earth's ozone layer.) Toxic nickel and other heavy metals are being eliminated or reduced in the manufacturing processes.

Of course, not all of these technologies and manufacturing processes are used for all microcomputers. But more and more of them are.

Personal Responsibility

Some of the things that you, as a computer user, can do to help protect the environment are the following:

- **Conserve:** The EPA estimates that 30 to 40 percent of computer users leave their machines running days, nights, and weekends. When through working for the day, turn off all computers and other energy-consuming devices. The EPA also estimates that 80 percent of the time a monitor is on, no one is looking at it. Use screen-saver programs that blank the computer screen after three to five minutes of inactivity.

- **Recycle:** U.S. businesses use an enormous amount of paper each year—a pile 48,900 miles high. Much of that, as well as the paper we throw out at home, can be recycled. Other recyclable items include computer boxes, packaging material, printer cartridges, and floppy disks.

- **Educate:** Be aware of and learn more about ecological dangers of all types. Make your concerns known to manufacturers and retail agencies. Support ecologically sound products.

CONCEPT CHECK

✔ What is a Green PC?

✔ What are the basic elements of the Green PC?

✔ What other actions can you take to help protect the environment?

A LOOK TO THE FUTURE

New legislation will be needed to define access to government files and to regulate government interference in free speech in the new electronic world.

Technology often has a way of outracing existing social and political institutions. For instance, citizens have a right to request government records under the Freedom of Information Act. But even in its most recent amendment, in 1986, the act does not mention the word *computer* or define the word *record*. Can the government therefore legally deny, as one agency did, a legitimate request for data on corporate compliance with occupational safety and health laws? *Access laws* lag behind even as the government collects more information than ever.

In addition, there has been a rise in computer-related crimes. These include bank and credit card fraud, viruses, and electronic break-ins of government and private computer systems. Law-enforcement agencies continue to crack down on these computer crimes. Yet they may also be jeopardizing the rights of computer users who are not breaking the law. Such users may be suffering illegal searches and violation of constitutional guarantees of free speech. However, it is unclear how the First Amendment protects speech and the Fourth Amendment protects against searches and seizures in this electronic world.

One professor of constitutional law has proposed a new amendment to the Constitution. This amendment would extend the other freedoms in the Bill of Rights, those on free speech and search and seizure restrictions. Under this amendment, all new technology and media for generating, storing, and altering information would be covered.

VISUAL SUMMARY — Privacy, Security, Ergonomics, and the Environment

PRIVACY

Computer ethics are guidelines for moral computer use. Four computer ethics issues are the following:

- **Privacy** concerns collection and use of data about individuals.
- **Accuracy** relates to the responsibility of those who collect data to ensure correctness.
- **Property** relates to who owns data and rights to software.
- **Access** relates to the responsibility of those who have data to control who is able to use it.

Large Databases

Large databases are constantly compiling information about us. A vast industry of data gatherers or "information resellers" collects data about us and sells it to direct marketers and others.

Private Networks

Some information networks have been used to eavesdrop on employees or to restrict members' messages.

Internet and the Web

All Web communications are subject to eavesdropping. Browsers record your activities in **history files. Cookies** deposited by Web sites collect information about you.

The Major Laws on Privacy

There are numerous federal laws governing privacy matters; however, each has drawbacks that make enforcement difficult.

SECURITY

Security is concerned with keeping hardware, software, data, and programs safe from unauthorized personnel.

Threats to Computer Security

Keeping information private depends on keeping computer systems safe from:

- Computer criminals—employees, outside users, **hackers/crackers**, and organized criminals.
- **Computer crime**—damage, theft, or manipulation.
- **Viruses**—programs that migrate through networks and operating systems attaching themselves to different programs and databases.
- Other hazards—natural forces, civil strife, terrorism, technological failures, and human errors.

Measures to Protect Computer Security

Security is concerned with protecting information, hardware, and software. Some measures are:

- **Encrypting** all messages.
- Restricting access through guards, passwords, dial-back systems, and **firewalls.**
- Anticipating disasters by providing **physical security** and **data security.**
- Backing up data frequently and storing the backups in safe locations.
- Securing microcomputers against extreme conditions and theft.

To be a competent end user you need to be aware of the potential impact of technology on people. You need to be sensitive to and knowledgeable about personal privacy, organizational security, ergonomics, and the environmental impact of technology.

ERGONOMICS

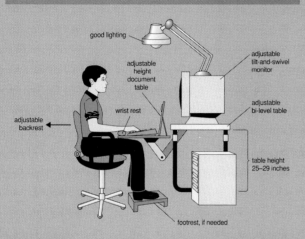

Ergonomics is the study of human factors related to things people use, including computers.

Physical Health

Physical health matters that have received the most attention can readily be avoided. Problems and their solutions are:

- Eyestrain and headache—take frequent breaks; avoid glare on the monitor.
- Back and neck pains—use adjustable chairs, tables, monitor stands, keyboards.
- Electromagnetic fields—sit 2 feet from the screen, 3 feet from adjacent computers.
- **Repetitive strain injury (RSI)**—including **carpal tunnel syndrome**—take frequent, short rest breaks; use good posture; maintain a healthy lifestyle; and use ergonomic keyboards.

Mental Health

Counterproductive mental irritations include:

- Noise from clattering printers and high-frequency squeal from monitors.
- Stress from excessive monitoring. **Technostress** is caused by unnaturally adapting to computers.

Design

Computers are being designed for easier and healthier use. There is a trend toward simplifying or stripping down features offered on new models.

THE ENVIRONMENT

Microcomputers are the greatest users of electricity in the workplace. The Environmental Protection Agency (EPA) has established the **Energy Star** program to encourage efficient use of energy in the computer industry. The industry has responded with the concept of the **Green PC.**

The Green PC

The basic elements of the Green PC include:

- System units with energy-saving microprocessors, sleep-mode capability, more efficient adapters, and no cooling fans.
- Display units that replace CRT displays with flat panels, use special power-down monitors, and use screen-saver software.
- Manufacturing that eliminates or reduces the use of harmful chemicals such as **chlorofluorocarbons (CFCs),** nickel, and other heavy metals.

Personal Responsibility

As a responsible computer user, you can help protect the environment by:

- Conserving energy by turning off computer systems at night and using screen-saver software.
- Recycling paper, computer boxes, packaging materials, printer cartridges, and floppy disks.
- Educating yourself and others about ecological dangers and using ecologically sound products.

Key Terms

access (236)
accuracy (236)
carpal tunnel syndrome (250)
chlorofluorocarbon (CFCs) (253)
cold site (247)
Computer Abuse Amendments Act of 1994 (239)
computer crime (241)
Computer Fraud and Abuse Act of 1986 (239)
Computer Matching and Privacy Protection Act of 1988 (239)
cookies (238)
cracker (241)
cumulative trauma disorder (250)
data security (247)
disaster recovery plan (247)
Electronic Communications Privacy Act of 1986 (239)

encrypting (246)
Energy Star (252)
ergonomics (249)
ethics (236)
Fair Credit Reporting Act of 1970 (238)
firewall (247)
Freedom of Information Act of 1970 (239)
Green PC (252)
hacker (241)
history file (238)
hot site (247)
National Information Infrastructure Protection Act of 1996 (239)
No Electronic Theft (NET) Act of 1997 (240)
password (247)
physical security (247)

privacy (236)
Privacy Act of 1974 (239)
property (236)
repetitive motion injury (250)
repetitive strain injury (RSI) (250)
Right to Financial Privacy Act of 1979 (239)
security (246)
snoopware (237)
Software Copyright Act of 1980 (243)
software piracy (243)
spike (243)
surge protector (246)
technostress (251)
Video Privacy Protection Act of 1988 (239)
virus (241)
voltage surge (243)
worm (241)

Chapter Review

LEVEL 1

Reviewing Facts and Terms
Matching

Match each numbered item with the most closely related lettered item. Write your answers in the spaces provided.

1. Concerns the collection and use of data about individuals. ____

2. Person who gains unauthorized access to a computer system for the fun and challenge of it. ____

3. Device that separates the computer from the power source, protecting it from power fluctuations. ____

4. Programs that migrate through networks and operating systems and attach themselves to different programs and databases. ____

5. Results from fast, repetitive work that can cause neck, wrist, hand, and arm pain. ____

6. Concerned with protecting information, hardware, and software. ____

7. Destructive program that fills a computer system with self-replicating information, clogging the system so that its operations are slowed or stopped. ____

8. Many Web sites have these specialized programs that record how often you visit a site, what you do there, and any other information that you provide. ____

9. Relates to who owns data and rights to software. ____

a. Access
b. Accuracy
c. Carpal tunnel syndrome
d. Computer crime
e. Cookies
f. Disaster recovery plan
g. Encrypting
h. Ergonomics
i. Ethics
j. Firewall
k. Freedom of Information Act of 1970
l. Green PC
m. Hacker
n. Privacy
o. Property
p. Repetitive strain injury
q. Security
r. Surge protector
s. Virus
t. Worm

10. Injury, found among heavy computer users, that consists of damage to nerves and tendons in the hands. ____

11. Process that codes messages sent over networks to protect the information from being viewed by unauthorized individuals. ____

12. The study of human factors related to things people use—concerned with fitting the job to the worker rather than vice versa. ____

13. Relates to the responsibility of those who collect data to ensure that the data is correct. ____

14. An illegal action in which the perpetrator uses special knowledge of computer technology. ____

15. Standards for moral conduct. ____

16. The computer industry's response to the EPA's goals for reducing waste associated with microcomputers. ____

17. Relates to the responsibility of those who have data to control who is able to use that data. ____

18. Plan describing ways, in the case of disaster, to continue operating until normal computer operations can be restored. ____

19. Gives you the right to look at data concerning you that is stored by the federal government. ____

20. Special hardware and software that act as a buffer between a corporation's private network and all external networks. ____

True/False

In the spaces provided, write T or F to indicate whether the statement is true or false.

1. Our legal system is the essential element used to control computers today. ____

2. Over 20 percent of businesses search through employees' electronic messages and computer files. ____

3. A worm keeps replicating itself until a computer system's operations are slowed or stopped. ____

4. Electromagnetic field (EMF) emissions can travel through a person's body. ____

5. Most people who use computers are midlevel managers. ____

Multiple Choice

Circle the letter of the correct answer.

1. The ethical issue that deals with the responsibility to control the availability of data:
   a. privacy   **c.** property   **e.** access
   b. accuracy   **d.** ownership

2. The largest category of computer criminals:
   a. students   **c.** outside users   **e.** database managers
   b. hackers   **d.** employees

3. The study of human factors related to things people use:
   a. data analysis   **c.** ergonomics   **e.** personal design
   b. human system performance **d.** expert analysis

4. A specific type of repetitive strain injury that causes damage to nerves and tendons in the hands:
   a. RSI   **c.** EMF   **e.** virus
   b. carpal tunnel syndrome **d.** hacker

5. The computer industry's response to the Energy Star program is the:

   a. Green PC   **c.** flat-panel display   **e.** network encryption

   b. multimedia PC   **d.** Fair Credit Reporting Act   standard

Completion

Complete each statement in the spaces provided.

1. _____ _____ is the unauthorized copying of programs for personal gain.

2. Computer _____ are guidelines for the morally acceptable use of computers in our society.

3. A common security measure for business is _____ or coding messages.

4. People who gain unauthorized access to a computer system for malicious purposes are called _____.

5. The new word _____ is used to describe harmful stress associated with computer use.

Reviewing Concepts

LEVEL 2

Open-Ended

On a separate sheet of paper, respond to each question and statement.

1. How are computers a threat to your privacy? Discuss what you can do to better ensure your privacy.

2. What are four types of computer criminals? In each case, give an example of how they can violate proper computer conduct.

3. What kind of activities can you perform to avoid computer-related eyestrain, headaches, and back and neck pain?

4. Describe some mental health problems associated with frequent computer use.

5. How are computers a threat to the environment? Discuss three things you can do to protect the environment.

Concept Mapping

On a separate sheet of paper, draw a concept map or a flowchart showing how the following terms are related. Show all relationships. Include any additional terms you can think of.

access	encrypting	property
accuracy	ergonomics	repetitive motion injury
carpal tunnel syndrome	ethics	repetitive strain injury
Computer Abuse Amendments Act of 1994	Fair Credit Reporting Act of 1970	Right to Financial Property Act of 1979
Computer Fraud and Abuse Act of 1986	firewall	security
Computer Matching and Privacy Protection Act of 1998	Freedom of Information Act of 1970	Software Copyright Act of 1980
computer crime	Green PC	software piracy
cookies	hacker	spike
cracker	history file	surge protector
data security	National Information Infrastructure Protection Act of 1996	Video Privacy Protection Act of 1988
disaster recovery plan	physical security	virus
Electronic Communications Privacy Act of 1986	privacy Privacy Act of 1974	worm

Critical Thinking Questions and Projects

LEVEL 3

Read each exercise and answer the related questions on a separate sheet of paper.

2001 2002

1. *Privacy:* Social security numbers have become a national identification number that, once given out, never goes away. It is often a person's student ID, tax number, military ID, and much more. Every time you provide the number on a credit or job application, it becomes available to thousands of people you don't know. Describe the worst scenario of what could happen to you because of the easy accessibility of your social security number.

2. *Virus Protection:* Several software suites are available to protect against viruses and other threats from the Internet. (See Making IT Work for You: Virus Protection on pages 244–45.) Organizations as well as individuals use this type of software. Discuss the security measures used by either your school or your workplace. Are they adequate? Why or why not? Do you own a computer system? If so, what security measures do you use? Are they adequate? Why or why not?

On the Web Exercises

1. Pretty Good Privacy

Pretty Good Privacy (PGP) is one of the most popular encryption software available. Learn more about your privacy rights and needs by visiting our Web site at http://

www.mhhe.com/it/oleary/exercise.mhtml to link to a site specializing in encryption software. Once connected to that site, explore the site and print out the Web page that discusses why you need PGP. Write a paragraph discussing individual privacy rights in the workplace and in regard to the government.

2. Crime on the Internet

Just like any community, the Internet community has crime. The best way to fight it is to know more about it. Visit the Infoseek site at http://infoseek.go. com, and search with

keywords such as "Internet crime," "e-mail fraud," and "Web police." Print a copy of the first page of the results of your search. Explore one or more of the sites and write a paragraph describing one type of computer crime and what we can do to fight against it.

3. Ergonomics

The right office equipment can go a long way in preventing stress and strain at work and at home. Visit our Web site at http://www.mhhe. com/it/oleary/

exercise.mhtml to link a site that specializes in ergonomic office products. Once connected to that site, explore it and select one or more products that would be helpful to you. Print out the Web page(s) describing the product(s) and write a paragraph describing why the product(s) would be helpful to you.

4. Computer Viruses

Computer viruses can strike anytime and can create serious problems. To learn more about them, visit our Web site at http://www. mhhe.com/it/oleary/ exercise.mhtml to link

to a site specializing in computer viruses. Once connected to that site, explore the site looking for a list of the most common viruses (try the *virus library* in the antivirus area). Print out the list. Select three of the viruses from the list, and write a paragraph describing what each could do and what effect it could have on a major corporation.

Databases

After you have read this chapter, you should be able to:

1. **Describe how data is organized: characters, fields, records, and files.**

2. **Understand the difference between batch processing and real-time processing.**

3. **Describe the difference between master files and transaction files.**

4. **Define and describe the three types of file organization: sequential, direct, and index sequential.**

5. **Describe the advantages of a database.**

6. **Describe the two essential parts of a database management system (DBMS).**

7. **Describe four ways of organizing a DBMS: hierarchical, network, relational, and object-oriented.**

8. **Distinguish among individual, company, distributed, and proprietary databases.**

9. **Discuss some issues of productivity and security.**

organized? What are files and databases, and why do you need to know anything about them? Perhaps the answer is this: To be a competent user of information in the Information Age, you have to know how to *find* that information.

At one time, it was not important for microcomputer users to know much about data files and databases. However, the recent arrival of very powerful microcomputers and

Like a library, secondary storage is designed to store information. How is such information

their connectivity to communications networks and the Internet has changed that. Communications lines

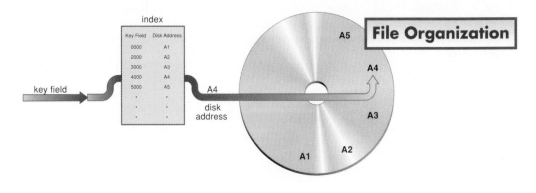

File Organization

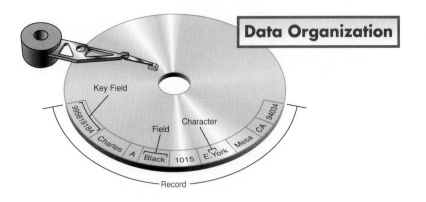

Data Organization

and the Internet extend the reach of your microcomputer well beyond the desktop.

Competent end users need to understand data fields, records, files, and databases. They need to be aware of the different ways in which databases can be organized and the different types of databases. Also, they need to know the most important database uses and issues.

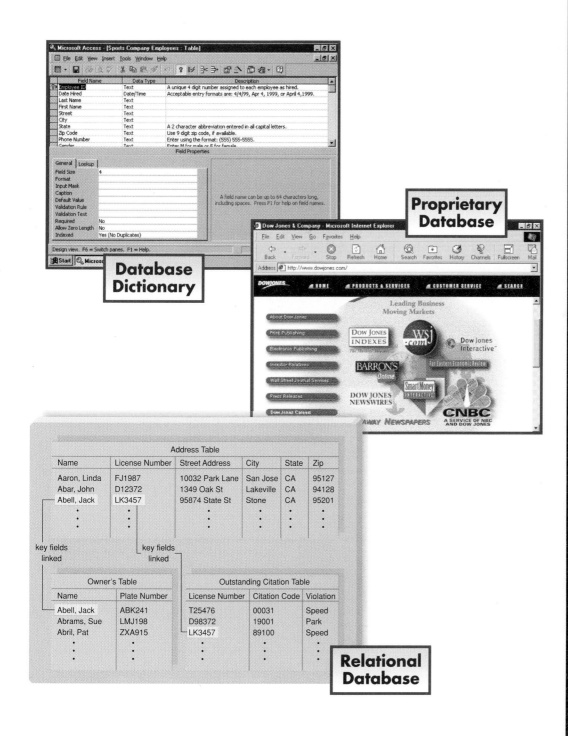

Database Dictionary

Proprietary Database

Relational Database

Data Files

Understanding how data files work means understanding data organization, key fields, batch versus real-time processing, master versus transaction files, and file organization.

Y ou want to know your final grades for the semester. You call your school's registrar after your last semester exams to find out your grade point average. Perhaps you are told, "Sorry, that's not in the computer yet." Why can't they tell you? How is the school's computer system different from, say, your bank's, where deposits and withdrawals seem to be recorded right away?

Data Organization

To be processed by the computer or stored in secondary storage, data is typically organized into groups or categories. Each group is more complex than the one before:

- **Character:** A **character** is a single letter, number, or special character such as a punctuation mark or $.
- **Field:** A **field** contains a set of related characters. On a college registration form or a driver's license, a person's first name is a field. Last name is another field, street address another field, city yet another field, and so on.
- **Record:** A **record** is a collection of related fields. Everything on a person's college registration form or driver's license, including identification number, is a record.
- **File:** A data **file** is a collection of related records. All the driver's licenses issued in one county could be a file.
- **Database:** A **database** is a collection of related files. All the driver's license files for the state could be a database.

An example of how data is organized is shown in Figure 11-1. Note that a student's name is not one field, but three: first name, middle initial, and last name.

Key Field

Figure 11-1 also shows the student's identification number. Is such a number really necessary? Certainly most people's names are different enough that at a small college, say, you might think identification numbers wouldn't be necessary. However, as anyone named John Smith or Linda Williams knows, there are plenty of other people around with the same name. Sometimes they even have the same middle initial. This is the reason for the student identification number: The number is unique, whereas the name may not be. This distinctive number is called a *key field*. A **key field** is the particular field of a record that is chosen to uniquely identify each record. Some common key fields are social security numbers, employee identification numbers, and part numbers.

Batch versus Real-Time Processing

Traditionally data is processed in two ways. These are *batch processing*, what we might call "later," and *real-time processing*, what we might call "now."

FIGURE 11-1
How data is organized

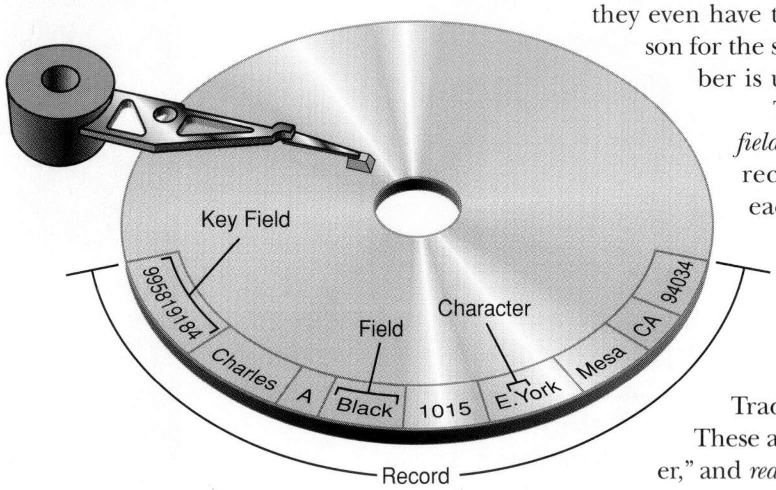

These two methods have been used to handle common record-keeping activities such as payroll and sales orders.

- **Batch processing:** In **batch processing,** data is collected over several hours, days, or even weeks. It is then processed all at once—as a "batch." If you have a bank credit card, your bill probably reflects batch processing. That is, during the month, you buy things and charge them to your credit card. Each time you charge something, a copy of the transaction is sent to the credit card company. At some point in the month, the company's data processing department puts all those transactions (and those of many other customers) together and processes them at one time. The company then sends you a single bill totaling the amount you owe.

- **Real-time processing:** Totaling up the sales charged to your bank credit card is an example of batch processing. You might use another kind of card—your bank's automated teller machine (ATM) card—for the second kind of processing. **Real-time processing** occurs when data is processed at the same time the transaction occurs. As you use your ATM card to withdraw cash, the system automatically computes the balance remaining in your account.

At one time, only tape storage, and therefore only sequential access storage (as we discussed in Chapter 6), was available. All processing then was batch processing and was done on mainframe computers. Even today, a great deal of mainframe time is dedicated to this kind of processing. Many smaller organizations, however, use microcomputers for this purpose.

Real-time processing is made possible by the availability of disk packs and direct access storage (as we described in Chapter 6). Direct access storage enables the user to quickly go directly to a particular record. (In sequential access storage, by contrast, the user must wait for the computer to scan several records one at a time. It continues scanning until it comes to the one that's needed.)

Not long ago, specialized terminals were used to enter data and perform real-time processing. Today, however, more and more microcomputers are being used for this purpose. As we have stated, microcomputers have become increasingly more powerful. Thus, smaller companies and departments of large companies use these machines by themselves for many real-time processing needs. That is, they use them without connecting to a mainframe.

Master versus Transaction Files

Two types of files are commonly used to update data—a *master file* and a *transaction file.*

- The **master file** is a complete file containing all records current up to the last update. An example is the data file used to prepare your last month's telephone bill or credit card bill.

- The **transaction file** contains *recent* changes to records that will be used to update the master file. An example could be a temporary "holding" file that accumulates telephone charges or credit card charges through the present month. At the end of the month, the transaction file containing all the new charges is used to update the master file. The result is a new master file that is current up to the end of the month.

File Organization

File organization may be of three types: *sequential, direct,* and *index sequential.*

- **Sequential file organization:** The simplest organization is **sequential file organization,** in which records are stored physically one after another in predetermined order. This order is determined by the key field on each record, such as a student identification number. (See Figure 11-2.)

This organization is very efficient whenever all or a large portion of the records need to be accessed—for example, when final grades are to be mailed out. There is also an equipment cost advantage because magnetic tapes and tape drives can be used. Both are less expensive than disks and disk drives. One disadvantage is that the records must be ordered in a specific way, and that can be time-consuming.

The major disadvantage, however, is that access to a particular record can be very slow. For example, to find the record of a particular student, the registrar's office would sequentially search through the records. It would search them one at a time until the student's number was found. If the number is 4315, the computer will start with record number 0000. It will go through 0001, 0002, and so on until it reaches the student's number.

FIGURE 11-2
Sequential file organization

- **Direct file organization:** To obtain particular records, **direct file organization** is much better. Records are not stored physically one after another. Rather, they are stored on a disk in a particular address or location that can be determined by their key field. This address is calculated by a technique known as **hashing.** Hashing programs use mathematical operations to convert the key field's numeric value to a particular storage address. These programs are used to initially store records and later to relocate them. (See Figure 11-3.)

  Unlike sequential access files, which are stored on either magnetic tape or disk, direct files can only be stored on disk. The primary advantage is that direct file organization can locate specific records very quickly. If your grades were stored in a direct file, the registrar could access them quickly using only your student identification number.

  The disadvantage of direct file organization is cost. It needs more storage space on disk. It also is not as good as sequential file organization for large numbers of updates or for listing large numbers of records.

- **Index sequential file organization: Index sequential file organization** is a compromise between sequential and direct file organizations. It stores records in a file in sequential order. However, an index sequential file

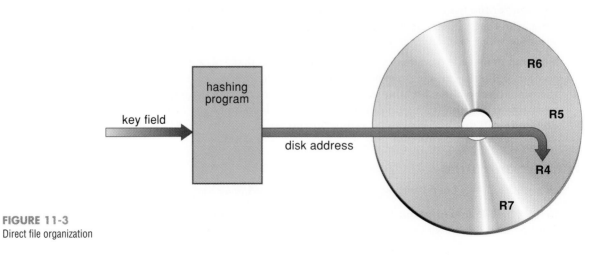

FIGURE 11-3
Direct file organization

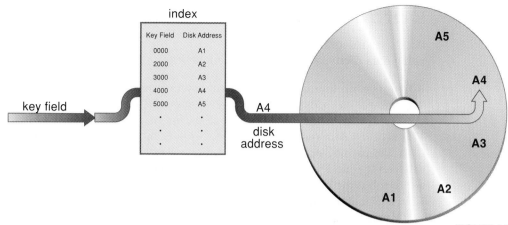

FIGURE 11-4
Index sequential file organization

also contains an index. The index lists the key to each group of records stored and the corresponding disk address for that group. When the user seeks a particular record, the computer starts searching sequentially by looking at the beginning of the record group. (See Figure 11-4.) Note: The disk addresses listed in Figure 11-4 represent the actual locations, which are identified by specific track and sector locations.

For example, the college registrar could index certain ranges of student identification numbers—0000 to 1999, 2000 to 2999, and so on. For the computer to find your number (for example, 4315), it would first go to the index. The index would give the location of the range in which your number appears on the disk (for example, 4000 to 4999). The computer would then search that range (A4) sequentially to find your number.

Index sequential file organization requires disks or other direct access storage devices. It is faster than sequential but not as fast as direct access. It's best used when large batches of transactions must occasionally be updated, yet users also want frequent, quick access to data. For example, every month a bank will update bank statements to send to its customers. However, customers and bank tellers need to be able to have up-to-the-minute information about checking accounts.

See Figure 11-5 for a summary of the advantages and disadvantages of each type of file organization.

FIGURE 11-5
Summary of the three types of file organization

Type	Advantages	Disadvantages
Sequential	Efficient access to all or large part of records, low cost	Slow access to specific records
Direct	Fast access to specific records	High cost
Index sequential	Faster than sequential, more efficient than direct	Not quite as efficient as sequential, not quite as fast as direct, high cost

CONCEPT CHECK

✔ List the categories into which data is stored.

✔ What are the two ways in which data is processed?

✔ Name the two file types commonly used to update data.

✔ File organization may be of these types: _____, _____, and _____.

Database

Databases integrate data. DBMSs create, modify, and access databases using data dictionaries and query languages.

Many organizations have multiple files on the same subject or person. For example, records for the same customer may appear in different files in the sales department, billing department, and credit department. This is called **data redundancy.** If the customer moves, then the address in each file must be updated. If one or more files are overlooked, problems will likely result. For example, a product ordered might be sent to the new address, but the bill might be sent to the old address. This situation results from a lack of **data integrity.**

Moreover, data spread around in different files is not as useful. The marketing department, for instance, might want to offer special promotions to customers who order large quantities of merchandise. However, they may be unable to do so because the information they need is in the billing department. A database can make the needed information available.

Need for Databases

For both individuals and organizations, there are many advantages to having databases:

- **Sharing:** In organizations, information from one department can be readily shared with others. Billing could let marketing know which customers ordered large quantities of merchandise.

- **Security:** Users are given passwords or access only to the kind of information they need. Thus, the payroll department may have access to employees' pay rates, but other departments would not.

- **Data redundancy:** With several departments having access to one file, there are fewer files. Excess storage is reduced. Microcomputers linked by a network to a file server, for example, could replace the hard disks located in several individual microcomputers.

- **Data integrity:** Older filing systems many times did not have "integrity." That is, a change made in the file in one department might not be made in the file in another department. As you might expect, this can cause serious problems and conflicts when data is used for important decisions affecting both departments.

Database Management

In order to create, modify, and gain access to the database, special software is required. This software is called a **database management system,** which is commonly abbreviated **DBMS.**

Some DBMSs, such as dBASE, are designed specifically for microcomputers. Other DBMSs are designed for minicomputers and mainframes. Once again, increased processing power and the wide use of communications networks linked to file servers are changing everything. Now microcomputer DBMSs have become more like the ones used for mainframes—and vice versa.

DBMS software is made up of a data dictionary and a query language.

- **Data dictionary:** A **data dictionary** contains a description of the structure of the data used in the database. For a particular item of data, it defines the names used for a particular field. It defines what type of data that field is (text, numeric, alphanumeric, date, time, or logic). It also specifies the number of characters in each field and whether that field is a key field. An example of a data dictionary appears in Figure 11-6.

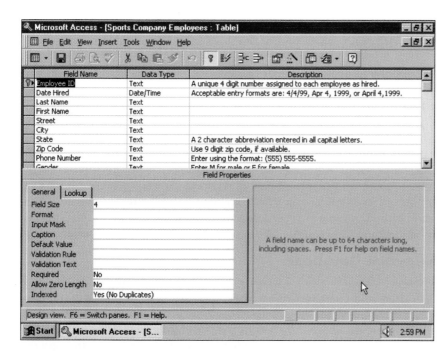

FIGURE 11-6
Data dictionary

- **Query language:** Access to most databases is accomplished with a **query language.** This is an easy-to-use language understandable to most users. The most widely used query language is **structured query language (SQL).**

Query languages have commands such as DISPLAY, ADD, COMPARE, LIST, SELECT, and UPDATE. For example, imagine you wanted the names of all salespeople in an organization whose sales were greater than their sales quotas. You might type the statement "DISPLAY ALL FOR SALES > QUOTA."

CONCEPT CHECK

✔ What are some of the advantages to having databases?

✔ What are the two components of DBMS software?

DBMS Organization

The four principal ways to organize DBMSs are hierarchical, network, relational, and object-oriented.

The purpose of a database is to integrate individual items of data—that is, to transform isolated facts into useful information. We saw that *files* can be organized in various ways (sequentially, for example) to best suit their use. Similarly, *databases* can also be organized in different ways to best fit their use. Although other arrangements have been tried, the four most common database formats are *hierarchical, network, relational,* and *object-oriented.*

Hierarchical Database

In a **hierarchical database,** fields or records are structured in **nodes.** Nodes are points connected like the branches of an upside-down tree. Each entry has one **parent node,** although a parent may have several **child nodes.** This is sometimes described as a *one-to-many relationship*. To find a particular field you have to start at the top with a parent and trace down the tree to a child.

The nodes farther down the system are subordinate to the ones above, like the hierarchy of managers in a corporation. An example of a hierarchical database for part of a nationwide airline reservations system is shown in Figure 11-7. The parent node is the "departure" city, Los Angeles. This parent has four children, labeled "arrival." New York, one of the children, has three children of its own. They are labeled "flight number." Flight 110 has three children, labeled "passenger."

The problem with a hierarchical database is that if one parent node is deleted, so are all the subordinate child nodes. Moreover, a child node cannot be added unless a parent node is added first. The most significant limitation is the rigid structure: one parent only per child, and no relationships or connections between the child nodes themselves.

Network Database

A **network database** also has a hierarchical arrangement of nodes. However, each child node may have more than one parent node. This is sometimes described as a *many-to-many relationship*. There are additional connections—called **pointers**—between parent nodes and child nodes. Thus, a node may be reached through more than one path. It may be traced down through different branches.

An example of the use of a network organization is that shown in our illustration for students taking courses. (See Figure 11-8.) If you trace through the logic of this organization, you can see that each student can have more than one teacher. Each teacher can also teach more than one course. Students may take more than a single course. This demonstrates how the network arrangement is more flexible and in many cases more efficient than the hierarchical arrangement.

Relational Database

The most flexible type of organization is the **relational database.** In this structure, there are no access paths down a hierarchy to an item of data.

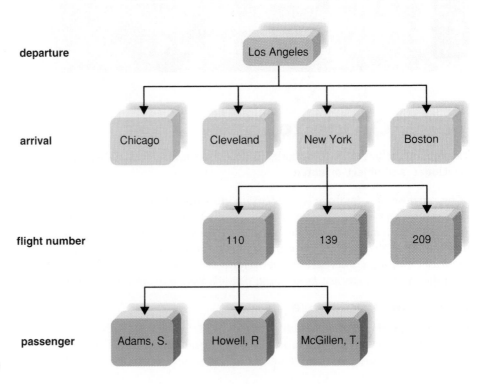

FIGURE 11-7
Example of a hierarchical database

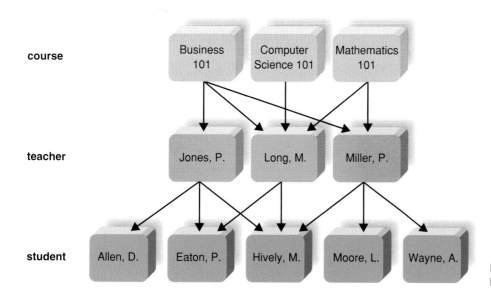

course

teacher

student

FIGURE 11-8
Example of a network database

Rather, the data elements are stored in different tables, each of which consists of rows and columns. A table is called a **relation.**

An example of a relational database is shown in Figure 11-9. The address table contains the names, driver's license numbers, and addresses for all registered drivers in a particular state. Within the table, a row resembles a record—for example, information about one driver. A column entry resembles a field. The driver's name is one field; the driver's license number is another field. All related tables must have a *common data item* (a key field). Thus, information stored on one table can be linked with information stored on another. One common field might be a person's name. Another might be a driver's license number.

Thus, police officers who stop a speeding car can radio the driver's license number and the car's license plate number to the department of motor vehicles. They can use the driver's license number as the key field. With

Address Table

Name	License Number	Street Address	City	State	Zip
Aaron, Linda	FJ1987	10032 Park Lane	San Jose	CA	95127
Abar, John	D12372	1349 Oak St	Lakeville	CA	94128
Abell, Jack	LK3457	95874 State St	Stone	CA	95201

key fields linked

key fields linked

Owner's Table

Name	Plate Number
Abell, Jack	ABK241
Abrams, Sue	LMJ198
Abril, Pat	ZXA915

Outstanding Citation Table

License Number	Citation Code	Violation
T25476	00031	Speed
D98372	19001	Park
LK3457	89100	Speed

FIGURE 11-9
Example of a relational database

it they can find out about any unpaid traffic violations (such as parking tickets). Also using the license plate number they can obtain the car owner's name and address. If the owner's name and address do not match the driver who has been stopped, the police officer may check further for a stolen vehicle.

The most valuable feature of relational databases is their simplicity. Entries can be easily added, deleted, and modified. The hierarchy and network databases are more rigid. The relational organization is common for microcomputer DBMSs, such as Access, Paradox, dBASE, and R:Base. Relational databases are also widely used for mainframe- and minicomputer-based systems.

Object-Oriented Database

The other data structures are designed to handle structured data such as names, addresses, pay rates, and so on. Object-oriented databases are more flexible and are also able to handle unstructured data such as photographs, graphics, audio, and video. **Object-oriented databases** keep track of objects, which are entities that contain both data and the action that can be taken on the data.

For example, a non-object-oriented health club database would only contain data about members, such as member name, address, monthly rate, and the like. An object-oriented health club database would consider members as objects. All of the data listed above would be associated with each member as well as unstructured data such as member photos. Plus the object-oriented database would include instructions for processing the data such as how to calculate and when to print monthly bills.

For a summary of the four types of DBMS organizations, see Figure 11-10.

CONCEPT CHECK

✔ List the four principal DBMS formats.

✔ Give a brief example of each type of DBMS format.

Type	Description
Hierarchical	Data structured in nodes organized like an upside-down tree; each parent node can have several children; each child node can only have one parent
Network	Like hierarchical except that each child can have several parents
Relational	Data stored in tables consisting of rows and columns
Object-oriented	Objects can include both data and actions to be taken on the data

FIGURE 11-10

Summary of the four types of DBMS organizations

Types of Databases

There are four kinds of databases: individual, company, distributed, and proprietary.

Databases may be small or large, limited in accessibility or widely accessible. Databases may be classified into four types: *individual, company* (or shared), *distributed,* and *proprietary.*

Individual

The **individual database** is also called a **microcomputer database.** It is a collection of integrated files primarily used by just one person. Typically, the data and the DBMS are under the direct control of the user. They are stored either on the user's hard-disk drive or on a LAN file server.

There may be many times in your life when you will find this kind of database valuable. If you are in sales, for instance, a microcomputer database can be used to keep track of your customers. If you are a sales manager, you can keep track of your salespeople and their performance. If you are an advertising account executive, you can keep track of what work and how many hours to charge each client.

Company or Shared

Companies, of course, create databases for their own use. The **company database** may be stored on a mainframe and managed by a computer professional (known as a *database administrator*). Users throughout the company have access to the database through their microcomputers linked to local area networks or wide area networks.

Company databases are of two types:

- The **common operational database** contains details about the operations of the company, such as inventory, production, and sales. It contains data describing the day-to-day operations of the organization.

- The **common user database** contains selected information both from the common operational database and from outside private (proprietary) databases. Managers can tap into this information on their microcomputers or terminals and use it for decision making.

As we will see in Chapter 12, company databases are the foundation for management information systems. For instance, a department store can record all sales transactions in the database. A sales manager can use this information to see which salespeople are selling the most products. The manager can then determine year-end sales bonuses. Or the store's buyer can learn which products are selling well or not selling and make adjustments when reordering. A top executive might combine overall store sales trends with information from outside databases about consumer and population trends. This information could be used to change the whole merchandising strategy of the store.

Distributed

Many times the data in a company is stored not in just one location but in several locations. It is made accessible through a variety of communications networks. The database, then, is a **distributed database.** That is, it is located in a place or places other than where users are located. Typically, database servers on a client/server network provide the link between users and the distant data.

For instance, some database information can be at regional offices. Some can be at company headquarters, some down the hall from you, and

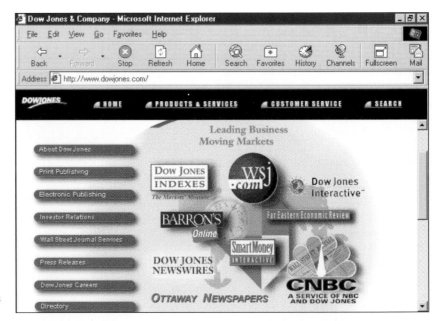

FIGURE 11-11
Proprietary database (Dow Jones
Interactive Publishing)

some even overseas. Sales figures for a chain of department stores, then, could be located at the various stores. But executives at district offices or at the chain's headquarters could have access to these figures.

Proprietary

A **proprietary database** is generally an enormous database that an organization develops to cover particular subjects. It offers access to this database to the public or selected outside individuals for a fee. Sometimes proprietary databases are also called *information utilities* or *data banks*. An example is Dow Jones Interactive Publishing, which offers a variety of financial services. (See Figure 11-11.)

Some important proprietary databases are the following:

- **CSi:** Offers consumer and business services, including electronic mail.
- **Dialog Information Services:** Offers business information, as well as technical and scientific information.
- **Dow Jones Interactive Publishing:** Provides world news and information on business, investments, and stocks.
- **Prodigy:** Offers news and information on business and economics, as well as leisure services.

Most of the proprietary databases are designed for organizational as well as individual use. Organizations typically pay a membership fee plus hourly use fees. Often, individuals are able to search the database to obtain a summary of available information without charge. They pay only for those items selected for further investigation.

See Figure 11-12 for a summary of the four types of databases.

CONCEPT CHECK

✔ What are the four types of databases?

✔ Give a brief example of each type of database.

Type	Description
Individual	Integrated files used by just one person
Company	Common operational or commonly used files shared in an organization
Distributed	Database spread geographically and accessed using database server
Proprietary	Information utilities or databanks available to users on a wide range of topics for a fee

FIGURE 11-12
Summary of the four types of databases

Database Uses and Issues

Databases help users keep current and plan for the future, but database security is important. Databases may be supervised by a database administrator.

Databases offer great opportunities for productivity. In fact, in corporate libraries, electronic databases are now considered more valuable than books and journals. However, maintaining databases means users must make constant efforts to keep them from being tampered with or misused.

Strategic Uses

Databases help users keep up to date and plan for the future. To support the needs of managers and other business professionals, many organizations collect data from a variety of internal and external databases. This data is then stored in a special type of database called a **data warehouse.** A technique called **data mining** is often used to search these databases to look for related information and patterns.

There are hundreds of databases available to help users with both general and specific business purposes, including:

- *Business directories* providing addresses, financial and marketing information, products, and trade and brand names.
- *Demographic data,* such as county and city statistics, current estimates on population and income, employment statistics, census data, and so on.
- *Business statistical information,* such as financial information on publicly traded companies, market potential of certain retail stores, and other business data and information.
- *Text databases* providing articles from business publications, press releases, reviews on companies and products, and so on.
- *Internet databases* covering a wide range of topics including all of the above. As mentioned earlier, Web search tools like Yahoo and HotBot maintain extensive databases of available Web sites.

Security

Precisely because databases are so valuable, their security has become a vital issue. As we discussed in Chapter 10, there are several database security concerns. One is that personal and private information about people stored in databases will be used for the wrong purposes. For instance, a person's credit

FIGURE 11-13
Security: Electronic fingerprint pads

history or medical records might be used to make hiring or promotion decisions. Another concern is with preventing unauthorized users from gaining access to a database. For example, there have been numerous instances in which a computer virus has been launched into a database or network.

Security may require putting guards in company computer rooms and checking the identification of everyone admitted. Some security systems electronically check fingerprints. (See Figure 11-13.) Security concerns are particularly important to organizations using WANs. Violations can occur without actually entering secured areas. As mentioned in Chapters 8 and 10, most major corporations today use special hardware and software called *firewalls* to control access to their internal networks.

Database Administrator

Librarians have had to be trained in the use of electronic databases so that they can help their corporate users. However, corporate databases of all sorts—not just those in the library—have become extremely important. Hence, many large organizations employ a **database administrator (DBA).** He or she helps determine the structure of the large databases and evaluates the performance of the DBMS. For shared databases, the DBA also determines which people have access to what kind of data; these are called **processing rights.** In addition, the DBA is concerned with such significant issues as security, privacy, and ethics.

CONCEPT CHECK

✔ What are some of the security concerns involved with databases?

✔ Why might an organization need a database administrator?

A LOOK TO THE FUTURE

The Internet on Wheels from Mercedes Benz is a leader in smart cars.

Did you see Bruce Willis's futuristic airborne taxi in *The Fifth Element*? Imagine a car that operates on its own, flies, has rocket boosters, monitors traffic conditions and weather, and entertains us as we travel. In the 21st century, smart cars may well look and act a lot like Bruce Willis's taxi . . . hopefully more sleek and efficient but just as smart.

Of course, none of today's cars can do all those things. But you can expect more cars that navigate themselves and surf the Internet in the near future. Several car makers offer global positioning systems (GPSs) that combine large databases of streets, highways, and interchanges with satellite communications. These GPSs precisely track a vehicle as it travels to its destination. You can obtain specific directions and navigational advice without ever leaving your car.

Daimler-Benz, maker of the Mercedes Benz cars, recently unveiled a concept car dubbed The Internet Multimedia on Wheels. This smart car allows occupants to access the Internet from most locations in Europe and North America. They foresee a day when you will be able to access Internet-based roadside services that will diagnose car problems remotely as well as deliver traffic advisories and headline news.

When will these smart cars be available? Don't expect to see any Internet applications in your sedan anytime soon. Most observers predict that it will be 10 years away. However, many of the other applications are already here, and more are on their way.

VISUAL SUMMARY Databases

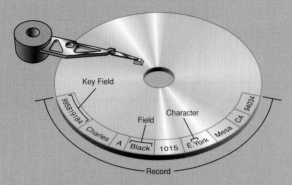

Understanding how data files are used requires an understanding of the following.

Data Organization

Data is organized by the following groups:
- **Character**—letter, number, special character.
- **Field**—set of related characters.
- **Record**—collection of related fields.
- **File**—collection of related fields.
- **Database**—collection of related files.

Key Field

A record is uniquely identified by its **key field.**

Batch versus Real-Time Processing

Two methods of processing are:
- **Batch processing**—transactions collected over time then processed all at once.
- **Real-time processing**—data processed at the same time transactions occur.

Master versus Transaction Files

- **Master file**—complete file, current to the last update.
- **Transaction file**—temporary "holding file" used to update the master file.

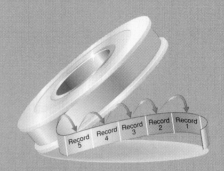

File Organization

File organization may be sequential, direct, or index sequential.
- **Sequential**—records are stored one after another in ascending or descending order; often used with magnetic tape.

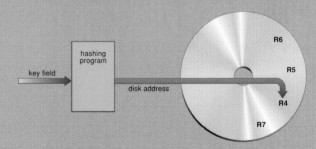

- **Direct**—records stored in order by a key field; often used with a magnetic disk.

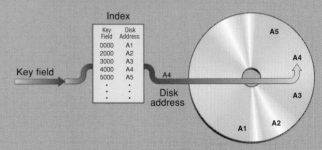

- **Index sequential**—records stored sequentially with an index for near-direct access; used with magnetic disk storage.

To be a competent end user you need to understand data fields, records, files, and databases. You need to be aware of the different ways in which databases can be organized and the different types of databases. Also, you need to know the most important database uses and issues.

DATABASE

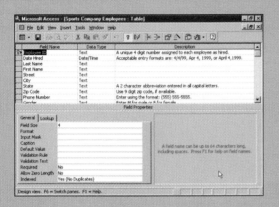

A **database** is a collection of integrated data—logically related files and records.

Need for Databases
Advantages of databases are:
- Sharing—users may share data with one another.
- Security—access is restricted to authorized people.
- **Data redundancy**—multiple files and data are avoided.
- **Data integrity**—changes in one file are made in other files as well.

Database Management
A **database management system (DBMS)** is the software for creating, modifying, and gaining access to the database. A DBMS consists of:
- **Data dictionary**—describes the structure of the database.
- **Query language**—language to access and use the database. **Structured Query Language (SQL)** is the most widely used.

DBMS ORGANIZATION

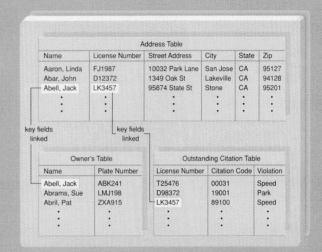

Four principal ways to organize DBMSs are *hierarchical, network, relational,* and *object-oriented.*

Hierarchical Database
In a **hierarchical database,** fields and records are structured in nodes, or points connected like tree branches. An entry may have a **parent node** with several **child nodes.**

Network Database
In a **network database**, nodes are arranged hierarchically, but a child node may have more than one parent. There are additional connections called **pointers.**

Relational Database
In a **relational database**, data is stored in **tables** consisting of rows and columns; related tables have a common **data element** (key field); data items are found by means of an **index.**

Object-Oriented Database
Unlike other types of organizations, **object-oriented databases** keep track of **objects** or entities that contain both data and the action that can be taken on the data.

TYPES OF DATABASES

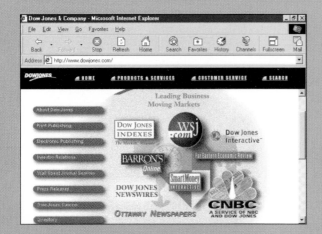

There are four types of databases: individual, company or shared, distributed, and proprietary.

Individual
The **individual database** is a collection of integrated files useful mainly to just one person.

Company or Shared
Two types of **company (shared) databases** are:
- **Common operational databases**—contains data on company operations.
- **Common user database**—contains selected data from common operational and from outside private databases.

Distributed
The **distributed database** is spread out geographically and is accessible by communications links.

Proprietary
A **proprietary database** is available by subscription to customers.

DATABASE USES AND ISSUES

Databases offer a great opportunity for increased productivity; however, security is always a concern.

Strategic Uses
Data warehousing is a new type of database that supports data mining. **Data mining** is a technique for searching and exploring for related information and patterns in data.

Databases available for business purposes:
- Business directories
- Demographic data
- Business statistical information
- Text databases
- Internet databases

Security
Two important security concerns are illegal use of data and unauthorized access.

Database Administrator
The **database administrator (DBA)** is a specialist who sets up and manages the database and determines **processing rights**—which people have access to what kind of data.

Key Terms

batch processing (265)
character (264)
child node (269)
common operational
   database (273)
common user database (273)
company database (273)
database (264)
database administrator
   (DBA) (276)
database management system
   (DBMS) (268)
data dictionary (268)
data integrity (268)
data mining (275)

data redundancy (268)
data warehouse (275)
direct file organization (266)
distributed database (273)
field (264)
file (264)
hashing (266)
hierarchical database (269)
index sequential file
   organization (266)
individual database (273)
key field (264)
master file (265)
microcomputer database (273)
network database (270)

nodes (269)
object-oriented database (272)
parent node (269)
pointer (270)
processing rights (277)
proprietary database (274)
query language (269)
real-time processing (265)
record (264)
relation (271)
relational database (270)
sequential file organization (265)
structured query language
   (SQL) (269)
transaction file (265)

Chapter Review

LEVEL 1

Reviewing Facts and Terms
Matching

Match each numbered item with the most closely related lettered item. Write your answers in the spaces provided.

1. Contains a set of related characters. ____

2. Person who helps determine the structure of the large databases and evaluates the performance of the DBMS. ____

3. The particular field of a record that is chosen to uniquely identify each record. ____

4. A complete file containing all records current up to the last update. ____

5. Means that files are accurate and consistent across departments within a company. ____

6. Fields or records are structured in these connected points. ____

7. Software that helps create, modify, and gain access to a database. ____

8. Access to most databases is accomplished with this. ____

9. Data is collected over several hours, days, weeks, etc. ____

10. Flexible, able to handle unstructured data such as photographs, graphics, audio, and video. ____

11. The most flexible type of database organization. ____

12. Contains a description of the structure of the data used in the database. ____

13. The most widely used query language. ____

14. A collection of related files. ____

a. Batch processing
b. Company database
c. Data dictionary
d. Data integrity
e. Data redundancy
f. Database
g. Database administrator
h. Database management system
i. Direct file organization
j. Field
k. Hashing
l. Key field
m. Master file
n. Network database
o. Nodes
p. Object-oriented database
q. Query language
r. Relational database
s. SQL
t. Transaction file

2001
2002

15. Users throughout a company have access to this database through microcomputers linked to a network. _____

16. Records are stored at a particular location, not physically one after another, and can be accessed by key field. _____

17. Contains recent changes to records that will be used to update the master file. _____

18. When records for the same customer appear in different files in multiple departments. _____

19. Hierarchical arrangement of nodes where each child node can have more than one parent node. _____

20. Process of converting a key field's numeric value to a particular storage address. _____

True/False

In the spaces provided, write T or F to indicate whether the statement is true or false.

1. A record is a collection of related files. _____

2. In batch processing, data is processed at the same time the transaction occurs. _____

3. A data dictionary describes the structure of the data in a database. _____

4. A distributed database contains data in more than one location. _____

5. Processing rights are typically determined by the database administrator to specify which people have access to what kind of data. _____

Multiple Choice

Circle the letter of the correct answer.

1. A collection of related fields:
   a. byte
   b. word
   c. character
   d. record
   e. file

2. A temporary file containing recent changes to records:
   a. master
   b. data
   c. transaction
   d. indexed
   e. batch

3. The database organization in which fields and records are structured in nodes, with each child node having only one parent:
   a. hierarchical
   b. network
   c. proprietary
   d. relational
   e. object-oriented

4. The type of database that is sometimes called an information utility or data bank:
   a. individual
   b. common operational
   c. common user
   d. distributed
   e. proprietary

5. The DBMS organization that can handle unstructured data such as photographs:
   a. hierarchical
   b. network
   c. data dictionary
   d. relational
   e. object-oriented

Completion

Complete each statement in the spaces provided.

1. The _____ field uniquely identifies each record.

2. _____ file organization is best for locating specific records.

3. A database is a collection of _____ data.

4. _____ databases are more flexible than hierarchical, network, and relational databases.

5. Large organizations employ database _____ to help determine database structures and evaluate database performance.

Reviewing Concepts

LEVEL
2

Open-Ended

On a separate sheet of paper, respond to each question and statement.

1. Describe how data is organized and give an example.

2. What are the differences between sequential, direct, and index sequential file organizations?

3. What are databases, and why are they needed?

4. Discuss the four principal ways of organizing a database.

5. Describe each of the four database types.

Concept Mapping

On a separate sheet of paper, draw a concept map or a flowchart showing how the following terms are related. Show all relationships. Include any additional terms you can think of.

batch processing	database management system	network database
character		nodes
child node	direct file organization	object-oriented database
common operational database	distributed database	parent node
common user database	field	proprietary database
company database	file	query language
data dictionary	hashing	real-time processing
data integrity	hierarchical database	record
data mining	index sequential file organization	relational database
data redundancy	individual database	sequential file organization
data warehouse	key field	Structured Query Language
database	master file	transaction file
database administrator	microcomputer database	

Critical Thinking Questions and Projects

2001
2002

LEVEL 3

Read each exercise and answer the related questions on a separate sheet of paper.

1. *Useful information utilities:* What is your major or prospective major? What kinds of information are you apt to be required to obtain for research papers, projects, and assignments? If you're in the health field, you may be required to learn about diet, exercise, drug recovery, and the like. If you're in marketing, you may need to know about sales forecasting and product marketing.

   Take a few minutes to list the areas of information required in your field. Then go to the library or use the Internet to look up which information utilities or data banks would be most valuable to you.

2. *Your school's database:* Almost all organizations keep information about their employees and customers in databases. Surely your college or university does also. Without conducting extensive research, respond to each of the following:

   a. What information about you is stored in your school's database?

   b. How do you suppose the data is organized? Specifically, give an example of a database, file, record, field, and character.

   c. Which type of file organization do you think is used, and why?

   d. Which type of DBMS organization is likely used, and why?

3. *Privacy:* After purchasing an appliance such as a stereo or a television, you are typically requested to complete a product warranty card that will list your name, address, telephone number, and purchase date. Often additional questions are included, such as: Where did you purchase the product? What is your age? or What is your income range? This information is used by the manufacturer to create and maintain a warranty database. If you didn't complete the warranty card and then had problems with the product, do you think it would still be under warranty? Many manufacturers use their warranty database to target prospective customers for other goods and services. Some organizations even sell this information to other companies. Do you see anything wrong with these activities?

On the Web Exercises

1. Financial Markets

There are several sites on the Web for financial information. These sites typically maintain a large, ever-expanding database that receives and organizes stock information.

Visit our Web site at http://www.mhhe.com/it/oleary/exercise.mhtml to link to one of these sites. Once connected to that site, look up the performance of Nike, Disney, and Coke. Print out information about each stock's price fluctuations. Write a paragraph discussing how businesses might use this database information.

2. Information Please

If you forget a telephone number, you can always consult the telephone directory or call for directory assistance. Or you can use the Web and get even more information. To see how this

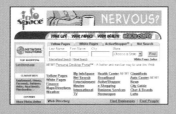

works, connect to our site at http://www.mhhe.com/it/oleary/exercise.mhtml to link to a site containing this type of information. Once connected to that site, use the white pages, and enter your family name, city, and state or province as indicated. Then use the map function to print out a map to your home. If you are not listed, try using a neighbor's or a friend's name and address. Use the directions option to print out directions for getting from your home to school. Write a brief paragraph describing who might use such a service and why.

3. Search Tool

Numerous search tools are available on the Web. Typically, they create, maintain, and use an enormous database of Web sites. When users specify a search topic, a search tool in-

vestigates its database and returns a ranked list of all related Web sites. Choose a topic of interest to you and determine a few key words that can be used to search for this topic. Then visit each of the following search tools using the same key words each time: http://www.webcrawler.com; http://infoseek.com; and http://www.lycos.com. Print out the first page of results for each search tool. Write a paragraph addressing the following questions: Are the lists identical? If not, why? Which search engine produced the best sites?

4. Music

The Web is full of sites that maintain large databases. For example, most large retail chains have sites containing databases listing all of their products. Visit our site at http://www.mhhe.com/

it/oleary/exercise.mhtml to link to a popular retail music chain. Once connected to that site, choose one of your favorite albums or artists. Print the information provided on the album or artist. Write a paragraph addressing the following questions: What type of file organization do you think is most likely used for this database? Why?

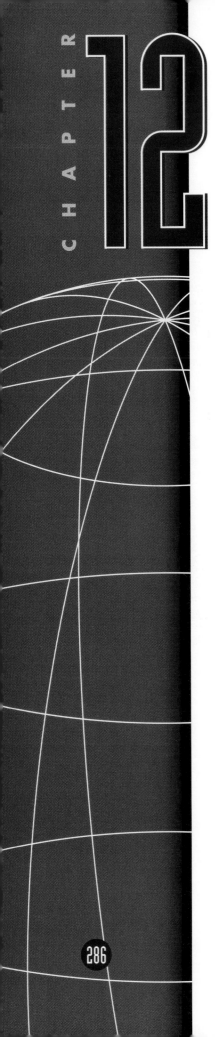

12

Information Systems

COMPETENCIES

After you have read this chapter, you should be able to:

1. Explain how organizations can be structured according to five functions and three management levels.

2. Describe how information flows in an organization.

3. Distinguish among a transaction processing system, a management information system, a decision support system, and an executive support system.

4. Distinguish between office automation systems and knowledge work systems.

5. Explain the difference between data workers and knowledge workers.

An **information system** (as we discussed in Chapter 1) is a collection of *people, procedures, software, hardware,* and *data.* They all work together to provide information essential to running an organization. This is information that will successfully produce a product or service and, for profit-oriented enterprises, derive a profit.

Why are computers used in organizations? No doubt you can easily state one reason: to keep records of events. However, another reason might be less obvious: to help make decisions. For example, point-of-sale terminals

Information Flow

Top managers:
long-range
planning

Middle managers:
control and planning

Supervisors:
control matters

Workers

record sales as well as recording which salesperson made each sale. This information can be used for decision making. For instance, it can help the sales manager decide which salespeople will get year-end bonuses for doing exceptional work.

The Internet, communications links, and databases connect you with information resources as well as information systems far beyond the surface of your desk. The microcomputer offers you access to a greater *quantity* of information than was possible a few years ago. In addition, you also have access to a better *quality* of information. As we show in this chapter, when you tap into a computer-based information system, you not only get information—you also get help in making decisions.

Competent end users need to understand how the information flows in an organization as it moves through the organization's different functional areas and management levels. They need to be aware of the different types of computer-based information systems, including transaction processing systems, management information systems, decision support systems, and executive support systems. They also need to understand the role and importance of databases to support each level or type of information system.

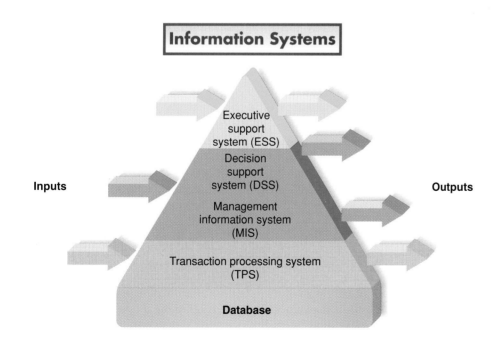

Organizational Information Flow

Information flows up and down among managers and sideways among departments.

In large and medium-sized organizations, computerized information systems don't just keep track of transactions and day-to-day business operations. They also support the flow of information within the organization. This information flows both vertically and horizontally. In order to understand this, we need to understand how an organization is structured. One way to examine an organization's structure is to view it from a functional perspective. That is, you can study the different basic functional areas in organizations and the different types of people within these functional areas.

As we describe these, you might consider how they apply to any organization you are familiar with. Or consider how they apply to a hypothetical manufacturer of sporting goods, the HealthWise Group. Think of this as a large company that manufactures equipment for sports and physical activities, including those that interest you. These goods range from every type of ball imaginable (from golf to tennis to soccer) to hockey pads, leotards, and exercise bicycles.

Functions

Depending on the services or products they provide, most organizations have departments that specialize in one of five basic functions. These are *accounting, marketing, human resources, production,* and *research.* (Even in very small organizations these functions must be performed—often by one person, the owner.) Whatever your job in an organization, it is likely to be in one of these functional areas.

To learn more about these functions, study Figure 12-1.

Management Levels

Most people who work in an organization are not managers, of course. At the base of the organizational pyramid are the assemblers, painters, welders, drivers, and so on. These people produce goods and services. Above them, however, are various levels of managers—people with titles such as supervisor, director, regional manager, and vice president. These are the people who do the planning, leading, organizing, and controlling necessary to see that the work gets done. At HealthWise, for example, the northwest district sales manager directs and coordinates all the salespeople in her area. Other job titles might be vice president of marketing, director of human resources, or production manager. In smaller organizations, these titles are often combined.

Management in many organizations is divided into three levels: supervisors, middle-level, and top-level. (See Figure 12-2.)

These levels may be described as follows:

- **Supervisors: Supervisors** manage and monitor the employees or workers, those who actually produce the goods and services. Thus, these managers have the responsibility relating to *operational matters.* They monitor day-to-day events and immediately take corrective action, if necessary. For example, at HealthWise, a production supervisor monitors the materials needed to build exercise bicycles. If parts begin to run low, the supervisor must take action immediately.

- **Middle management: Middle-level managers** deal with *control, planning* (also called *tactical planning*), and *decision making.* They implement the long-term goals of the organization. For example, the HealthWise regional sales manager for the Northwest sets sales goals for district sales managers in Washington, Oregon, and Idaho. She or

Organizational Function

Human resources finds and hires people and handles matters such as sick leave and retirement benefits. In addition, it is concerned with evaluation, compensation, and professional development. As you might imagine, HealthWise has rather good health benefits.

Accounting tracks all financial activity. At HealthWise, this department records bills and other financial transactions with sporting goods stores. It also produces financial statements including budgets and forecasts of financial performance.

Research conducts basic research and relates new discoveries to the firm's current or new products department. Research people at HealthWise explore new ideas from exercise physiologists about muscle development. They use this knowledge to design new physical fitness machines.

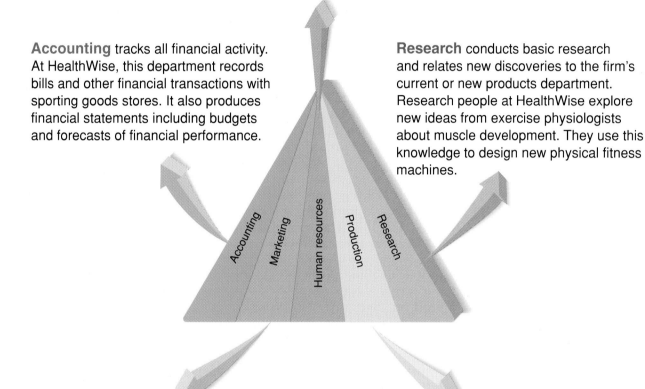

Marketing handles planning, pricing, promoting, selling, and distributing goods and services to customers. At HealthWise, they even get involved with creating a customer newsletter that is distributed via the corporate Web page.

Production takes in raw materials and people to work to turn out finished goods (or services). It may be a manufacturing activity or—in the case of a retail store—an operations activity. At HealthWise, this department purchases steel and aluminum to be used in weight-lifting and exercise machines.

FIGURE 12-1
The five functions of an organization

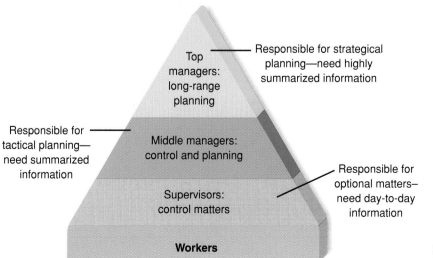

FIGURE 12-2
Three levels of management

he also monitors their sales performance and takes corrective action if necessary.

- **Top management: Top-level managers** are concerned with *long-range planning* (also called *strategic planning*). They need information that will help them to plan the future growth and direction of the organization. For example, the HealthWise vice president of marketing might need to determine the demand and the sales strategy for a new product. Such a product might be a stationary exercise bicycle with a biometric feedback mechanism.

Information Flow

Each level of management has different information needs. Top-level managers need information that is summarized in capsule form to reveal the overall condition of the business. They also need information from outside the organization, because top-level managers need to forecast and plan for long-range events. Middle-level managers need summarized information— weekly or monthly reports. They need to develop budget projections as well as to evaluate the performance of supervisors. Supervisors need detailed, very current day-to-day information on their units so that they can keep operations running smoothly.

To support these different needs, information *flows* in different directions. (See Figure 12-3.) The top-level managers, such as the chief executive officer (CEO), need information from below and from all departments. They also need information from outside the organization. For example, at HealthWise, they are deciding whether to introduce a line of hockey equipment in the southwestern United States. The vice president of marketing must look at relevant data. Such data might include availability of ice rinks and census data about the number of young people. It might also include sales histories on related cold-weather sports equipment.

For middle-level managers, the information flow is both vertical and horizontal across functional lines within the organization. For example, the regional sales managers at HealthWise set their sales goals by coordinating with middle managers in the production department. They are able to tell sales managers what products will be produced, how many, and when. An example of a product might be an exercise bicycle. The regional sales managers also must coordinate with the strategic goals set by the top managers. They must set and monitor the sales goals for the supervisors beneath them.

For supervisory managers, information flow is primarily vertical. That is, supervisors communicate mainly with their middle managers and with the

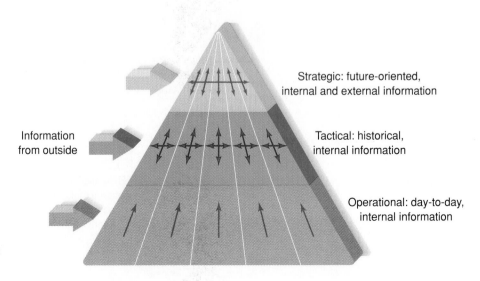

FIGURE 12-3
Information flow within an organization

workers beneath them. For instance, at HealthWise, production supervisors rarely communicate with people in the accounting department. However, they are constantly communicating with production-line workers and with their own managers.

Now we know how a large organization is usually structured and how information flows within the organization. But how is a computer-based information system likely to be set up to support its needs? And what do you, as a microcomputer user, need to know to use it?

CONCEPT CHECK

✔ What are the basic functions of most organizations?

✔ What are the levels of management within an organization?

✔ Describe how information flows within an organization.

Computer-Based Information Systems

There are four kinds of computer-based information systems: transaction processing system, management information system, decision support system, and executive support system.

All large organizations have computer-based information systems. Some systems record routine activities: employees hired, materials purchased, products produced, and the like. Such recorded events are called **transactions.** Other systems use these recorded events to help managerial planning and control. The systems form a pyramid, each primarily (but not exclusively) supporting one level of management. (See Figure 12-4.)

- **Transaction processing system:** The **transaction processing system (TPS)** records day-to-day transactions such as customer orders, bills, inventory levels, and production output. The TPS helps supervisors by

FIGURE 12-4
The three levels of information systems

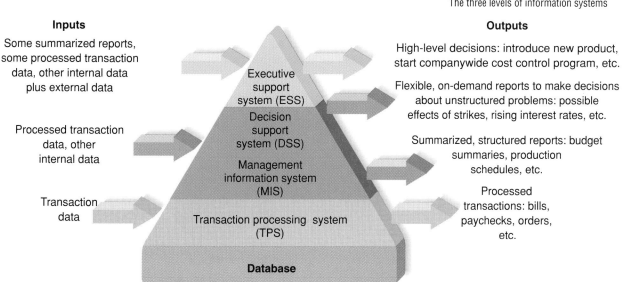

Inputs		Outputs
Some summarized reports, some processed transaction data, other internal data plus external data	Executive support system (ESS)	High-level decisions: introduce new product, start companywide cost control program, etc.
	Decision support system (DSS)	Flexible, on-demand reports to make decisions about unstructured problems: possible effects of strikes, rising interest rates, etc.
Processed transaction data, other internal data	Management information system (MIS)	Summarized, structured reports: budget summaries, production schedules, etc.
Transaction data	Transaction processing system (TPS)	Processed transactions: bills, paychecks, orders, etc.

Database

generating databases that act as the foundation for the other information systems.

- **Management information system:** The **management information system (MIS)** summarizes the detailed data of the transaction processing system in standard reports for middle-level managers. Such reports might include production schedules and budget summaries.

- **Decision support system:** The **decision support system (DSS)** provides a flexible tool for analysis. The DSS helps middle-level managers and others in the organization analyze a wide range of problems, such as the effect of events and trends outside the organization. Like the MIS, the DSS draws on the detailed data of the transaction processing system.

- **Executive support system:** The **executive support system (ESS),** also known as the **executive information system (EIS),** is an easy-to-use system that presents information in a very highly summarized form. It helps top-level managers oversee the company's operations and develop strategic plans. The ESS combines the internal data from TPS and MIS with external data.

Let us describe these information systems in more detail.

CONCEPT CHECK

✔ What are the four kinds of computer-based information systems?

Transaction Processing Systems

A transaction processing system records routine operations.

A *transaction processing system (TPS)* helps an organization keep track of routine operations and records these events in a database. (See Figure 12-5.) For this reason, some firms call this the **data processing system (DPS).** The data from operations—for example, customer orders for HealthWise's products—makes up a database that records the transactions of the company. This database of transactions is used to support an MIS, DSS, and ESS.

One of the most essential transaction processing systems for any organization is in the accounting area. (See Figure 12-6.) Every accounting department handles six basic activities. Five of these are sales order processing,

FIGURE 12-5
Transaction processing system database.

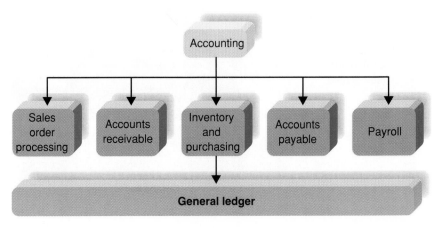

FIGURE 12-6
Transaction processing system for accounting

accounts receivable, inventory and purchasing, accounts payable, and payroll. All of these are recorded in the general ledger, the sixth activity.

Let us take a look at these six activities. They will make up the basis of the accounting system for almost any office you might work in.

FIGURE 12-7
Sales order processing for Canadian Airlines International

- The **sales order processing** activity records the customer requests for the company's products or services. (See Figure 12-7.) When an order comes in—a request for a set of barbells, for example—the warehouse is alerted to ship a product.

- The **accounts receivable** activity records money received from or owed by customers. HealthWise keeps track of bills paid by sporting goods stores and by gyms and health clubs to which it sells directly.

- The parts and finished goods that the company has in stock are called **inventory**—all exercise machines in the warehouse, for example. An *inventory control system* keeps records of the number of each kind of part or finished good in the warehouse. **Purchasing** is the buying of materials and services. Often a *purchase order* is used. This is a form that shows the name of the company supplying the material or service and what is being purchased.

- **Accounts payable** refers to money the company owes its suppliers for materials and services it has received—steel and aluminum, for example.

- The **payroll** activity is concerned with calculating employee paychecks. Amounts are generally determined by the kind of job, hours worked, and kinds of deductions (such as taxes, social security, medical insurance). Paychecks may be calculated from employee time cards or, in some cases, supervisors' time sheets.

- The **general ledger** keeps track of all summaries of all the foregoing transactions. A typical general ledger system can produce income statements and balance sheets. *Income statements* show a company's financial performance—income, expenses, and the difference between them for a specific time period. *Balance sheets* list the overall financial condition of an organization. They include assets (for example, buildings and property owned), liabilities (debts), and how much of the organization (the equity) is owned by the owners.

CONCEPT CHECK

✔ What is the purpose of a transaction processing system?

✔ Name the six activities that make up the accounting system for a typical organization?

294

Management Information Systems

A management information system produces summarized, structured reports.

A *management information system (MIS)* is a computer-based information system that produces standardized reports in summarized, structured form. (See Figure 12-8.) It is used to support middle managers. An MIS differs from a transaction processing system in a significant way. Whereas a transaction processing system *creates* databases, an MIS *uses* databases. Indeed, an MIS can draw from the databases of *several* departments. Thus, an MIS requires a *database management system* that integrates the databases of the different departments. Middle managers need summary data often drawn from across different functional areas.

An MIS produces reports that are *predetermined*. That is, they follow a predetermined format and always show the same kinds of content. Although reports may differ from one industry to another, there are three common categories of reports: periodic, exception, and demand.

The Sports Company
Regional Sales Report

Region	Actual Sales	Target	Difference
Central	$166,430	$175,000	($8,570)
Northern	137,228	130,000	7,228
Southern	137,772	135,000	2,772
Eastern	152,289	155,000	(2,711)
Western	167,017	160,000	7,017

FIGURE 12-8
Management information system report

- **Periodic reports** are produced at regular intervals—weekly, monthly, or quarterly, for instance. Examples are HealthWise's monthly sales and production reports. The sales reports from district sales managers are combined into a monthly report for the regional sales managers. For comparison purposes, a regional manager is also able to see the sales reports of other regional managers.

- **Exception reports** call attention to unusual events. An example is a sales report that shows that certain items are selling significantly above or below marketing department forecasts. For instance, if fewer exercise bicycles are selling than were predicted for the northwest sales region, the regional manager will receive an exception report. That report may be used to alert the district managers and salespeople to give this product more attention.

- The opposite of a periodic report, a **demand report** is produced on request. An example is a report on the numbers and types of jobs held by women and minorities. Such a report is not needed periodically, but it may be required when requested by the U.S. government. At HealthWise, many government contracts require this information. It is used to certify that HealthWise is achieving certain government equal-opportunity guidelines.

> **CONCEPT CHECK**
> ✔ What is the purpose of a management information system?
> ✔ Give the three common categories of MIS reports.

Decision Support Systems

A DSS helps decision makers analyze unanticipated situations.

Managers often must deal with unanticipated questions. For example, the HealthWise manager in charge of manufacturing might ask how a strike would affect production schedules. A *decision support system (DSS)* enables managers to get answers to such unexpected and generally nonrecurring kinds of problems. Frequently, a team is formed to address large problems. A **group decision support system (GDSS)** is then used to support this collective work.

A DSS, then, is quite different from a transaction processing system, which simply records data. It is also different from a management information system, which summarizes data in predetermined reports. A DSS is used to *analyze* data. Moreover, it produces reports that do not have a fixed format. This makes the DSS a flexible tool for analysis.

Many DSSs are designed for large computer systems. However, microcomputers, with their increased power and sophisticated software, such as spreadsheet and database programs, are being used for DSS. Users of a DSS are managers, not computer programmers. Thus, a DSS must be easy to use—or most likely it will not be used at all. A HealthWise marketing manager might want to know which territories are not meeting their monthly sales quotas. To find out, the executive could query the database for all "SALES < QUOTA." (See Figure 12-9.)

How does a decision support system work? Essentially, it consists of four parts: the user, system software, data, and decision models.

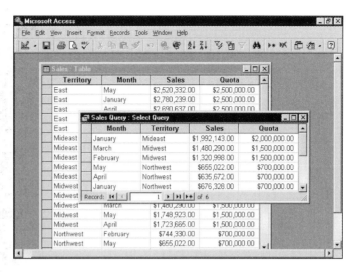

FIGURE 12-9
Decision support system query results for SALES < QUOTA

- The **user** could be you. In general, the user is someone who has to make decisions—a manager, often a middle-level manager.

- **System software** is essentially the operating system—programs designed to work behind the scenes to handle detailed operating procedures. In order to give the user a good, comfortable interface, the software typically is menu- or icon-driven. That is, the screen presents easily understood lists of commands or icons, giving the user several options.

- **Data** in a DSS is typically stored in a database and consists of two kinds. *Internal data*—data from within the organization—consists principally of transactions from the transaction processing system. *External data* is data gathered from outside the organization. Examples are data provided by marketing research firms, trade associations, and the U.S. government (such as customer profiles, census data, and economic forecasts).

- **Decision models** give the DSS its analytical capabilities. There are three basic types of models: strategic, tactical, and operational. *Strategic models* assist top-level managers in long-range planning, such as stating company objectives or planning plant locations. *Tactical models* help middle-level managers control the work of the organization, such as financial planning and sales promotion planning. *Operational models* help lower-level managers accomplish the organization's day-to-day activities, such as evaluating and maintaining quality control.

CONCEPT CHECK

✔ What is the purpose of a decision support system?

✔ Describe the four parts of a decision support system.

Executive Support Systems

Executive support systems are specially designed, simplified systems for top executives.

Using a DSS requires some training. Many top managers have other people in their offices running DSSs and reporting their findings. Top-level executives also want something more concise than an MIS—something that produces very focused reports.

Executive support systems (ESSs) consist of sophisticated software that, like an MIS or a DSS, can draw together data from an organization's databases in meaningful patterns. However, an ESS is specifically designed to be easy to use. This is so that a top executive with little spare time, for example, can obtain essential information without extensive training. Thus, information is often displayed in very condensed form and in bold graphics.

Consider an executive support system used by the president of HealthWise. It is available on his microcomputer. The first thing each morning, the president calls up the ESS on his display screen, as shown in the left-hand illustration. (See Figure 12-10.) Note that the screen gives a condensed account of activities in the five different areas of the company. (These are Accounting, Marketing, Production, Human Resources, and Research.) On this particular morning, the ESS shows business in four areas proceeding smoothly. However, in the first area, Accounting, the percentage of late-paying customers—past due accounts—has increased by 3 percent. Three percent may not seem like much, but HealthWise has had a history of problems with late payers, which has left the company at times strapped for cash. The president decides to find out the details. To do so, he presses *1* (corresponding to Accounting) on his keyboard.

The right-hand screen shows information about past due accounts expressed in graphic form. The status of today's late payers is shown in blue. The status of late payers at this time a year ago is shown in yellow. The differences between today and a year ago are very clearly presented. The differences are significant. For example, approximately $70,000 was late 1 to 10 days last year. This year, over $80,000 was late. The president knows that he must take some action to speed up customer payments. (For example, he or she might call this to the attention of the vice president of accounting. The vice president might decide to offer discounts to early payers or charge more interest to late payers.)

ESSs permit a firm's top executives to gain direct access to information about the company's performance. Some also have electronic mail setups that

FIGURE 12-10
An executive information system

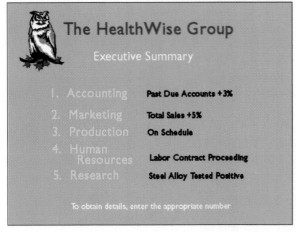

Information in condensed text form

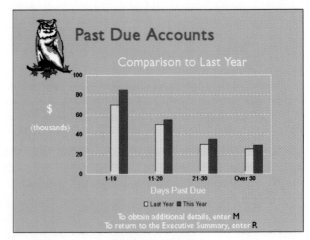

Details in graphic form

Type	Description
TPS	Tracks routine operations and records events in databases, also known as data processing systems
MIS	Produces standardized reports (periodic, exception, and demand) using databases created by TPS
DSS	Analyzes unanticipated situations using data (internal and external) and decision models (strategic, tactical, and operational)
ESS	Presents summary information in a flexible, easy to use, graphical format designed for top executives

FIGURE 12-11
Summary of information systems

allow managers to communicate directly with other executives. Some systems even have structured forms to help managers streamline their thoughts before sending electronic memos. In addition, an ESS may be organized to retrieve information from databases outside the company, such as business-news services. This enables a firm to watch for stories on competitors and stay current on relevant news events that could affect its business. For example, news of increased sports injuries caused by running and aerobic dancing, and the consequent lessened interest by people in these activities, might cause HealthWise to alter its sales and production goals for its line of fitness-related shoes.

For a summary of the different types of information systems, see Figure 12-11.

CONCEPT CHECK

✔ What is the purpose of an executive support system?

Other Information Systems

Information workers use office automation systems and knowledge work systems.

We have discussed only four information systems: TPS to support lower-level managers, MIS and DSS to support middle-level managers, and ESS to support top-level managers. There are many other information systems to support different individuals and functions. The fastest-growing are information systems designed to support information workers.

Information workers create, distribute, and communicate information. They are the organization's secretaries, clerks, engineers, and scientists, to name a few. Some are involved with distribution and communication of information (like the secretaries and clerks). They are called *data workers*. Others are involved with the creation of information (like the engineers and scientists). They are called *knowledge workers*.

Two systems to support information workers are:

- **Office automation systems: Office automation systems (OASs)** are designed primarily to support data workers. These systems focus on managing documents, communicating, and scheduling. Documents are managed using word processing, desktop publishing, and other image technologies. As we discussed in Chapter 7, communications include e-mail, voice-messaging, and videoconferencing. (See Figure 12-12.)

- **Knowledge work systems:** Knowledge workers use OAS systems. Additionally, they use specialized information systems called **knowledge work systems (KWSs)** to create information in their areas of expertise. For example, engineers involved in product design and manufacturing use **computer-aided design/computer-aided manufacturing (CAD/CAM)** systems. (See Figure 12-13.) These KWSs consist of powerful microcomputers running special programs that integrate the design and manufacture activities. CAD/CAM is widely used in the manufacture of automobiles and other products.

CONCEPT CHECK

✔ Who are information workers? data workers? knowledge workers?

✔ Name two systems to support information workers.

FIGURE 12-12
Videoconferencing: individuals and groups can see and share information

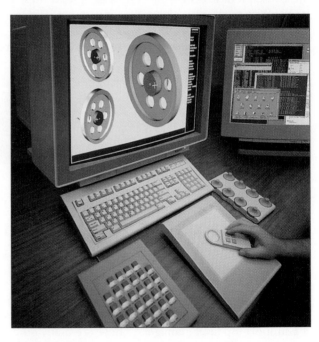

FIGURE 12-13
CAD/CAM: Knowledge work systems used by design and manufacturing engineers

A LOOK TO THE FUTURE

Technology can cause information overload.
E-mail is the major source of information overload.

Have you ever questioned the value of technology? Has it really helped us and made us more productive? Is it possible that the various devices intended to increase our productivity have actually had the opposite effect?

E-mail, cell phones, notebook computers, and the Web are great. They allow us to communicate, work almost anywhere, and have access to vast amounts of data. However, unless we are careful, they can create information overload and have a negative effect on our ability to get work done.

Several studies have found that e-mail is the major source of information overload. It was recently reported that a typical knowledge worker in a large corporation sends and receives over 100 e-mail messages a day. Furthermore, the study concluded that the majority of these messages are not necessary. Here are some tips to control e-mail overload:

- *Be selective.* Look first at the subject line in an e-mail—read only those of direct and immediate interest to you; look next at the sender line—read only those from people important to you; postpone or ignore the others.

- *Remove.* After reading an e-mail, respond if necessary, then either file it in the appropriate folder or delete it.

- *Protect.* Limit your e-mail by giving your address only to those who need it.

- *Be brief.* When responding, be concise and direct.

- *Stop spam.* Spam is unwanted e-mail advertisements—avoid mailing lists, complain to those who send spam, and ask to have your name removed from their mailing list.

- *Don't respond.* You do not have to respond to an e-mail; be selective, respond only to those worthy of your time.

Is information overload part of your future? If you are like today's busy executives, it probably will be.

VISUAL SUMMARY Information Systems

HOW INFORMATION FLOWS

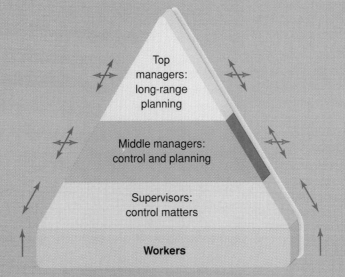

Information flows in an **organization** through functional areas and between management levels.

Functions

Most organizations have separate departments to perform five functions:

- **Accounting**—tracks all financial activities and generates periodic financial statements.
- **Marketing**—advertises, promotes, and sells the product (or service).
- **Production**—makes the product (or service) using raw materials and people to turn out finished goods.
- **Human resources**—finds and hires people, handles such matters as sick leave, retirement benefits, evaluation, compensation, and professional development.
- **Research and/or development**—conducts product research and development, monitors and troubleshoots new products.

Management Levels

The three basic management levels are:

- **Top-level**—long-range planning and forecasting.
- **Middle-level**—control, planning, and implementing long-term goals.
- **Supervisors**—control of operational matters, monitoring day-to-day events, and supervising workers.

Information Flow

Information flows within an organization in different directions.

- For **top-level managers** the informational flow is primarily up within the organization and into the organization from the outside.
- For **middle-level managers** the information flow is horizontally across and vertically within departments.
- For **supervisors** the information flow is primarily vertical.

To be a competent end user you need to understand how information flows through functional areas and management levels. You need to be aware of the different types of computer-based information systems including transaction processing systems, management information systems, decision support systems, and executive support systems.

TYPES OF INFORMATION SYSTEMS

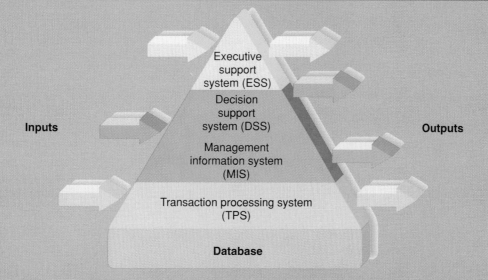

Inputs — Outputs

Executive support system (ESS)
Decision support system (DSS)
Management information system (MIS)
Transaction processing system (TPS)

Database

All organizations have **computer-based information systems,** including the following.

Transaction Processing Systems

Transaction processing systems (TPSs) record day-to-day transactions. An example is in accounting, which handles six activities: **sales order processing, accounts receivable, inventory and purchasing, accounts payable, payroll,** and **general ledger.**

Management Information Systems

Management information systems (MISs) produce predetermined **periodic, exception,** and **demand reports.** Management information systems use database management systems to integrate the database of different departments.

Decision Support Systems

Decision support systems (DSSs) enable managers to get answers for unanticipated questions. Teams formed to address large problems use **group decision support systems (GDSSs).** A decision support system consists of the user, software system, data, and decision models. Three types of decision models are *strategic, tactical,* and *operational.*

Executive Support Systems

Executive support systems (ESSs) assist top-level executives. An executive support system is similar to MIS or DSS but easier to use. ESSs are designed specifically for top-level decision makers.

Other Systems

- **Office automation systems (OASs)** support *data workers* who are involved with distribution and communication of information.
- **Knowledge work systems (KWSs)** support *knowledge workers,* who create information.

Key Terms

accounts payable (293)
accounts receivable (293)
computer-aided design/computer-aided manufacturing (CAD/CAM) (298)
data (295)
data processing system (DPS) (292)
decision models (295)
decision support system (DSS) (292)
demand report (294)
exception report (294)
executive information system (EIS) (292)

executive support system (ESS) (292)
general ledger (293)
group decision support system (GDSS) (295)
information system (286)
information worker (297)
inventory (293)
knowledge work system (KWS) (298)
management information system (MIS) (292)
middle-level manager (288)

office automation system (OAS) (297)
payroll (293)
periodic report (294)
purchasing (293)
sales order processing (293)
supervisor (288)
system software (295)
top-level manager (290)
transaction (291)
transaction processing system (TPS) (291)
user (295)

Chapter Review

LEVEL 1

Reviewing Facts and Terms
Matching

Match each numbered item with the most closely related lettered item. Write your answers in the spaces provided.

1. Summarizes the detailed data of the transaction processing system in standard reports for middle-level managers. _____

2. An easy-to-use system that helps top-level managers oversee the company's operations and develop strategic plans. _____

3. Systems that focus on managing documents, communicating, and scheduling. _____

4. People who create, distribute, and communicate information. _____

5. People who deal with control, planning, and decision making. _____

6. Helps an organization keep track of routine operations and records these events in a database. _____

7. A collection of people, procedures, software, hardware and data. _____

8. People who manage and monitor the employees or workers. _____

9. Specialized information systems used to create information in an area of expertise. _____

10. Enables managers to get answers to unexpected and generally nonrecurring kinds of problems. _____

11. Refers to the money a company owes its suppliers for materials and services it has received. _____

a. Accounts payable
b. Accounts receivable
c. CAD/CAM
d. Data processing system
e. Decision models
f. Decision support system
g. Demand report
h. Executive support system
i. Information system
j. Information workers
k. Inventory
l. Knowledge work systems
m. Management information system
n. Middle-level managers
o. Office automation system
p. Supervisors
q. System software
r. Top-level managers
s. Transactions
t. User

12. In a DSS, this is essentially the operating system—handling the detailed operating procedures. ____

13. Opposite of a periodic report, this is produced on request. ____

14. Routine activities such as employees hired, materials purchased, products produced, etc. ____

15. Knowledge work systems consisting of powerful microcomputers running special programs that integrate design and manufacture activities. ____

16. Give a decision support system its analytical capabilities. ____

17. People concerned with long-range planning. ____

18. The parts and finished goods that a company has in stock. ____

19. In a DSS, the person who has to make decisions—maybe a manager. ____

20. Refers to the money received from or owed by customers. ____

True/False

In the spaces provided, write T or F to indicate whether the statement is true or false.

1. For small companies, keeping accurate records and making good decisions are extremely important. ____

2. The production department takes in raw materials and puts people to work to turn out finished goods (or services). ____

3. In smaller organizations such titles as vice president of marketing and director of human resources are often combined. ____

4. CEO stands for chief executive officer. ____

5. Decision support systems summarize the detailed data of the transaction processing system in standard reports. ____

Multiple Choice

Circle the letter of the correct answer.

1. An information system is a collection of hardware, software, procedures, data, and:
   a. people
   b. functions
   c. managers
   d. accounting systems
   e. information

2. This department finds and hires people and handles such matters as sick leave and retirement benefits:
   a. accounting
   b. production
   c. marketing
   d. human resources
   e. research

3. The level of manager who deals with control and planning:
   a. executive
   b. top manager
   c. middle manager
   d. supervisor
   e. vice president

4. The level of manager whose information flow is primarily vertical:
   a. supervisor
   b. top manager
   c. middle manager
   d. executive
   e. vice president

5. The information system that provides a flexible tool for analysis:
   a. database information system
   b. transaction processing system
   c. management information system
   d. executive information system
   e. decision support system

Completion

Complete each statement in the spaces provided.

1. The _____ department relates new discoveries and is responsible for product development.

2. _____ managers are concerned with long-range planning.

3. _____ _____ systems record day-to-day activities such as customer orders and inventory levels.

4. _____ information systems are specially designed, simplified systems for top-level executives.

5. _____ workers use office automation and knowledge work systems.

Reviewing Concepts

LEVEL 2

Open-Ended

On a separate sheet of paper, respond to each question and statement.

1. What are the five departments found in medium-sized and large organizations? Discuss the function of each department.

2. What are the three levels of management? What are the responsibilities of managers at each level?

3. Explain the differences between the three types of reports produced by a management information system.

4. Discuss the three types of decision models used in a DSS.

5. Explain what an executive support system is. Give an example.

Concept Mapping

On a separate sheet of paper, draw a concept map or a flowchart showing how the following terms are related. Show all relationships. Include any additional terms you can think of.

accounting
accounts payable
accounts receivable
CAD/CAM
data
data processing system
decision models
decision support system
demand report
exception report
executive information
   system
executive support system

group decision support
   system
human resources
information system
information worker
inventory
knowledge work system
management information
   system
marketing
middle-level manager
office automation system
payroll

periodic report
production
purchasing
research
sales order processing
supervisor
system software
top-level manager
transaction
transaction processing
   system
user

Critical Thinking Questions and Projects

LEVEL 3

Read each exercise and answer the related questions on a separate sheet of paper.

1. *Functions and transactions:* What is the equivalent of "production" in a hotel or "marketing" in a college? The five functions or departments of an organization—accounting, marketing, production, human resources, and research—clearly exist in a for-profit organization such as an apparel manufacturer. However, they would not be found—at least not in the same form—in an employment agency, a department store, or a hospital. Nevertheless, these organizations do offer products or services, and they probably (if large enough) have three management levels.

   Choose an organization, such as your college, and interview someone to find out what might constitute departments and management layers. Summarize your findings in a drawing identifying the departments and the levels of management. See if you can discover the transactions that go into the database of one of the departments—for example, registrar, housing, fund-raising, financial aid, or alumni affairs.

2. *Data and information:* Almost all organizations collect and use data. Large organizations typically have formal names for the systems designed to collect and use the data. Although different organizations may use different names, the most common names are transaction processing, management information, decision support, and executive information systems.

   Choose an organization, such as your college, and consider the data and information needs. Describe:

   a. the transaction processing system—identify the data collected and briefly describe the six basic activities of the accounting department.

   b. the management information system—describe a typical periodic report, exception report, and demand report.

   c. the decision support system—discuss a problem that might be addressed.

   d. the executive support system—present a typical display the president would see first thing each morning.

3. *Computers and productivity:* The basic reason for installing and learning computers is that they help us increase our productivity. But do they really? Research has shown some of the following problems: (1) People in organizations spend time fine-tuning their word processing documents when such polishing is not necessary. (2) They devote time to unnecessarily putting spreadsheet information into graphic form. (3) They fill up ordinary memos with unnecessary facts gleaned from expensive computer searches. (4) They create unnecessary pressure by transmitting messages via modem or fax when regular mail would do. (5) They spend unproductive hours surfing the Internet and browsing the Web.

    These are serious problems that you will run across in the workplace. Discuss the following:

  a. Who or what creates these problems?

  b. What would you do to prevent these problems if you were a supervisor in an office?

  c. What can you do to avoid contributing to these problems yourself?

4. *Security:* All organizations are concerned about controlling and limiting access to their information systems and corporate databases. Financial institutions such as credit card companies are particularly sensitive to security issues. Consider what might happen if your credit card number and credit history fell into the wrong hands. Who do you suppose would be responsible for any financial losses? Most observers agree that it is the joint responsibility of the financial institution and the individual cardholder to ensure security. List three measures that you feel your credit card company should take to protect you. List three measures you should take to protect yourself.

On the Web Exercises

1. CAD

Computer-aided drawing (CAD) programs are often used to build precise models of buildings or products on the computer. Visit our

site at http://www.mhhe.com/it/oleary/exercise.mhtml to link to a CAD site. Once connected explore the site and learn about CAD systems. Print out the page you find the most interesting. Write a brief paragraph describing how and why many businesses use CAD programs.

2. Intranets

Within corporate computing, intranets are rapidly growing in popularity. To learn more about this important topic, visit the Yahoo site at http://www.

yahoo.com. Once connected to that site try using the keyword "intranet" or the subject heading "Computers and Internet: Communications and Networking: Intranet." Print the Web page you find the most informative. Write a paragraph on what kinds of companies use intranets, how intranets are helpful, and what the downsides are in using intranets in the workplace.

3. Global Positioning Systems

Global positioning systems (GPSs) use a series of satellites that ring the world. Using these satellites, a special receiver can locate your exact location anywhere in the

world. To learn more, visit our Web site at http://www. mhhe.com/it/oleary/exercise.mhtml to link to a GPS service provider. Once connected to that site, print the opening Web page. Write a paragraph describing the GPS product and discussing its accuracy.

4. News

Current and late-breaking news often hits the Web before it is printed and sometimes before it makes the television news. Visit our Web site at http://www.

mhhe.com/it/oleary/exercise.mhtml to link to a site specializing in news. Once connected, locate an article of interest to you. Print the article and write a brief paragraph summarizing it.

2001
2002

13

Systems Analysis and Design

COMPETENCIES

After you have read this chapter, you should be able to:

1. Describe the six phases of the systems life cycle.
2. Discuss how problems or needs are identified during Phase 1, preliminary investigation.
3. Explain how the current system is studied and new requirements are specified in Phase 2, systems analysis.
4. Describe how a new or alternative information system is designed in Phase 3, systems design.
5. Explain how new hardware and software are acquired, developed, and tested in Phase 4, systems development.
6. Discuss how a new information system is installed and users are trained in Phase 5, systems implementation.
7. Describe Phase 6, systems maintenance, the systems audit and ongoing evaluation, to see if a new system is doing what it's supposed to.
8. Describe prototyping and RAD.

Most people in an organization are involved with an information system of some kind, as we saw in the previous chapter. For an organization to *create* a system and for users to make it truly useful require considerable thought and effort. Fortunately, there is a six-step problem-solving process for accomplishing this. It is known as *systems analysis and design*.

Big organizations can make big mistakes. For example, General Motors spent $40 billion putting in factory robots and other high technology in its automaking plants. It then removed much of this equipment and reinstalled that basic part of the assembly line, the conveyor belt. Why did the high-tech production systems fail? The probable reason was that GM didn't devote

1. Preliminary investigation

2. Systems analysis

3. Systems design

4. Systems development

5. Systems implementation

6. Systems maintenance

enough energy to training its workforce in using the new systems.

The government also can make big mistakes. In one year, the Internal Revenue Service computer system was so overwhelmed it could not deliver tax refunds on time. The reason? Despite extensive testing of much of the system, not all testing was completed. Thus, when the new system was phased in, the IRS found it could not process tax returns as quickly as it had hoped.

Both of these examples show the necessity for thorough planning—especially when an organization is trying to implement a new kind of system. Despite the spectacular failures just mentioned, there is a way to reduce the chances for such mistakes. It is called *systems analysis and design.*

Competent end users need to understand the importance of systems analysis and design. They need to be aware of the relationship of an organization's chart to its managerial structure. Additionally, they need to know the six phases of the systems development life cycle: preliminary investigation, systems analysis, systems design, systems development, systems implementation, and systems maintenance.

Organizational Chart

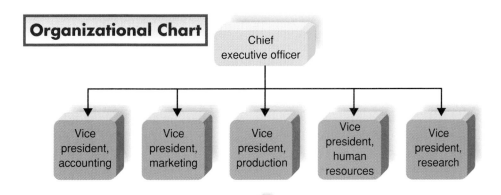

Management Pyramid

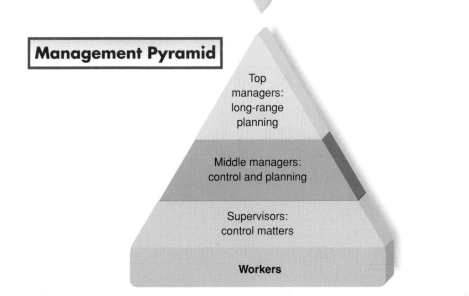

Systems Analysis and Design

Systems analysis and design is a six-phase problem-solving procedure for examining and improving an information system.

W e described different types of information systems in the last chapter. Now let us consider: What, exactly, is a **system?** We can define it as a collection of activities and elements organized to accomplish a goal. As we saw in Chapter 12, an *information system* is a collection of hardware, software, people, procedures, and data. These work together to provide information essential to running an organization. This information helps to produce a product or service and, for profit-oriented businesses, derive a profit.

Information—about orders received, products shipped, money owed, and so on—flows into an organization from the outside. Information—about what supplies have been received, which customers have paid their bills, and so on—also flows inside the organization. In order to avoid confusion, these flows of information must follow some system. However, from time to time, organizations need to change their information systems. Reasons include organizational growth, mergers and acquisitions, new marketing opportunities, revisions in governmental regulations, and availability of new technology.

Systems analysis and design is a six-phase problem-solving procedure for examining and improving an information system.

The six phases make up the **systems life cycle.** (See Figure 13-1.) The phases are as follows:

1. *Preliminary investigation:* The information problems or needs are identified.
2. *Systems analysis:* The present system is studied in depth. New requirements are specified.
3. *Systems design:* A new or alternative information system is designed.
4. *Systems development:* New hardware and software are acquired, developed, and tested.

FIGURE 13-1
The six-phase systems life cycle

5. *Systems implementation:* The new information system is installed and adapted to the new system, and people are trained to use it.

6. *Systems maintenance:* In this ongoing phase, the system is periodically evaluated and updated as needed.

In organizations, this six-phase systems life cycle is used by computer professionals known as **systems analysts.** These people study an organization's systems to determine what actions to take and how to use computer technology to assist them. A recent survey by *Money* magazine compared salary, prestige, and security of 100 widely held jobs. The top job classification was computer engineer, followed by computer systems analyst.

You may well find yourself working with a systems analyst. Or you may even become one yourself. It's important that you understand how the six phases work. After all, you better than anyone should understand what is needed in your part of the organization. And you should be the one best able to express that need. Developing a large computer-based information system requires the close collaboration of end users and systems analysts.

The procedure is also one that *you* as an end user can perform, working alone or with a systems analyst. In fact, you may *have* to use the procedure. More and more end users are developing their own information systems. This is because in many organizations there is a three-year backlog of work for systems analysts. For instance, suppose you recognize that there is a need for certain information within your organization. Obtaining this information will require the introduction of new hardware and software. You go to seek expert help from systems analysts in studying these information needs. At that point you discover they are so overworked it will take them three years to get to your request! You can see, then, why many managers are learning to do these activities themselves. In any case, learning the six steps described in this chapter will raise your computer competency. It will also give you skills to solve a wide range of problems. These skills can make you more valuable to an organization.

CONCEPT CHECK

✔ What is a system?

✔ Name the six phases of the systems life cycle.

Phase 1: Preliminary Investigation

In the preliminary investigation phase, the problems are briefly identified and a few solutions are suggested.

The first phase is a **preliminary investigation** of a proposed project to determine the need for a new information system. This usually is requested by an end user or a manager who wants something done that is not presently being done. For example, suppose you work for Advantage Advertising, a fast-growing advertising agency. Advantage Advertising produces a variety of different ads for a wide range of different clients. The agency employs both regular staff workers and on-call freelancers. One of your responsibilities is keeping track of the work performed for each client and the employees who performed the work. In addition, you are responsible for tabulating the final bill for each project.

312

Step	Description
1	Define problem
2	Suggest alternatives
3	Write preliminary investigation report

FIGURE 13-2
Phase 1: Preliminary investigation

How do you figure out how to charge which clients for which work done by which employees? This kind of problem is common to many service organizations (such as lawyers' and contractors' offices). Indeed, it is a problem in any organization where people charge for their "time" and clients need proof of hours worked.

In Phase 1, the systems analyst—or the end user—is concerned with three tasks. These are (1) briefly defining the problem, (2) suggesting alternative solutions, and (3) preparing a short report. (See Figure 13-2.) This report will help management decide whether to pursue the project further. (If you are an end user employing this procedure for yourself, you may not produce a written report. Rather, you would report your findings directly to your supervisor.)

Defining the Problem

Defining the problem means examining whatever current information system is in use. Determining what information is needed, by whom, when, and why, is accomplished by interviewing and making observations. (See Figure 13-3.) If the information system is large, this survey is done by a systems analyst. If the system is small, the survey can be done by the end user.

For example, suppose Advantage Advertising account executives, copywriters, and graphic artists currently just record the time spent on different jobs on their desk calendars. (Examples might be "Client A, telephone conference, 15 minutes"; "Client B, design layout, 2 hours.") This approach is somewhat helter-skelter. Written calendar entries are too unprofessional to be shown to clients. Moreover, a large job often has many people working on it. It is difficult to pull together all their notations to make up a bill for the client. Some freelancers work at home, and their time slips are not readily available. These matters constitute a statement of the problem: The company has a manual time-and-billing system that is slow and difficult to implement.

As an end user, you might experience difficulties with this system yourself. You're in someone else's office, and a telephone call comes in for you from a client. Your desk calendar is back in your own office. You have two choices. You can always carry your calendar with you. As an alternative, you can remember to note the time you spent on various tasks when you return to your office. The secretary to the account executive is continually after you (and everyone else at Advantage) to provide photocopies of your calendar. This is so that various clients can be billed for the work done on various jobs. Surely, you think, there must be a better way to handle time and billing.

FIGURE 13-3
Determining information requirements

Suggesting Alternative Systems

This step is simply to suggest some possible plans as alternatives to the present arrangement. For instance, Advantage could hire more secretaries to collect the information from everyone's calendars (including telephoning those working at home). Or it could use the existing system of network-linked microcomputers that staffers and freelancers presently use. Perhaps, you think, there is already some off-the-shelf packaged software available that could be used for a time-and-billing system. At least there might be one that would make your own job easier.

Preparing a Short Report

For large projects, the systems analyst writes a report summarizing the results of the preliminary investigation and suggesting alternative systems. (See Figure 13-4.) The report may also include schedules for further development of the project. This document is presented to higher management, along with a recommendation to continue or discontinue the project. Management then decides whether to finance the second phase, the systems analysis.

For Advantage Advertising, your report might point out that billing is frequently delayed. It could say that some tasks may even "slip through the cracks" and not get charged at all. Thus, as the analyst has noted, you suggest the project might pay for itself merely by eliminating lost or forgotten charges.

FIGURE 13-4
Preliminary investigation report

CONCEPT CHECK

✔ What is the purpose of the preliminary investigation phase?

✔ What are the three tasks the system analyst is concerned with during this phase?

Phase 2: Analysis

In the systems analysis phase, the present system is studied in depth, and new requirements are specified.

In Phase 2, **systems analysis,** data is collected about the present system. This data is then analyzed, and new requirements are determined. We are not concerned with a new design here, only with determining the *requirements* for a new system. Systems analysis is concerned with gathering and analyzing the data. It usually is completed by documenting the analysis in a report. (See Figure 13-5.)

Gathering Data

When gathering data, the systems analyst—or the end user doing systems analysis—expands on the data gathered during Phase 1. He or she adds details about how the current system works. Data is obtained from observation and interviews. It is also obtained from studying documents that describe the formal lines of authority and standard operating procedures. One document is the **organization chart,** which shows levels of management and formal lines of authority. (See Figure 13-6, top.) You might note that an organization chart resembles the hierarchy of three levels of management we described in Chapter 12. The levels are top managers, middle managers, and supervisors. (See Figure 13-6, bottom.) In addition, data may be obtained from questionnaires given to people using the system.

Step	Description
1	Gather data
2	Analyze data
3	Write system analysis report

FIGURE 13-5
Phase 2: Analysis

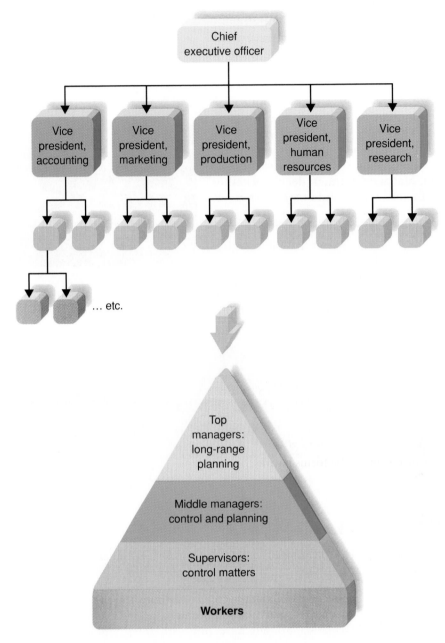

FIGURE 13-6
Example of an organization chart—
and how it corresponds to the
management pyramid

Note in our illustration that we have preserved the department labeled "Production." (Refer to Figure 13-6, top.) However, the name in an advertising agency might be something like "Creative Services." Obviously, the products an advertising agency produces are ads: radio and television commercials, magazine and newspaper ads, billboard ads, and so on. In any case, if the agency is working on a major advertising campaign, people from several departments might be involved. There might also be people from different management levels within the departments. Their time charges will vary, depending on how much they are paid.

Analyzing the Data

In the step of analyzing the data, the idea is to learn how information currently flows and to pinpoint why it isn't flowing appropriately. The whole point of this step is to apply *logic* to the existing arrangement to see how workable it is. Many times the current system is not operating correctly because prescribed procedures are not being followed. That is, the system may not really need to be redesigned. Rather, the people in it may need to be shown how to follow correct procedures.

Many different tools are available to assist systems analysts and end users in the analysis phase. Some of the principal ones are as follows:

- **Checklists:** Numerous checklists are available to assist in this stage. A **checklist** is a list of questions. It is helpful in guiding the systems analyst and end user through key issues for the present system.

  For example, one question might be "Can reports be easily prepared from the files and documents currently in use?" Another might be "How easily can the present time-and-billing system adapt to change and growth?"

- **Top-down analysis method:** The **top-down analysis method** is used to identify the top-level components of a complex system. Each component is then broken down into smaller and smaller components. This approach makes each component easier to analyze and deal with.

  For instance, the systems analyst might look at the present kind of bill submitted to a client for a complex advertising campaign. The analyst might note the categories of costs—employee salaries, telephone and mailing charges, travel, supplies, and so on.

- **Grid charts:** A **grid chart** shows the relationship between input and output documents. An example is shown in Figure 13-7 that indicates the relationship between the data input and the outputs.

  For instance, a time sheet is one of many inputs that produces a particular report, such as a client's bill. (Other inputs might be forms having to do with telephone conferences and travel expenses.) Horizontal rows represent inputs, such as time sheet forms. Vertical columns represent output documents, such as different clients' bills. A checkmark at the intersection of a row and column means that the input document is used to create the output document.

Forms (input)	Reports (output)		
	Client billing	Personnel expense	Support cost
Time sheet	✓	✓	
Telephone log	✓		✓
Travel log	✓		✓

- **Decision tables:** A **decision table** shows the decision rules that apply when certain conditions occur. Figure 13-8 shows a decision table to evaluate whether to accept a client's proposed advertising project. The first decision rule applies if both conditions are met. If the project is less than $10,000 and if the client has a good credit history, the firm will accept the project without requiring a deposit.

Conditions	Decision rules			
	1	2	3	4
1. Project less than $10,000	Y	Y	N	N
2. Good credit history	Y	N	Y	N
Actions	1	2	3	4
1. Accept project	✓	✓	✓	
2. Require deposit			✓	✓
3. Reject project				✓

- **System flowcharts:** System **flowcharts** show the flow of input data to processing and finally to

FIGURE 13-9
Example of a system flowchart

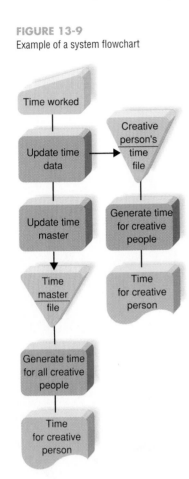

FIGURE 13-9
Example of a system flowchart

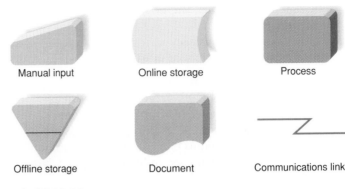

FIGURE 13-10
System flowchart symbols

output, or distribution of information. An example of a system flowchart keeping track of time for advertising "creative people" is shown in Figure 13-9. The explanation of the symbols used (and others not used) appears in Figure 13-10. Note that this describes the present manual, or noncomputerized, system. (A *system* flowchart is not the same as a *program* flowchart, which is very detailed. Program flowcharts are discussed in Chapter 14.)

- **Data flow diagrams: Data flow diagrams** show the data or information flow within an information system. The data is traced from its origination through processing, storage, and output. An example of a data flow diagram is shown in Figure 13-11. The explanation of the symbols used appears in Figure 13-12.

- **Automated design tools: Automated design tools** are software packages that evaluate hardware and software alternatives according to requirements given by the systems analyst. They are also called **computer-aided software engineering (CASE) tools.** These tools are not limited to system analysis. They are used in system design and development as well. CASE tools relieve the system analysts of many repetitive tasks, develop clear documentation, and, for larger projects, coordinate team-member activities.

For a summary of the analysis tools, see Figure 13-13.

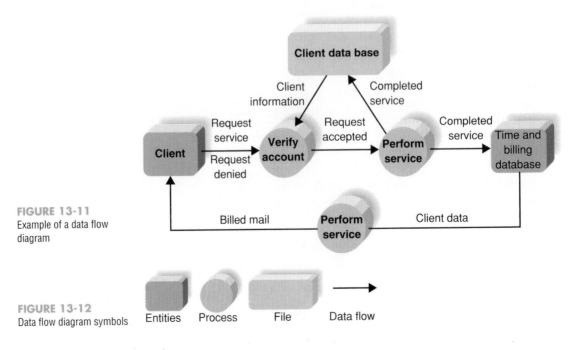

FIGURE 13-11
Example of a data flow diagram

FIGURE 13-12
Data flow diagram symbols

Tool	Description
Checklist	Provides a list of questions about key issues
Top-down analysis	Divides a complex system into components, beginning at the top
Grid chart	Shows relationships between inputs and outputs
Decision table	Specifies decision rules and circumstances when specific rules are to be applied
System flowchart	Shows movement of input data, processing, and output or distribution of information
Data flow diagram	Shows data flow within an organization or application
CASE	Automates the analysis, design, and development of information systems

FIGURE 13-13
Summary of analysis tools

Documenting Systems Analysis

In larger organizations, the systems analysis stage is typically documented in a report for higher management. The systems analysis report describes the current information system, the requirements for a new system, and a possible development schedule. For example, at Advantage Advertising, the system flowcharts show the present flow of information in a manual time-and-billing system. Some boxes in the system flowchart might be replaced with symbols showing where a computerized information system could work better. For example, in our flowchart, the offline storage symbol ("time master file") might be replaced by an online storage symbol. (Refer back to Figure 13-9.) That is, the information in the file would be instantly accessible.

Management studies the report and decides whether to continue with the project. Let us assume your boss and higher management have decided to continue. You now move on to Phase 3, systems design.

CONCEPT CHECK

✔ What is the purpose of the analysis phase?

✔ List some common analysis tools.

FIGURE 13-14
Phase 3: Design

Phase 3: Design

In the systems design phase, a new or alternative information system is designed.

Phase 3 is **systems design.** It consists of three tasks: (1) designing alternative systems, (2) selecting the best system, and (3) writing a systems design report. (See Figure 13-14.)

Step	Description
1	Design alternative systems
2	Select best alternative
3	Write system design report

Designing Alternative Systems

In almost all instances, more than one design can be developed to meet the information needs. Systems designers evaluate each alternative system for feasibility. By *feasibility* we mean three things:

- **Economic feasibility:** Will the costs of the new system be justified by the benefits it promises?
- **Technical feasibility:** Are reliable hardware, software, and training available to make the system work?
- **Operational feasibility:** Can the system actually be made to operate in the organization, or will people—employees, managers, clients—resist it?

Selecting the Best System

When choosing the best design, managers must consider these four questions. (1) Will the system fit in with the organization's overall information system? (2) Will the system be flexible enough so it can be modified in the future? (3) Can it be made secure against unauthorized use? (4) Are the benefits worth the costs?

For example, one aspect you have to consider at Advantage Advertising is security. Should freelancers and outside vendors enter data directly into a computerized time-and-billing system, or should they keep submitting time slips manually? In allowing these outside people to directly input information, are you also allowing them access to files they should not see? Do these files contain confidential information, perhaps information of value to rival advertising agencies?

FIGURE 13-15
Comparing costs and benefits

Writing the Systems Design Report

The systems design report is prepared for higher management and describes the alternative designs. It presents the costs versus the benefits and outlines the effect of alternative designs on the organization. (See Figure 13-15.) It usually concludes by recommending one of the alternatives.

CONCEPT CHECK

✔ What is the purpose of the design phase?

✔ Identify the factors that need to be considered when choosing the best systems design.

Phase 4: Development

In the systems development phase, new hardware and software are developed, acquired, and tested.

Phase 4 is **systems development.** It has three steps: (1) developing software, (2) acquiring hardware, and (3) testing the new system. (See Figure 13-16.)

Developing Software

Application software for the new information system can be obtained in two ways. It can be purchased as off-the-shelf packaged software and possibly modified, or it can be custom designed. If any of the software is being specially created, the steps we will outline on programming (in Chapter 14) should be followed.

With the systems analyst's help, you have looked at time-and-billing packaged software designed for service organizations. Such organizations might include advertising agencies, law firms, and building contractors. The systems analyst points out the importance of time-and-billing data's being collected in an appropriate manner so that it can be used for a variety of purposes. Such a system will not only help supervisory and middle managers do their jobs but also help top managers make decisions.

Unfortunately, you find that none of the packaged software will do. Most of the packages seem to work well for one person (you). However, none seem to be designed for many people working together. It appears, then, that software will have to be custom designed. (We discuss the process of developing software in Chapter 14, on programming.)

Acquiring Hardware

Some new systems may not require new computer equipment, but others will. The kinds needed and the places they are to be installed must be determined. This is a very critical area. Switching or upgrading equipment can be a tremendously expensive proposition. Will a microcomputer system be sufficient as a company grows? Are networks expandable? Will people have to undergo costly training?

The systems analyst tells you that there are several different makes and models of microcomputers currently in use at Advantage Advertising. (See Figure 13-17.) Fortunately, all are connected by a local area network to a file server that can hold the time-and-billing data. To maintain security, the systems analyst suggests that an electronic mailbox can be installed for free-lancers and others outside the company. They can use this electronic mailbox to post their time charges. Thus, it appears that existing hardware will work just fine.

Step	Description
1	Develop software
2	Acquire hardware
3	Test system

FIGURE 13-16
Phase 4: Development

FIGURE 13-17
Acquiring hardware

FIGURE 13-18
Testing a new system

Testing the New System

After the software and equipment have been installed, the system should be tested. Sample data is fed into the system. The processed information is then evaluated to see whether results are correct. Testing may take several months if the new system is complex.

For this step, you take some time and expense charges from an ad campaign that Advantage ran the previous year. You then ask some people in Creative Services to test it on the system. (See Figure 13-18.) You observe that time is often charged in fractions of minutes and that the software ignores these fractions of time. You also see that some of the people in Creative Services have problems knowing where to enter their times. To solve the first problem, you must see that the software is corrected to allow for fractional minutes. To solve the second problem, you must see that the software is modified so that it displays an improved user entry screen. After the system has been thoroughly tested and revised as necessary, you are ready to put it into use.

CONCEPT CHECK

✔ What is the purpose of the development phase?

✔ What are the ways by which application software can be obtained?

Phase 5: Implementation

In the systems implementation phase, the new information system is installed, and people are trained to use it.

Step	Description
1	Select conversion type
2	Train users

FIGURE 13-19
Phase 5: Implementation

Another name for Phase 5, **systems implementation,** is **conversion.** It is the process of changing—converting—from the old system to the new and training people to use the new system. (See Figure 13-19.)

Types of Conversion

There are four approaches to conversion: *direct, parallel, pilot,* and *phased.*

- In the **direct approach,** the conversion is done simply by abandoning the old and starting up the new. This can be risky. If anything is still wrong with the new system, the old system is no longer available to fall back on.

  The direct approach is not recommended precisely because it is so risky. Problems, big or small, invariably crop up in a new system. In a large system, a problem might just mean catastrophe.

- In the **parallel approach,** old and new systems are operated side by side until the new one has proved to be reliable.

  This approach is low-risk. If the new system fails, the organization can just switch to the old system to keep going. However, keeping enough equipment and people active to manage two systems at the same time can be very expensive. Thus, the parallel approach is used only in cases in which the cost of failure or of interrupted operation is great.

- In the **pilot approach,** the new system is tried out in only one part of the organization. Once the system is working smoothly in that part, it is implemented throughout the rest of the organization.

    The pilot approach is certainly less expensive than the parallel approach. It also is somewhat riskier. However, the risks can be controlled because problems will be confined to only certain areas of the organization. Difficulties will not affect the entire organization.

- In the **phased approach,** the new system is implemented gradually over a period of time.

    The entire implementation process is broken down into parts or phases. Implementation begins with the first phase and once it is successfully implemented, the second phase begins. This process continues until all phases are operating smoothly. This is an expensive proposition, because the implementation is done slowly. However, it is certainly one of the least risky approaches.

In general, the pilot and phased approaches are the most favored methods. Pilot is preferred when there are many people in an organization performing similar operations—for instance, all sales clerks in a department store. Phased is more appropriate for organizations in which people are performing different operations. For a summary of the different types of conversions, see Figure 13-20.

You and the systems analyst succeed in convincing the top managers of Advantage Advertising to take a pilot approach. The reason is that it is easy to select one trial group—the group of which you are a member. Moreover, this group is eager to try the new system. Thus, the new time-and-billing system is tried first with a handful of people in your particular department. (See Figure 13-21.)

Training

Training people is important, of course. Unfortunately, it is one of the most commonly overlooked activities. Some people may begin training early, even before the equipment is delivered, so that they can adjust more easily. In some cases, a professional software trainer may be brought in to show people how to operate the system. However, at Advantage Advertising the time-and-billing software is simple enough that the systems analyst can act as the trainer.

Type	Description	Discussion
Direct	Abandon the old	Very risky; not recommended
Parallel	Run old and new side by side	Very low risk; however, very expensive; not generally recommended
Pilot	Convert part of organization first	Less expensive but riskier than parallel conversion; recommended for situations with many people performing similar operations
Phased	Implement gradually	Less risky but more expensive than parallel conversion; recommended for situations with many people performing different operations

FIGURE 13-20
Types of conversion

FIGURE 13-21
Trial or pilot group for new system

CONCEPT CHECK

✔ What is the goal of the implementation phase?

✔ Briefly describe the four approaches to conversion.

Step	Description
1	Perform system audit
2	Conduct periodic evaluations

FIGURE 13-22
Phase 6: Maintenance

Phase 6: Maintenance

Systems maintenance is first a systems audit and then an ongoing evaluation to see whether a system is performing productively.

After implementation comes **systems maintenance,** the last step in the systems life cycle. This phase is a very important, ongoing activity. Most organizations spend more time and money on this phase than any of the others. Maintenance has two parts—a *systems audit* and a *periodic evaluation*. (See Figure 13-22.)

In the **systems audit,** the system's performance is compared to the original design specifications. This is to determine whether the new procedures are actually furthering productivity. If they are not, some further redesign may be necessary.

After the systems audit, the new information system is periodically evaluated and further modified, if necessary. All systems should be evaluated from time to time to see if they are meeting the goals and providing the service they are supposed to.

For example, over time the transaction database at Advantage Advertising is expanded. After a year or two, the systems analyst might suggest that the time-and-billing part of it be reevaluated. For instance, the analyst might discover that telephone and mailing charges need to be separated. This might be because, with more people using the electronic mailbox and more people sharing data, telephone charges are now higher. The six-step systems life cycle is summarized in Figure 13-23.

Phase	Activity
1. Preliminary investigation	Define problem, suggest alternatives, prepare short report
2. Systems analysis	Gather data, analyze data, document
3. Systems design	Design alternatives, select best alternative, write report
4. Systems development	Develop software, acquire hardware, test system
5. Systems implementation	Convert, train
6. Systems maintenance	Perform system audit, evaluate periodically

FIGURE 13-23
Summary of systems life cycle

CONCEPT CHECK

✔ What is the purpose of the maintenance phase?

✔ Name the two parts of the maintenance phase.

Prototyping and RAD

Prototyping and RAD are two alternatives to the systems development life cycle approach.

Is it necessary to follow every phase of the systems life cycle? It may be desirable, but often there is no time to do so. For instance, hardware may change so fast that there is no opportunity for evaluation, design, and testing as just described. Two alternative approaches that require much less time are prototyping and rapid applications development (RAD).

Prototyping

Prototyping means to build a *model* or *prototype* that can be modified before the actual system is installed. For instance, the systems analyst for Advantage Advertising might develop a proposed or prototype menu as a possible screen display for the time-and-billing system. Users would try it out and provide feedback to the systems analyst. The systems analyst would revise the prototype until the users felt it was ready to put into place. Typically, the development time for prototyping is shorter; however, it is sometimes more difficult to manage the project and to control costs.

RAD

Rapid applications development (RAD) involves the use of powerful development software; small, specialized teams; and highly trained personnel. For example, the systems analyst for Advantage Advertising would use specialized development software like CASE, form small teams consisting of select users and managers, and obtain assistance from other highly qualified

FIGURE 13-24
Rapid applications development (RAD)

analysts. (See Figure 13-24.) Although the resulting time-and-billing system would likely cost more, the time for development would be shorter and the quality of the completed system development time would be better.

CONCEPT CHECK

✔ What is meant by prototyping?

✔ What is involved in RAD?

A LOOK TO THE FUTURE

The systems life cycle will be shortened using a method called rapid applications development.

Most observers firmly believe that the pace of business is now faster than ever before. The time to develop a product and bring it to market in many cases is now months rather than years. Internet technologies, in particular, have provided tools to support the rapid introduction of new products and services.

To stay competitive, corporations must integrate these new technologies into their existing way of doing business. Existing systems are being modified, and entirely new systems are being developed. In most cases, the traditional systems life cycle approach takes too long—sometimes years to develop a system. In the future, we will see increasing use of prototyping and RAD approaches and increased end user involvement in the development process.

VISUAL SUMMARY Systems Analysis and Design

PHASE 1: PRELIMINARY INVESTIGATION

The **preliminary investigation** determines the need for a new information system. It is typically requested by an end user or a manager.

Three tasks of this phase are the following.

Defining the Problem
The current information system is examined to determine who needs what information, when the information is needed, and why.

If the existing information system is large, then a **systems analyst** conducts the survey. Otherwise, the end user conducts the survey.

Suggesting Alternative Systems
Some possible alternative systems are suggested. Based on interviews and observations made in defining the problem, alternative information systems are identified.

Preparing a Short Report
To document and to communicate the findings of Phase 1, preliminary investigation, a short report is prepared and presented to management.

PHASE 2: ANALYSIS

Summary of Analysis Tools

Tool	Description
Checklist	Provides a list of questions about key issues
Top-down analysis	Divides a complex system into components, beginning at the top
Grid chart	Shows relationships between inputs and outputs
Decision table	Specifies decision rules and circumstances when specific rules are to be applied
System flowchart	Shows movement of input data, processing, and output or distribution of information
Data flow diagram	Shows data flow within an organization or application
CASE	Automates the analysis, design, and development of information systems

In **systems analysis** data is collected about the present system. The focus is on determining the requirements for a new system.

Three tasks of this phase are the following.

Gathering Data
Data is gathered by observation, interviews, questionnaires, and looking at documents. One helpful document is the **organization chart,** which shows a company's functions and levels of management.

Analyzing the Data
There are several tools for the analysis of data, including checklists, top-down analysis, grid charts, and decision tables.

Documenting Systems Analysis
To document and to communicate the findings of Phase 2, analysis, a report is prepared for higher management.

To be a competent end user you need to understand the importance of systems analysis and design and the relationship of an organization's chart to its managerial structure. Additionally, you need to know the six phases of the systems development life cycle including preliminary investigation, analysis, design, development, implementation, and maintenance.

PHASE 3: DESIGN

Systems design consists of the following three tasks.

Designing Alternative Systems
Alternative information systems are designed. Each alternative is evaluated for **economic, technical, and operational feasibility.**

Selecting the Best System
Four questions considered when selecting the best system:
- Will the system fit into an overall information system?
- Will the system be flexible enough to be modified as needed in the future?
- Will it be secure against unauthorized use?
- Will the system's benefits exceed its costs?

Writing the System Design Report
To document and to communicate the findings of Phase 3, design, a report is prepared for higher management.

PHASE 4: DEVELOPMENT

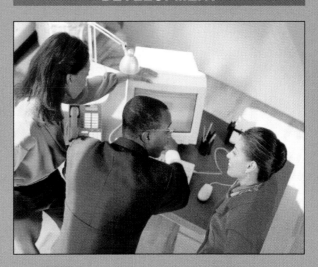

The **systems development** phase has three steps: *developing software, acquiring hardware,* and *testing.*

Developing Software
Two ways to acquire software are:
- Purchase—buying off-the-shelf packaged software to be modified if necessary.
- Custom designed—create programs following programming steps presented in Chapter 14.

Acquiring Hardware
Acquiring hardware is very critical and involves consideration for future company growth, existing networks, communication capabilities, and training.

Testing the System
Using sample data, the new system is tested. Can take several months for a complex system.

PHASE 5: IMPLEMENTATION

Type	Description	Discussion
Direct	Abandon the old	Very risky; not recommended
Parallel	Run old and new side by side	Very low risk; however, very expensive; not generally recommended
Pilot	Convert part of organization first	Less expensive but riskier than parallel conversion; recommended for situations with many people performing similar operations
Phased	Implement gradually	Less risky but more expensive than parallel conversion; recommended for situations with many people performing different operations

Systems implementation (conversion) is the process of changing to the new system and training people.

Types of Conversion

Four ways to convert are:

- **Direct approach**—abandoning the old and starting up the new; can be very risky and not recommended.
- **Parallel approach**—operating the old and new side by side until the new one proves its worth; low risk but expensive.
- **Pilot approach**—trying out the new system in only one part of an organization. Compared to parallel, pilot is riskier and less expensive.
- **Phased approach**—implementing the new system gradually; low risk but expensive.

Training

A software trainer may be used to train end users in the new system.

PHASE 6: MAINTENANCE

Systems maintenance is the final phase.

System Audit

Once the system is operational, the systems analyst compares it to the original design specifications. If the system does not meet these specifications, some further redesign of the system may be required.

Periodic Evaluation

The new system is periodically evaluated to ensure that it is operating efficiently. If it is not, some redesign may be required.

PROTOTYPING AND RAD

Due to time pressures, the system life cycle may not always be feasible. Two alternatives that require less time are *prototyping* and *RAD*.

Prototyping

Prototyping uses a model or prototype that is modified.

RAD

Rapid application development (RAD) uses powerful development software, specialized teams, and highly trained personnel.

2001
2002

Key Terms

automated design tool (316)
checklist (315)
computer-aided software engineering
    (CASE) tool (316)
conversion (320)
data flow diagram (316)
decision table (315)
direct approach (320)
economic feasibility (318)
grid chart (315)
operational feasibility (318)
organization chart (313)
parallel approach (320)
phased approach (321)
pilot approach (321)
preliminary investigation (311)

prototyping (323)
rapid applications development (RAD) (323)
system (310)
system flowchart (315)
systems analysis (313)
systems analysis and design (310)
systems analyst (310)
systems audit (322)
systems design (317)
systems development (318)
systems implementation (320)
systems life cycle (310)
systems maintenance (322)
technical feasibility (318)
top-down analysis method (315)

Chapter Review

LEVEL 1

Reviewing Facts and Terms
Matching

Match each numbered item with the most closely related lettered item. Write your answers in the spaces provided.

1. To build a model that can be modified before the actual system is installed. _____

2. A collection of activities and elements organized to accomplish a goal. _____

3. Are reliable hardware, software, and training available to make the system work? _____

4. A six-phase problem-solving procedure for examining and improving an information system. _____

5. The new system is tried out in only one part of the organization. _____

6. Shows the data or information flow within an information system. _____

7. Process of converting from the old system to the new and training people to use the new system. _____

8. Can the system actually be made to operate in the organization, or will people resist it? _____

9. The system's performance is compared to the original design specifications. _____

10. Involves the use of powerful development software; small, specialized teams; and highly trained personnel. _____

11. Shows the decision rules that apply when certain conditions occur. _____

a. Automated design tool
b. Data flow diagram
c. Decision table
d. Economic feasibility
e. Operational feasibility
f. Parallel approach
g. Phased approach
h. Pilot approach
i. Preliminary investigation
j. Prototyping
k. RAD
l. System
m. System flowchart
n. Systems life cycle
o. Systems analyst
p. Systems audit
q. Systems design
r. Systems implementation
s. Systems maintenance
t. Technical feasibility

2001
2002

12. The phase in which problems or needs are identified. ____

13. Old and new systems are operated side by side until the new one has proved to be reliable. ____

14. Person who studies an organization's systems to determine what actions to take and how to use computer technology to assist them. ____

15. Will the costs of the new system be justified by the benefits it promises? ____

16. A very important and ongoing activity, evaluating a system—the last phase of the system life cycle. ____

17. Software packages that evaluate hardware and software alternatives. ____

18. Consists of three tasks: designing alternative systems, selecting the best system, and writing a systems design report. ____

19. Shows the flow of input data to processing and finally to output, or distribution of information. ____

20. The new system is implemented gradually over a period of time. ____

True/False

In the spaces provided, write T or F to indicate whether the statement is true or false.

1. Systems analysis and design is a way to reduce the chance of creating an ineffective information system. ____

2. In large organizations, the person who uses the systems life cycle the most is called a database administrator. ____

3. Defining the problem is a task in Phase 2, design. ____

4. Software is either purchased or developed in the implementation phase. ____

5. In pilot conversion, one part of an organization initially tries out the new system. ____

Multiple Choice

Circle the letter of the correct answer.

1. A collection of hardware, software, people, procedures, and data:
   a. analysis and design
   b. system
   c. network
   d. microcomputer
   e. design

2. This phase in the systems life cycle focuses on evaluating and determining the need for a new information system:
   a. preliminary investigation
   b. systems design
   c. systems development
   d. systems implementation
   e. systems maintenance

3. Phase 2, analysis, involves gathering data, analyzing the data, and:
   a. designing the new system
   b. creating programs
   c. auditing the existing system
   d. documenting the systems analysis stage
   e. training

4. The evaluation of economic, technical, and operational feasibility is made during this phase:
   a. preliminary investigation
   b. systems design
   c. systems development
   d. systems implementation
   e. systems maintenance

5. The final step in Phase 4, development, is:
   a. designing alternative systems
   b. selecting the best system
   c. developing software
   d. acquiring hardware
   e. testing the new system

Completion

Complete each statement in the spaces provided.

1. The six-phase problem-solving procedure for systems analysis and design is the _____ _____ _____.

2. Defining the problem, suggesting alternative systems, and preparing a short report are all parts of _____ _____.

3. _____ _____ diagrams trace data from its origination through processing, storage, and output.

4. Once the new system is operational, a _____ _____ is performed to compare the original design specifications with the actual system.

5. A _____ is a model of a system.

Reviewing Concepts

Open-Ended

On a separate sheet of paper, respond to each question and statement.

1. What is the purpose of systems analysis and design? Who is involved with this process?

2. What are the six phases in the systems life cycle? Briefly describe each phase.

3. Describe top-down analysis method.

4. Explain the three steps in Phase 4, systems development.

5. Describe the four possible ways of implementing a system. Which one would you recommend?

Concept Mapping

On a separate piece of paper, draw a concept map or a flowchart showing how the following terms are related. Show all relationships. Include any additional terms you can think of.

automated design tool	organization chart	systems analyst
CASE tool	parallel approach	systems audit
conversion	phased approach	systems design
data flow diagram	pilot approach	systems development
decision table	preliminary investigation	systems implementation
direct approach	prototyping	systems life cycle
economic feasibility	RAD system	systems maintenance
grid chart	system flowchart	technical feasibility
operational feasibility	systems analysis	top-down analysis method

Critical Thinking Questions and Projects

LEVEL 3

Read each exercise and answer the related questions on a separate sheet of paper.

1. *Systems analysis:* Apply the systems life cycle to an activity at school or work that you find inefficient or irritating: parking problems, preregistration, financial aid applications, or some other activity. Apply Phase 1, Preliminary Investigation, by briefly identifying the problem and suggesting possible solutions. Write a short report to the college administration or to higher management to help them decide whether to go ahead with Phase 2.

2. *Systems analysts:* Interview a systems analyst, perhaps one working at the computer center or information systems department at your college or at work. Ask them to describe their work and to discuss typical projects. With whom do they normally work? Do they work on more than one project at a time? What types of jobs do they like best? What aspects of their job do they like best? Ask if job opportunities for systems analysts are plentiful. Inquire about job security and satisfaction. Do they use the systems life cycle approach? Why or why not?

3. *Security:* Organizations are constantly evaluating the effectiveness of their security systems to protect their information systems and corporate databases. Whenever a new information system is being developed or an existing one is being revised, security is a critical issue. Review the Advantage Advertising case presented in this chapter. Identify potential security risks and suggest procedures to control those risks.

On the (Web) Exercises

1. Health and Fitness

One of the best resources for the latest information about health and fitness is the Web. Visit our Web site at http://www.mhhe.com/it/oleary/exercise.mhtml to link to one of these sites. Once connected to

that site, explore and learn more about health and fitness. Print out the page you find most informative. Write a paragraph on the ways that you can improve your nutrition.

2. Job Search

You can find almost anything on the Web—even a job. Visit our Web site at http://www.mhhe.com/it/oleary/exercise.mhtmlto link to one of these job-search sites. Once connected to that

site, explore the job opportunities and find one that interests you. Print out the job description and write a paragraph describing the value of the job-search service and who would most likely use it.

3. Management Information Systems

Search the Internet for recent information on management information systems (MIS). Visit the Yahoo site at http://www.yahoo.com and investigate the cat-

egory of "Business and Economy: Management Information Systems." Select the most interesting Web site. Print out the first page of the site and write a paragraph summarizing its content.

4. Astrology

There are numerous Web sites that provide horoscopes, including romance predictions and explanations of astrological terms. Visit our Web site at http://www.mhhe.com/it/oleary/exercise.mhtml to link to one of

these sites. Once connected to that site, print out your horoscope. Write a paragraph describing how the horoscope's prediction could relate to you.

Programming and Languages

COMPETENCIES

After you have read this chapter, you should be able to:

1. Understand the six steps of programming.

2. Describe Step 1, program specification.

3. Describe Step 2, program design, and the program-design tools of top-down program design, pseudocode, flowcharts, and logic structures.

4. Explain Step 3, coding the program.

5. Describe Step 4, testing the program, and the tools for finding and removing errors.

6. Discuss Step 5, documenting the program.

7. Discuss Step 6, maintaining the program.

8. Describe CASE tools and object-oriented software development.

9. Explain the five generations of programming languages.

How do you go about getting a job? You look through newspaper classified ads, check with employment services, write to prospective employers, and so on. In other words, you do some *general problem solving* to come up with a broad plan. This is similar to what you do in systems analysis and design. Once you have determined a *particular* job you would like to have, you then do some *specific problem solving*. That is what you do in programming. In this chapter, we describe programming in two parts. They are (1) the steps in the programming process and (2) some of the programming languages available.

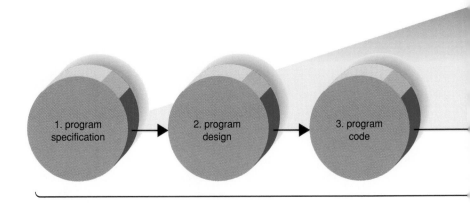

Why should you need to know anything about programming? The answer is simple. You might need to deal with programmers in the course of your work. You may also be required to do some programming yourself in the future. A new field has emerged known as *end-user application development*. In this field, users like you create their own business application programs, without the assistance of a programmer. Thus, organizations avoid paying high software development costs. You and other end users avoid waiting months for programmers to get around to projects important to you.

In Chapter 13, we described the six phases of the systems life cycle. Programming is part of Phase 4, systems development. Competent end users need to understand the relationship between the systems life cycle's Systems Development and programming. Additionally, they need to know the six steps of programming including program specification, program design, program code, program test, program documentation, and program maintenance.

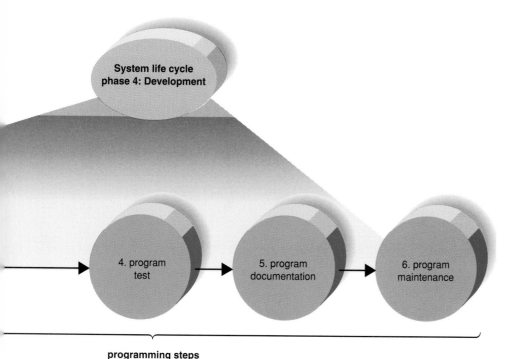

System life cycle phase 4: Development

4. program test

5. program documentation

6. program maintenance

programming steps

Programs and Programming

Programming is a six-step procedure for producing a program—a list of instructions—for the computer.

What exactly *is* programming? Many people think of it as simply typing words into a computer. That may be part of it—but certainly not all of it. Programming, as we've hinted before, is actually a *problem-solving procedure*.

What Is a Program?

To see how programming works, think about what a program is. A **program** is a list of instructions for the computer to follow to accomplish the task of processing data into information. The instructions are made up of statements used in a programming language, such as BASIC, C, or Java.

Application software or *application programs,* as we said in Chapter 1, are the kind of programs that do "end-user work." These are things such as word processing and accounting tasks. *System software,* we said, is concerned with "background" tasks such as housekeeping chores involving computer operations. In this chapter we are concerned with application programs.

You are probably most familiar with one kind of application program—the *prewritten* or *packaged* programs. These are so-called off-the-shelf programs such as word processors, spreadsheets, and database managers. However, application programs may also be *created* or *custom-made*—either by a professional programmer or by you, the end user. In Chapter 13, we saw that the systems analyst looked into the availability of time-and-billing software for Advantage Advertising. Will off-the-shelf software do the job, or should it be custom written? This is one of the first things that needs to be decided in programming.

What Is Programming?

A program is a list of instructions for the computer to follow to process data. **Programming,** also known as **software development,** is a six-step procedure for creating that list of instructions. Only *one* of those steps consists of typing (keying) statements into a computer.

The six steps are as follows:

1. Program specification
2. Program design
3. Program code
4. Program test
5. Program documentation
6. Program maintenance

CONCEPT CHECK

✔ Define *programming.* What are its six steps?

✔ What is a program?

Step 1: Program Specification

In the program specification step, the objectives, outputs, inputs, and processing requirements are determined.

Program specification is also called **program definition** or **program analysis**. It requires that the programmer—or you, the end user, if you are following this procedure—specify five tasks. They are (1) the program's objectives, (2) the desired output, (3) the input data required, (4) the processing requirements, and (5) the documentation. (See Figure 14-1.)

Program Objectives

You solve all kinds of problems every day. A problem might be deciding how to commute to school or work or which homework or report to do first. Thus, every day you determine your *objectives*—the problems you are trying to solve. Programming is the same. You need to make a clear statement of the problem you are trying to solve. (See Figure 14-2.) An example would be "I want a time-and-billing system to record the time I spend on different jobs for different clients of Advantage Advertising."

Desired Output

It is best always to specify outputs before inputs. That is, you need to list what you want to *get out* of the computer system. Then you should determine what will *go into* it. The best way to do this is to draw a picture. You—the end user, not the programmer—should sketch or write how you want the output to look when it's done. It might be printed out or displayed on the monitor.

For example, if you want a time-and-billing report, you might write or draw something like Figure 14-3. Another form of output from the program might be bills to clients.

Input Data

Once you know the output you want, you can determine the input data and the source of this data. For example, for a time-and-billing report, you can specify that one source of data to be processed should be time cards. These are usually logs or statements of hours worked submitted on paper forms. The log shown in Figure 14-4 is an example of the kind of input data used in Advantage Advertising's manual system. Note that military time is used. For example, instead of writing "5:45 P.M.," people would write "1745."

Task	Description
1	Specify objectives
2	Specify output
3	Determine required input
4	Define processing requirements
5	Document specification

FIGURE 14-1
Step 1: Program specification

FIGURE 14-2
Problem definition: make a clear statement of the problem

Client name: Allen Realty		Month and year: Jan'00		
Date	Worker	Regular Hours & Rate	Overtime Hours & Rate	Bill
1/2	M. Jones	5 @ $10	1 @ $15	$65.00
	K. Williams	4 @ $30	2 @ $45	$210.00

FIGURE 14-3
End user's sketch of desired output

Daily Log			
Worker:			
Date:			
Client	Job	Time in	Time out
A	TV commercial	800	915
B	Billboard ad	935	1200
C	Brochure	1315	1545
D	Magazine ad	1600	1745

FIGURE 14-4
Example of statement of hours worked—manual system; hours are expressed in military time

Processing Requirements

Here you define the processing tasks that must happen for input data to be processed into output. For Advantage, one of the tasks for the program will be to add the hours worked for different jobs for different clients.

Program Specifications Document

As in the systems life cycle, ongoing documentation is essential. You should record program objectives, desired outputs, needed inputs, and required processing. This leads to the next step, program design.

CONCEPT CHECK

✔ What are the five tasks of the program specification phase?

Step 2: Program Design

In the program-design step, a solution is created using programming techniques such as top-down program design, pseudocode, flowcharts, and logic structures.

After program specification, you begin **program design.** (See Figure 14-5.) Here you plan a solution, preferably using **structured programming techniques.** These techniques consist of the following: (1) top-down program design, (2) pseudocode, (3) flowcharts, and (4) logic structures.

Top-Down Program Design

First you determine the outputs and inputs of the computer program you will create. Then you can use **top-down program design** to identify the program's processing steps. Such steps are called **program modules** (or just **modules**). Each module is made up of logically related program statements.

An example of a top-down program design for a time-and-billing report is shown in Figure 14-6. Each of the boxes shown is a module. Under the rules of top-down design, each module should have a single function. The program must pass in sequence from one module to the next until all modules have been

Task	Description
1	Plan a solution using structured programming techniques
2	Document solution

FIGURE 14-5
Step 2: Program design

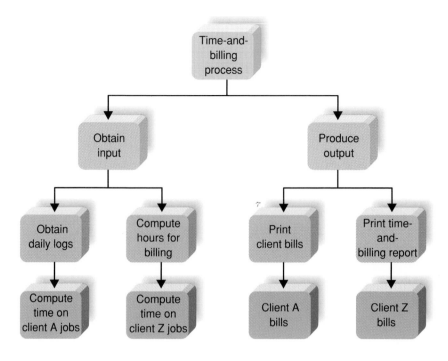

FIGURE 14-6
Example of top-down program design

processed by the computer. Three of the boxes—"obtain input," "compute hours for billing," and "produce output"—correspond to the three principal computer system operations. These operations are *input, process,* and *output.*

Pseudocode

Pseudocode (pronounced "*soo*-doh-code") is an outline of the logic of the program you will write. It is like doing a summary of the program before it is written. Figure 14-7 shows the pseudocode you might write for one module in the time-and-billing program. This shows the reasoning behind determining hours—including overtime hours—worked for different jobs for one client, Client A. Again, note this expresses the *logic* of what you want the program to do.

Flowcharts

We mentioned system flowcharts in the previous chapter. Here we are concerned with **program flowcharts.** These graphically present the detailed sequence of steps needed to solve a programming problem. Figure 14-8 presents the standard flowcharting symbols. An example of a program flowchart

Processing

Input/output

Decision

Connector

Terminal

Compute time for Client A

Set total regular hours and total overtime hours to zero.

Get time in and time out for a job.

If worked past 1700 hours, then compute overtime hours.

Compute regular hours.

Add regular hours to total regular hours.

Add overtime hours to total overtime hours.

If there are more jobs for that client, go back and compute for that job as well.

FIGURE 14-7
Example of pseudocode

FIGURE 14-8
Flowchart symbols

is presented in Figure 14-9. This flowchart expresses all the logic for just *one* module—"Compute time on Client A jobs"—in the top-down program design.

Perhaps you can see from this flowchart why a computer is a computer, and not just a fancy adding machine. A computer does more than arithmetic. It also *makes comparisons*—whether something is greater than or less than, equal to or not equal to.

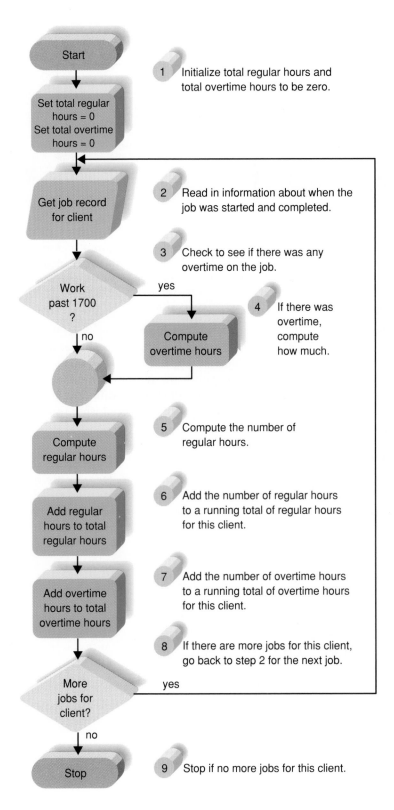

FIGURE 14-9
Flowchart example

But have we skipped something? How do we *know* which kind of twists and turns to put in a flowchart so that it will logically work? The answer is based on the use of logic structures, as we will explain.

Logic Structures

How do you link the various parts of the flowchart? The best way is a combination of three **logic structures** called *sequence, selection,* and *loop.* Using these arrangements enables you to write so-called *structured programs,* which take much of the guesswork out of programming. Let us look at the logic structures.

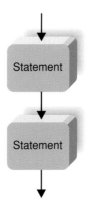

- In the **sequence structure,** one program statement follows another. (See Figure 14-10.)

  Consider, for example, the "compute time" flowchart. (Refer back to Figure 14-9.) The two "add" boxes are "Add regular hours to total regular hours" and "Add overtime hours to total overtime hours." They logically follow each other. There is no question of "yes" or "no," of a decision suggesting other consequences.

FIGURE 14-10
Sequence logic structure

- The **selection structure** occurs when a decision must be made. The outcome of the decision determines which of two paths to follow. (See Figure 14-11.) This structure is also known as an **IF-THEN-ELSE structure,** because that is how you can formulate the decision.

  Consider, for example, the selection structure in the "compute time" flowchart, which is concerned about computing overtime hours. (Refer back to Figure 14-9.) It might be expressed in detail as follows:

  IF hour finished for this job is later than or equal to 1700 hours (5:00 P.M.),

  THEN overtime hours equal the number of hours past 1700 hours,

  ELSE overtime hours equal zero.

- The **loop structure** describes a process that may be repeated as long as a certain condition remains true. The structure is called a "loop" or "iteration" because the program loops around (iterates or repeats) again and again.

  The loop structure has two variations: *DO UNTIL* and *DO WHILE.* (See Figure 14-12.) The **DO UNTIL structure** is the most used form. An example is as follows.

  DO read in job information UNTIL there are no more jobs.

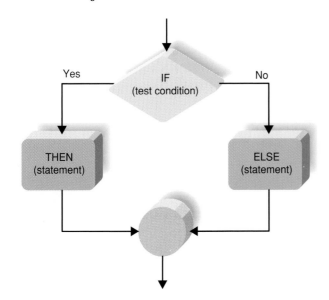

FIGURE 14-11
Selection (IF-THEN-ELSE) logic structure

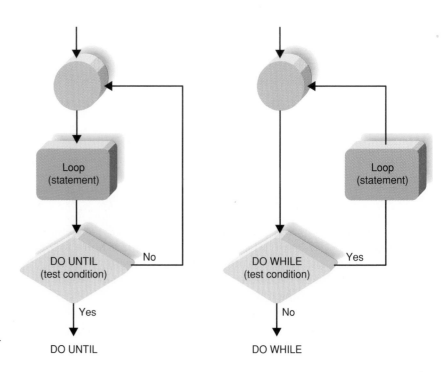

FIGURE 14-12
Loop logic structures: DO UNTIL and DO WHILE

An example of the **DO WHILE structure** is:

DO read in job information WHILE (that is, as long as) there are more jobs.

There is a difference between the two loop structures. You may have several statements that need to be repeated. If so, the decision when to *stop* repeating them can appear at the *beginning* of the loop (DO WHILE). Or, it can appear at the *end* of the loop (DO UNTIL). The DO UNTIL loop means that the loop statements will be executed at least once. This is because the loop statements are executed *before* you are asked whether to stop.

A summary of the structured programming techniques is presented in Figure 14-13.

The last thing to do before leaving the program design step is to document the logic of the design. This report typically includes pseudocode, flowcharts, and logic structures. Now you are ready for the next step, program code.

Technique	Description
Top-down design	Major processing steps, called program modules, are identified
Pseudocode	A narrative expression of the logic of the program is written
Program flowcharts	Graphic representation of the steps needed to solve the programming problem is drawn
Logic structures	Three arrangements are used in program flowcharts to write structured programs

FIGURE 14-13
Summary of structured programming techniques

CONCEPT CHECK

✔ Define the goal of the program design step.

✔ What are the techniques employed during program design?

Step 3: Program Code

"Coding" is the actual writing of the program, using a programming language.

Writing the program is called **coding.** Here you use the logic you developed in the program design step to actually write the program. (See Figure 14-14.) That is, you write out—using pencil and paper or typing on a computer keyboard—the letters, numbers, and symbols that make up the program. An example of the handwritten code for the "compute time" module is shown in Figure 14-15. This is the "program code" that instructs the computer what to do. Coding is what many people think of when they think of programming. As we've pointed out, however, it is only one of the six steps in the programming process.

The Good Program

What are the qualities of a good program? Above all, it should be reliable—that is, it should work under most conditions. It should catch obvious and common input errors. It should also be well documented and understandable by programmers other than the person who wrote it. After all, someone may need to make changes in the program in the future. The best way to code effective programs is to write so-called *structured programs,* using the logic structures described in Step 2.

Which Language?

An important decision is the selection of the programming language. There are hundreds of programming languages. The most popular for microcomputers have been C++ and Visual Basic. We describe programming languages later in this chapter. First you must determine the program's logic. Then you can write (code) it in whatever language you choose that is appropriate and available on your computer. A C++ program for the "compute time" module is illustrated in Figure 14-16. The next step is testing, or debugging, the program.

Task	Description
1	Define program logic
2	Select programming language
3	Code or write the program

FIGURE 14-14
Step 3: Program code

```
total_regular = 0;
total_overtime = 0;
while (input_file != NULL
{
    input_file >> hour_in >> minute_in >> hour_out >> minute_out;
    if (hour_out >= 17)
        overtime = (hour_out - 17) + (minute_out/60);
    else
        overtime = 0;
        regular = (hour_out - hour_in) + (minute_out - minute_in)/60) - overtime;
    total_regular = total_regular + regular;
    total_overtime = total_overtime - overtime;
}
```

FIGURE 14-15
Example of handwritten code for computing the time worked

```
#include <fstream.h>

void main (void)
{
    ifstream input_file;

    float total_regular, total_overtime, regular, overtime;
    int hour_in, minute_in, hour_out, minute_out;
    input_file.open("time.txt",ios::in);

    total_regular = 0;
    total_overtime = 0;

    while (input_file != NULL)
    {
        input_file >> hour_in >> minute_in >> hour_out >> minute_out;

        if (hour_out >= 17)
            overtime = (hour_out-17) +(minute_out/(float)60);
        else
            overtime = 0;
            regular = ((hour_out - hour_in) +(minute_out
                        - minute_in)/(float)60)    - overtime;
        total_regular += regular;
        total_overtime += overtime;
    }

    cout <<"Regular: " << total_regular <<endl;
    cout <<"Overtime " << total_overtime <<endl;
}
```

FIGURE 14-16
The "compute time"
program written in C++

CONCEPT CHECK

✔ What is accomplished during the program coding step?

✔ What makes a good program?

Step 4: Program Test

Debugging is testing a program and correcting syntax and logic errors.

Debugging is a programmer's word for testing and then *eliminating* errors ("getting the bugs out"). It means running the program on a computer and then fixing the parts that do not work. (See Figure 14-17.) Programming errors are of two types: *syntax errors* and *logic errors*.

Syntax Errors

A **syntax error** is a violation of the rules of the programming language. For example, in C++, each statement must end with a semicolon (;). If the semicolon is omitted, the program will not run due to a syntax error.

Logic Errors

A **logic error** occurs when the programmer uses an incorrect calculation or leaves out a programming procedure. For example, a payroll program that did not compute overtime hours would have a logic error.

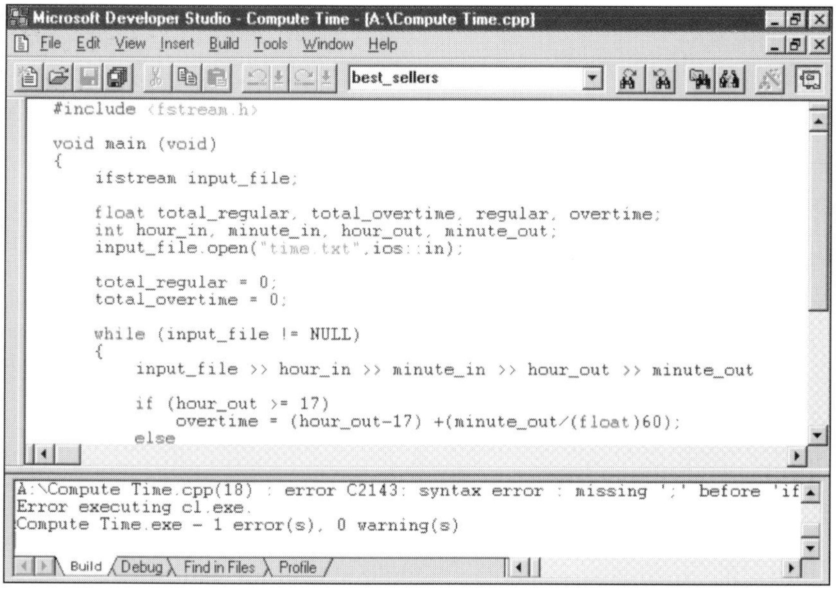

FIGURE 14-17
Syntax error identified

Testing Process

Several methods have been devised for finding and removing both types of errors:

- **Desk checking:** In **desk checking,** a programmer sitting at a desk checks (proofreads) a printout of the program. The programmer goes through the listing line by line looking for syntax and logic errors.

- **Manual testing with sample data:** Both correct and incorrect data are run through the program—manually, not with a computer—to test for correct processing results.

- **Attempt at translation:** The program is run through a computer, using a translator program. The translator attempts to translate the written program from the programming language (such as C++) into the machine language. Before the program will run, it must be free of syntax errors. Such errors will be identified by the translating program. (See Figure 14-17.)

- **Testing sample data on the computer:** After all syntax errors have been corrected, the program is tested for logic errors. Sample data is used to test the correct execution of each program statement.

- **Testing by a select group of potential users:** This is sometimes called *beta testing.* It is usually the final step in testing a program. Potential users try out the program and provide feedback.

For a summary of Step 4: Program Test, see Figure 14-18.

Task	Description
1	Desk-check for syntax and logic errors
2	Manually test with sample data
3	Translate program to identify syntax errors
4	Run program with sample data
5	Beta-test with potential users

FIGURE 14-18
Step 4: Program test

CONCEPT CHECK

✔ What is the purpose of the program testing phase?

✔ Briefly describe methods used to find and remove program errors.

Step 5: Program Documentation

"Documenting" means writing a description of the purpose and process of the program.

Task	Description
1	Review prior documentation
2	Write final documentation
3	Distribute to users, operators, and programmers

FIGURE 14-19
Step 5: Program documentation

Documentation consists of written descriptions and procedures about a program and how to use it. It is not something done just at the end of the programming process. Documentation is carried on throughout all the programming steps. This documentation is typically within the program itself and in printed documents. In this step, all the prior documentation is reviewed and finalized. Documentation is important for people who may be involved with the program in the future. (See Figure 14-19.) These people may include the following:

- **Users:** Users need to know how to use the software. Some organizations may offer training courses to guide users through the program. However, other organizations may expect users to be able to learn a package just from the written documentation. Two examples of this sort of documentation are the manuals that accompany the software and the help option within most microcomputer applications.

- **Operators:** Documentation must be provided for computer operators. If the program sends them error messages, for instance, they need to know what to do about them.

- **Programmers:** As time passes, even the creator of the original program may not remember much about it. Other programmers wishing to update and modify it—that is, perform program maintenance— may find themselves frustrated without adequate documentation. This kind of documentation should include text and program flowcharts, program listings, and sample output. It might also include system flowcharts to show how the particular program relates to other programs within an information system.

CONCEPT CHECK

✔ What is meant by *documentation?*

✔ Who is affected by documentation?

Task	Description
1	Search for and correct operational errors
2	Evaluate and improve program's ease of use
3	Standardize software
4	Evaluate changing needs

FIGURE 14-20
Step 6: Program maintenance

Step 6: Program Maintenance

Programmers update software to correct errors, improve usability, standardize, and adjust to organizational changes.

The final step is **program maintenance.** (See Figure 14-20.) As much as 75 percent of the total lifetime cost for an application program is for maintenance. This activity is so commonplace that a special job title, *maintenance programmer,* exists.

The purpose of program maintenance is to ensure that current programs are operating error free, efficiently, and effectively. Activities in this area fall into two categories: operations and changing needs.

Operations

Operations activities concern locating and correcting operational errors, making programs easier to use, and standardizing software using structured

Step	Primary Activity
1. Program specification	Determine program objectives, desired output, required input, and processing requirements
2. Program design	Use structured programming techniques
3. Program code	Select programming language; write the program
4. Program test	Perform desk check and manual checks; attempt translation; test using sample data; beta-test with potential users
5. Program documentation	Write procedure for users, operators, and programmers
6. Program maintenance	Adjust for errors, inefficient or ineffective operations, nonstandard code, and changes over time

FIGURE 14-21
Summary of six steps in programming

programming techniques. For properly designed programs these activities should be minimal.

Changing Needs

The category of changing needs is unavoidable. All organizations change over time, and their programs must change with them. Programs need to be adjusted for a variety of reasons, including new tax laws, new information needs, and new company policies.

Figure 14-21 summarizes the six steps of the programming process.

CONCEPT CHECK

✔ What is the purpose of program maintenance?

✔ Name the two categories of program maintenance.

CASE and OOP

CASE tools automate the development process. Object-oriented software development changes the approach.

You hear about efficiency and productivity everywhere. They are particularly important for software development. Two resources that promise to help are *CASE tools* and *object-oriented software development*.

CASE Tools

Professional programmers are constantly looking for ways to make their work easier, faster, and more reliable. One tool we mentioned in Chapter 13, CASE, is meeting this need. **Computer-aided software engineering (CASE) tools** provide some automation and assistance in program design, coding, and testing. (See Figure 14-22.)

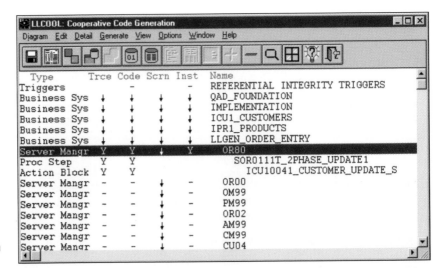

FIGURE 14-22
CASE tool: providing code-generation assistance

Object-Oriented Software Development

Traditional systems development is a careful, step-by-step approach focusing on the procedures needed to complete a certain objective. **Object-oriented software development** focuses less on the procedures and more on defining the relationships between previously defined procedures or objects. **Object-oriented programming (OOP)** is a process by which a program is organized into objects. Each object contains both the data and processing operations necessary to perform a task. Let's explain what this means.

In the past, programs were developed as giant entities, from the first line of code to the last. This has been compared to building a car from scratch. Object-oriented programming is like building a car from prefabricated parts—carburetor, generator, fenders, and so on. Object-oriented programs use *modules* called objects. Objects are reusable, self-contained components. Programs built with these objects assume that certain functions are the same. For example, many programs, from spreadsheets to database managers, have an instruction that will sort lists of names in alphabetical order. A programmer might use this module or object for alphabetizing in many other programs. There is no need to invent this activity anew every time. C++ is one of the most widely used object-oriented programming languages.

Some expect that object-oriented software development will improve productivity and efficiency tenfold. However, this will depend on how many available objects are clearly defined, programmed, documented, and saved in libraries for later use.

CONCEPT CHECK

✔ What are CASE tools?

✔ What is meant by object-oriented programming?

Generations of Programming Languages

Languages are described as occurring in "generations," from machine languages to natural languages.

Computer professionals talk about **levels** or **generations of programming languages,** ranging from "low" to "high." Programming languages are called *lower level* when they are closer to the language the computer itself uses. The computer understands the 0s and 1s that make up bits and bytes. Programming languages are called *higher level* when they are closer to the language humans use—that is, for English speakers, more like English.

There are five generations of programming languages. These are (1) machine languages, (2) assembly languages, (3) procedural languages, (4) problem-oriented languages, and (5) natural languages.

Machine Languages: The First Generation

We mentioned in Chapter 4 that a byte is made up of *bits,* consisting of 1s and 0s. These 1s and 0s may correspond to electricity's being on or off in the computer. They may also correspond to a magnetic charge's being present or absent on storage media such as disk or tape. From this two-state system have been built coding schemes that allow us to construct letters, numbers, punctuation marks, and other special characters. Examples of these coding schemes, as we saw, are ASCII and EBCDIC.

Data represented in 1s and 0s is said to be written in **machine language.** To see how hard this is to understand, imagine if you had to code this:

> 1111001001110011110100100001000001110000000101011

Machine languages also vary according to make of computer—another characteristic that makes them hard to work with.

Assembly Languages: The Second Generation

Assembly languages have a clear advantage over the 1s and 0s of machine language because they use abbreviations or mnemonics. These are easier for human beings to remember. The machine language code we gave earlier could be expressed in assembly language as

> **ADD 210(8,13),02B(4,7)**

This is still pretty obscure, of course, and so assembly language is also considered low-level.

Assembly languages also vary from computer to computer. With the third generation, we advance to high-level languages, many of which are considered **portable languages.** That is, they can be run on more than one kind of computer—they are "portable" from one machine to another.

High-Level Procedural Languages: The Third Generation

People are able to understand languages that are more like their own (e.g., English) than machine languages or assembly languages. These more English-like programming languages are called "high-level" languages. However, most people still require some training in order to use higher-level languages. This is particularly true of procedural languages.

Procedural languages are programming languages with names like BASIC, Pascal, C, COBOL, and FORTRAN. They are called procedural because they are designed to express the logic—the procedures—that can solve general problems. Procedural languages, then, are intended to solve *general* problems. COBOL, for instance, is used in all kinds of business applications, such as payroll and inventory control. It is fourth-generation languages, discussed next, that are intended to solve *specific* problems.

For a procedural language to work on a computer, it must be translated into machine language so that the computer understands it. Depending on the language, this translation is performed by either a *compiler* or an *interpreter.*

- A **compiler** converts the programmer's procedural language program, called the *source code,* into a machine language code, called the *object code.* This object code can then be saved and run later. Examples of procedural languages using compilers are the standard versions of Pascal, COBOL, and FORTRAN.

- An **interpreter** converts the procedural language one statement at a time into machine code just before it is to be executed. No object code is saved. An example of a procedural language using an interpreter is the standard version of BASIC.

What is the difference between using a compiler and using an interpreter? When a program is run, the compiler requires two steps. The first step is to convert the entire program's source code to object code. The second step is to run the object code. The interpreter, in contrast, converts and runs the program one line at a time. The advantage of a compiler language is that once the object code has been obtained, the program executes faster. The advantage of an interpreter language is that programs are easier to develop.

The principal procedural languages with which you may come in contact are as follows:

- **Ada:** **Ada** is named after Augusta Ada, the English Countess of Lovelace, who is regarded as the first programmer. Ada was developed under the sponsorship of the U.S. Department of Defense. Originally designed for weapons systems, it has commercial uses as well. Because of its structured design, modules (sections) of a large program can be written, compiled, and tested separately—before the entire program is put together.

- **BASIC:** Short for *B*eginner's *A*ll-purpose *S*ymbolic *I*nstruction *C*ode, **BASIC** is a popular microcomputer language. Widely used on microcomputers and easy to learn, it is suited to both beginning and experienced programmers. It is also interactive—you and the computer communicate with each other directly during the writing and running of programs.

  Another version created by the Microsoft Corporation is **Visual BASIC,** which has been hailed as a programming breakthrough. Visual BASIC makes it easier for novice programmers, as well as professionals, to develop customized applications for Windows. It has become quite popular for corporate, in-house development.

- **C/C++:** **C** is a general-purpose language that also works well with microcomputers. It is useful for writing operating systems, spreadsheet programs, database programs, and some scientific applications. Programs are portable: They can be run without change on a variety of computers. **C++** is a version of C that incorporates object-oriented technologies. It is widely used and has effectively increased programmer productivity.

- **COBOL:** COBOL—which stands for *CO*mmon *B*usiness-*O*riented *L*anguage—was one of the most frequently used programming languages in business. Though harder to learn than BASIC, its logic is easier for a person who is not a trained programmer to understand. Writing a COBOL program is sort of like writing the outline for a term paper. The program is divided into four divisions. The divisions in turn are divided into sections, which are divided into paragraphs, then into statements.

- **FORTRAN:** Short for *FOR*mula *TRAN*slation, **FORTRAN** is a widely used scientific and mathematical language. It is very useful for processing complex formulas. Thus, many scientific and engineering programs have been written in this language.

- **HTML:** Short for *Hyper*Text *M*arkup *L*anguage, **HTML** is not strictly a programming language like BASIC or C++. It consists of statements or tags that are saved in document files. Browsers interpret these HTML documents to display Web pages and provide links to other Web pages and to related audio, video, and graphic files.

- **Java: Java** is one of the newest and most exciting programming languages. As we noted in Chapter 8, Java programs called **applets** are widely used on the Internet to add animation and interest to Web pages. Unlike the programs written with other languages, a program written with Java can run on any system: Windows, Mac, or Unix. Recently Java has been used to develop general-application software. Some experts predict that Java will become the most widely used programming language within the next few years.

- **Pascal:** Another language that is widely used on microcomputers and easy to learn is **Pascal.** It is named after Blaise Pascal, a 17th-century French mathematician. This language has become quite popular in computer science educational programs. One advantage of Pascal is that it encourages programmers to follow structured coding procedures. It also works well for graphics.

Problem-Oriented Languages: The Fourth Generation

Third-generation languages are valuable, but they require training in programming. Problem-oriented languages, also known as *very high level languages,* require little special training on the part of the user.

Unlike general-purpose languages, **problem-oriented languages** are designed to solve specific problems. Some of these fourth-generation languages are used for very specific applications. For example, IFPS (interactive financial planning system) is used to develop financial models. Many consider Lotus 1-2-3 and dBASE to be flexible fourth-generation languages. This group also includes query languages and application generators:

- **Query languages: Query languages** enable nonprogrammers to use certain easily understood commands to search and generate reports from a database. An example is the commands used on an airline reservations system by clerks needing flight information.

- **Application generators:** An **application generator** contains a number of modules—logically related program statements—that have been preprogrammed to accomplish various tasks. An example would be a module that calculates overtime pay. The programmer can simply state which task is needed for a particular application. The application generator creates the program code by selecting the appropriate modules.

Generation	Sample Statement
First: Machine	10010001
Second: Assembly	ADD 210(8, 13),02B(4, 7)
Third: Procedural	Overtime: = 0
Fourth: Problem	FIND NAME = "JONES"
Fifth: Natural	IF patient is dizzy, THEN check temperature and blood pressure

FIGURE 14-23
Summary of five programming generations

Natural Languages: The Fifth Generation

Natural languages are still being developed. They are designed to give people a more human ("natural") connection with computers. The languages are human languages: English, French, Japanese, or whatever. Researchers also hope that natural languages will enable a computer to *learn*—to "remember" information, as people do, and to improve upon it. Clearly, this area is extremely challenging.

The five generations of programming languages are summarized in Figure 14-23.

CONCEPT CHECK

✔ Outline the five generations of programming languages.

✔ What distinguishes "lower level" from "higher level" programming languages?

A LOOK TO THE FUTURE

The Year 2000 Problem could cost billions to fix and trillions in legal fees for the year 2002!

The second millennium has just started. Many looked at the year 2000 as the beginning of exciting new and challenging opportunities. In programming and information systems circles, it represented a significant challenge. In fact, they have given this challenge a name—The Year 2000 Problem or Y2K.

Some thought that the year 2000 would begin with chaos. They predicted that millions of miscalculations on finance, scheduling, insurance, and accounting applications would be commonplace. Your credit cards, school loans, and other financial aid would suddenly be canceled. Social security checks would be stopped, federal assistance programs discontinued, and millions of applications programs fail to execute.

The source of this potential problem began 20 years ago when programmers first standardized how computers would handle dates. Only two digits were typically used to represent years. For example, the year 1980 was recorded simply as 80. Since that time thousands of databases and millions of programs have been created using that standard. Many of them are still widely used today, and they work fine as long as the year's first two digits are 19.

Does this sound like a difficult problem to fix? The experts thought so. One well-respected industry source estimated that the costs to avoid this problem would reach $600 billion. The American Bar Association predicted legal costs of $2 trillion for related lawsuits. The federal government, state governments, and every major corporation had teams of programmers reviewing program code, assessing potential problems, and reprogramming.

The year 2000 did not begin with chaos. Some experts, however, warn that we are not out of the woods yet. They point out that Y2K problems may surface in the future when seldom used programs are run for the first time in this millennium.

VISUAL SUMMARY Programming and Languages

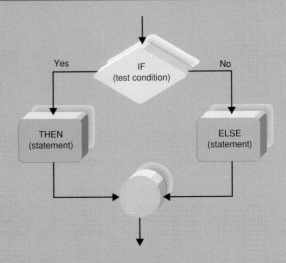

STEP 1: PROGRAM SPECIFICATION

Program specification, also called **program definition** or **program analysis,** consists of specifying five tasks related to objectives, outputs, inputs, requirements, and documentation.

Program Objectives
The first task is to clearly define the problem to solve in the form of program objectives.

Desired Output
Next, focus on the desired output before considering the required inputs.

Input Data
Once outputs are defined, determine the necessary input data and the source of the data.

Processing Requirements
Next, determine the steps necessary (processing requirements) to use input to produce output.

Program Specifications Document
Finally, document this step's program objectives, outputs, inputs, and processing requirements.

STEP 2 PROGRAM DESIGN

In **program design,** a solution is designed using, preferably, **structured programming techniques,** including the following.

Top-Down Design
In **top-down design,** major processing steps, called **program modules,** are identified.

Pseudocode
Pseudocode is an outline of the logic of the program you will write.

Flowcharts
Program flowcharts are graphic representations of the steps necessary to solve a programming problem.

Logic Structures
Logic structures are arrangements of programming statements. Three types are:
- **Sequence**—one program statement followed by another.
- **Selection** (or **IF-THEN-ELSE**)—when a decision must be made.
- **Loop**—when process is repeated as long as a condition is true.

To be a competent end user, you need to understand the relationship between the systems life cycle's Phase 4: Systems Development and programming. Additionally, you need to know the six steps of programming including program specification, program design, program coding, program test, program documentation, and program maintenance.

STEP 3: PROGRAM CODE

```
total_regular = 0;
total_overtime = 0;
while (input_file != NULL
{
    input_file >> hour_in >> minute_in >> hour_out >> minute_out;
    if (hour_out >= 17)
        overtime = (hour_out - 17) + (minute_out/60);
    else
        overtime = 0;
        regular = (hour_out - hour_in) + (minute_out - minute_in)/60) - overtime;
    total_regular = total_regular + regular;
    total_overtime = total_overtime - overtime;
}
```

Coding is writing a program. There are several important aspects of writing a program. Two are writing good programs and selecting which language to use.

The Good Program
Good programs are designed using the three basic logic structures. Programs coded in this manner are called *structured programs.*

Which Language
With hundreds of programming languages available, the most widely used for microcomputers include C++ and Visual Basic.

STEP 4: PROGRAM TEST

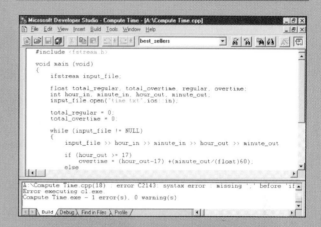

Debugging is a process of testing and eliminating errors in a program. *Syntax* and *logic* are two types of programming errors.

Syntax Errors
Syntax errors are violations in the rules of a programming language.

Logic Errors
Logic errors are incorrect calculations or procedures.

Testing Process
Five methods for testing for syntax and logic errors:
- **Desk checking**—careful reading of a printout of the program.
- **Manual testing**—using sample data to test for correct results.
- **Attempt at translation**—running program using a translator program.
- **Testing sample data**—testing the program for logic errors using sample data.
- **Testing by users**—*beta testing* by selected users is the final step.

STEP 5: PROGRAM DOCUMENTATION

Task	Description
1	Review prior documentation
2	Write final documentation
3	Distribute to users, operators, and programmers

Documentation consists of a written description of the program and the procedures for running it. People who use documentation include:

- Users, who need to know how to use the program.
- Operators, who need to know how to execute the program and how to recognize and correct errors.
- Programmers, who may need to update and maintain the program in the future.

STEP 6: PROGRAM MAINTENANCE

Task	Description
1	Search for and correct operational errors
2	Evaluate and improve program's ease of use
3	Standardize software
4	Evaluate changing needs

Maintenance is designed to ensure that the program operates correctly, efficiently, and effectively. Two categories of activities are the following.

Operations

Operations activities include locating and correcting errors, improving usability, and standardizing software.

Changing Needs

Organizations change over time and their programs must change with them.

CASE AND OOP

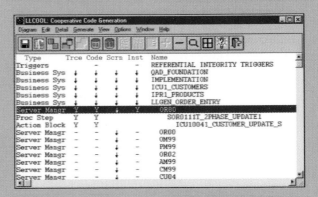

Computer-aided software engineering (CASE) provides automation and assistance in program design, coding, and testing. **Object-oriented software** development focuses on defining relationships between previously defined procedures or objects.

PROGRAMMING GENERATIONS

Generation	Sample Statement
First: Machine	10010001
Second: Assembly	ADD 210(8, 13),02B(4, 7)
Third: Procedural	Overtime: = 0
Fourth: Problem	FIND NAME = "JONES"
Fifth: Natural	IF patient is dizzy, THEN check temperature and blood pressure

Programming languages have **levels** or **generations** ranging from "low" to "high." The higher the language the closer it is to the language of humans.

Key Terms

Ada (350)
applets (351)
application generator (351)
assembly language (349)
BASIC (350)
C (350)
C++ (350)
COBOL (351)
coding (343)
compiler (350)
computer-aided software engineering (CASE) tool (347)
debugging (344)
desk checking (345)
documentation (346)
DO UNTIL structure (341)
DO WHILE structure (342)
FORTRAN (351)
generations of programming
languages (349)

HTML (351)
IF-THEN-ELSE structure (341)
interpreter (350)
Java (351)
levels of programming
languages (349)
logic error (344)
logic structure (341)
loop structure (341)
machine language (349)
module (338)
natural languages (352)
object-oriented programming
(OOP) (348)
object-oriented software
development (348)
Pascal (351)
portable language (349)
problem-oriented language
(351)

procedural language (350)
program (326)
program analysis (327)
program definition (327)
program design (328)
program flowchart (329)
program maintenance (346)
programming (326)
program module (328)
program specification (327)
pseudocode (329)
query language (351)
selection structure (341)
sequence structure (341)
software development (336)
structured programming
techniques (338)
syntax error (344)
top-down program design (338)
Visual BASIC (350)

Chapter Review

LEVEL 1

Reviewing Facts and Terms
Matching

Match each numbered item with the most closely related lettered item. Write your answers in the spaces provided.

1. Converts the procedural language, one statement at a time, into machine code just before it is to be executed. ____

2. Process of testing then eliminating errors in a program. ____

3. Using these arrangements allows you to write structured programs. ____

4. In this step objectives, outputs, inputs, and processing requirements are determined. ____

5. A violation of the rules of the programming language. ____

6. Data represented in 1s and 0s. ____

7. Enable nonprogrammers to use certain easily understood commands to search and generate reports from a database. ____

8. Program design techniques that consist of top-down program design, pseudocode, flowcharts, and logic structures. ____

9. Widely used on the Internet to add animation and interest to Web pages. ____

a. Applets
b. Assembly languages
c. BASIC
d. Coding
e. Compiler
f. Debugging
g. Interpreter
h. Logic error
i. Logic structures
j. Loop
k. Machine language
l. Natural languages
m. Procedural languages
n. Program specification
o. Program design
p. Pseudocode
q. Query language
r. Sequence structure
s. Structured programming techniques
t. Syntax error

10. A logic structure where one program statement follows another. ____

11. Converts the programmer's procedural language program into machine language object code. ____

12. In this step you plan a solution. ____

13. Beginner's All-purpose Symbolic Instruction Code. ____

14. An outline of the logic for the program you will write. ____

15. Occurs when the programmer uses an incorrect calculation or leaves out a programming procedure. ____

16. The actual process of writing a program. ____

17. Languages that are designed to express the logic that can solve general problems. ____

18. Languages designed to give people a more human connection with computers. ____

19. Programming languages that use abbreviations and mnemonics. ____

20. A process that is repeated as long as a condition is true. ____

True/False

In the spaces provided, write T or F to indicate whether the statement is true or false.

1. A program is a list of instructions for the computer to follow to accomplish the task of processing data into information. ____

2. All documentation is performed in the last step of programming. ____

3. CASE stands for Computer Association of Scientific Engineers. ____

4. A translator converts a programming language into machine language. ____

5. Assembly languages represent the first generation of programming languages. ____

Multiple Choice

Circle the letter of the correct answer.

1. Another name for application software is:
  a. application programs
  b. system software
  c. programming languages
  d. package software
  e. custom-made programs

2. The structured programming technique that graphically presents the detailed steps needed to solve the problem:
  a. top-down design
  b. pseudocode
  c. flowcharts
  d. logic structures
  e. object-oriented programming

3. The last thing to do before leaving the program design step:
  a. determine outputs
  b. document
  c. test
  d. select programming language
  e. code

4. The last step in programming:
   a. design
   b. analysis
   c. test
   d. maintenance
   e. coding

5. Translator that converts procedural languages one statement at a time into machine code before it is executed:
   a. BASIC
   b. interpreter
   c. Pascal
   d. query language
   e. computer

Completion

Complete each statement in the spaces provided.

1. In program specification, you should specify _____ before inputs.

2. The logic structure _____ is used when a decision must be made.

3. _____ is another name for testing the program.

4. Unlike traditional software development, object-oriented development focuses on defining the _____ between procedures or objects.

5. _____ is the most frequently used language in business programming.

Reviewing Concepts

Open-Ended

On a separate sheet of paper, respond to each question and statement.

1. Describe the six steps involved in programming.

2. Identify and give an example for each of the three logic structures.

3. List and discuss the five generations of programming languages.

4. Explain the difference between a computer and an interpreter.

5. Discuss the principal advantage(s) of natural languages, if researchers are successful in developing them.

Concept Mapping

On a separate sheet of paper, draw a concept map or a flowchart showing how the following terms are related. Show all relationships. Include any additional terms you can think of.

Ada	compiler	HTML
applets	debugging	IF-THEN-ELSE structure
assembly language	desk checking	interpreter
BASIC	DO UNTIL structure	Java
C	DO WHILE structure	levels of programming
C++	documentation	languages
CASE tool	FORTRAN	logic error
COBOL	generations of program-	loop structure
coding	ming language	machine language

module
natural languages
object-oriented program-
  ming
object-oriented software
  development
Pascal
portable language
problem-oriented language
procedural language

program
program analysis
program definition
program design
program flowchart
program maintenance
programming
program module
program specification
pseudocode

query language
selection structure
sequence structure
software development
structured programming
  techniques
syntax error
top-down program design
Visual BASIC

Critical Thinking Questions and Projects

LEVEL 3

Read each exercise and answer the related questions on a separate sheet of paper.

1. *Pencil-and-paper programming:* Suppose you are the manager of a clothing store. Using just pencil and paper, see if you can devise the steps in a program that will do the following:

   a. Pay a sales bonus of 1 percent of total sales to salespeople who sell $50,000 worth of clothes.

   b. Pay a sales bonus of 5 percent of total sales to salespeople who sell $250,000 worth of clothes.

   c. Pay a sales bonus of 10 percent of total sales to salespeople who sell $500,000 worth of clothes.

2. *Programming the beginning of your day:* Flowcharts can be used to graphically present the sequence of steps involved in almost any activity. Different shapes and symbols are used to represent different kinds of activities: processing, input and output of data, decision making, and starting and stopping. The lines and arrows connecting the shapes show the sequence of steps used to complete the activity. Create a flowchart that describes what you do each morning to get ready for the day.

3. *Ethics:* Most people will agree that it is wrong to copy someone else's work. It is generally recognized as unethical as well as illegal to make and distribute unauthorized copies of programs. But what about copying the logic used in a program? Let's say you are going to design a new Web page using HTML. One of the most common ways to start is to explore a variety of Web sites, select the best ones, and examine their HTML codes. Do you think this approach is unethical? Why or why not? What if you copied just part of the code? What if you copied all the code but made some changes? If you were going to create your own spreadsheet application program and started by examining the code used for Excel, would your earlier answers apply? Why or why not?

On the Web Exercises

1. Java

Java has been touted as the next great programming language, but it won't get far without real-world business applications. Find out what Java can do for you in the busi-  ness world by visiting our Web site at http://www.mhhe. com/it/oleary/exercise.mhtml to link to a site specializing in Java. Once connected to that site, explore the site and print out the most informative Web page. Write a paragraph discussing the advantages and disadvantages of Java for business applications.

2. Active X

Microsoft developed Active X as a response to Java. Microsoft has incorporated Active X into its Windows DNA product. To learn more about  Windows DNA, visit our Web site at http://www.mhhe.com/ it/oleary/exercise.mhtml to link to a site specializing in Windows DNA. Once connected to that site, print out the most informative Web page and write a paragraph describing Windows DNA.

3. The Year 2000 Problem

The year 2000 may have created a problem for years to come. Many programs were designed only for dates with years in the 1900s. To learn more about this potential problem,  connect to the HotBot search tool at http://news.hotbot. com/index.html and use the keywords "year 2000 problem." Print the first page of results, and write a paragraph clearly defining the problem and describing the programs that have been (or will be) affected and how.

4. Halting Problem

Programs can be written to solve many of today's real-world business problems. But did you know that some problems cannot be solved by any computer? To learn about a classic  example of a problem that is "fundamentally unsolvable", connect to the online search engine Ask.com, and ask for the definition of the *halting problem*. Print out the page that describes this problem. Using your own words, write a paragraph describing the problem and summarizing why it cannot be solved by a computer.

Your Future and Information Technology

COMPETENCIES

After you have read this chapter, you should be able to:

1. **Explain why it's important to have an individual strategy in order to be a "winner" in the information age.**

2. **Describe how technology is changing the nature of competition.**

3. **Discuss three ways people may react to new technology.**

4. **Describe how you can use your computer competence to stay current and to take charge of your career.**

5. **Discuss what computer trainers, database administrators, network managers, programmers, systems analysts, technical writers, and Webmasters do.**

Throughout this book, we have emphasized practical subjects that are useful to you now or will be very soon. Accordingly, this final chapter is not about the far future of, say, 10 years from now. Rather, it is about today and the near future—about developments whose outlines we can already see. It is about how organizations adapt to technological change. It is also about what you as an individual can do to keep your computer competency up to date.

Are the times changing any faster now than they ever have? It's hard to say. People who were alive when radios, cars, and airplanes were being introduced certainly lived through

Technology and Organizations

Technology and People

some dramatic changes. Has technology made our own times even more dynamic? Whatever the answer, it is clear we live in a fast-paced age. The challenge for you as an individual is to devise ways to stay current and to use technology to your advantage. For example, you can use the Web to locate job opportunities.

To stay competent, end users need to recognize the impact of technological change on organizations and people. They need to know how to use change to their advantage and to be winners. Although end users do not need to be specialists in information technology, they should be aware of career opportunities in the area.

Careers in Information Systems

Title	Description
Systems analyst	Analyzes, designs, and implements information systems
Webmaster	Designs, creates, monitors, and evaluates corporate Web sites
Database administrator	Structures, coordinates, links, and maintains databases
Programmer	Revises existing software and creates new software
Network manager	Monitors existing networks and implements new networks
Technical writer	Creates user manuals and documentation for information systems
Computer trainer	Provides classes and support for computer users

Be a Winner

Changing Times

To be a winner in the information revolution, you need an *individual* strategy.

Most businesses have become aware that they must adapt to changing technology or be left behind. Many organizations are now making formal plans to keep track of technology and implement it in their competitive strategies. For example, banks have found that automated teller machines (ATMs) are vital to retail banking. (See Figure 15-1.) Not only do they require fewer human tellers, but they can also be made available 24 hours a day. More and more banks are also trying to go electronic, doing away with paper transactions wherever possible. Thus, ATM cards can now be used in many places to buy gas or groceries. Many banks are also trying to popularize home banking, so that customers can use microcomputers for certain financial tasks. In addition, banks are exploring the use of some very sophisticated application programs. These programs will accept cursive writing (the handwriting on checks) directly as input, verify check signatures, and process the check without human intervention.

Clearly, such changes do away with some jobs—those of many bank tellers and cashiers, for example. However, they create opportunities for other people. New technology requires people who are truly capable of working with it. These are not the people who think every piece of equipment is so simple they can just turn it on and use it. Nor are they those who think each new machine is a potential disaster. In other words, new technology needs people who are not afraid to learn it and are able to manage it. The real issue, then, is not how to make technology better. Rather, it is how to integrate the technology with people.

FIGURE 15-1
Automated teller machines are examples of technology used in business strategy

You are in a very favorable position compared with many other people in industry today. After reading the previous chapters, you have learned more than just the basics of hardware, software, connectivity, and the Internet. You have learned the most *current* technology. You are therefore able to use these tools to your advantage—to be a winner.

How do you become and stay a winner? In brief, the answer is: You must form your own individual strategy for dealing with change. First let us look at how businesses are handling technological change. Then let's look at how people are reacting to these changes. Finally, we will offer a few suggestions that will enable you to keep up with—and profit from—the information revolution.

CONCEPT CHECK

✔ Cite examples of ways computers are changing the business world.

Technology and Organizations

Technology changes the nature of competition by introducing new products, new enterprises, and new relationships among customers and suppliers.

Technology can introduce new ways for businesses to compete with each other. Some of the principal changes are as follows.

New Products

Technology creates products that operate faster, are priced cheaper, are often of better quality, or are wholly new. Indeed, new products can be individually tailored to a particular customer's needs. For example, financial services company Merrill Lynch took advantage of technology to launch a cash management account. This account combines information on a person's checking, savings, credit card, and securities accounts into a single monthly statement. It automatically sets aside "idle" funds into interest-bearing money market funds. Customers can access their accounts on the Web and get a complete picture of their financial condition at any time. However, even if they don't pay much attention to their statements, their surplus funds are invested automatically.

New Enterprises

Information technology can build entire new businesses. Two examples are Internet service providers and Web site development companies.

- Just a few years ago, the only computer connectivity options available to individuals were through online service providers like America Online and through colleges and universities. Now, hundreds of national service providers and thousands of local service providers are available.

- Thousands of small companies specializing in Web site development have sprung up in the past three years. These companies help small- to medium-sized organizations by providing assistance in evaluating, creating, and maintaining Web sites. Large organizations employ specialists called *Webmasters* to provide this type of support.

New Customer and Supplier Relationships

Businesses that make their information systems easily available may make their customers less likely to take their business elsewhere. For instance, Federal Express, the overnight package delivery service, does everything possible to make its customers dependent on it. Upon request, customers receive airbills with their name, address, and account number preprinted on them, making shipping and billing easier. Package numbers are scanned into the company's information system, so that they can be tracked from pickup point to destination. (See Figure 15-2.) Thus, apprehensive customers can be informed very quickly of the exact location of their package as it travels toward its destination.

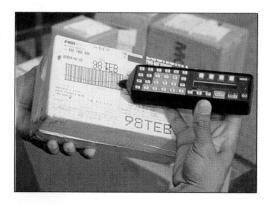

FIGURE 15-2

Federal Express couriers scan bar codes on every package, transferring customer and delivery data to a worldwide network that can be closely monitored by customer service agents

CONCEPT CHECK

✔ How does technology change the nature of competition among businesses?

Technology and People

People may be cynical, naïve, or frustrated in response to technology.

Clearly, recent technological changes, and those sure to come in the near future, will produce some upheavals in the years ahead. How should we be prepared for them?

People have different coping styles when it comes to technology. It has been suggested, for instance, that people react to the notion of microcomputers in business in three ways. These ways are *cynicism, naïveté,* and *frustration.*

Cynicism

The cynic feels that, for a manager at least, the idea of using a microcomputer is overrated. (See Figure 15-3.) Learning and using it take too much time, time that could be delegated to someone else. Doing spreadsheets and word processing, according to the cynic, are tasks that managers should understand. However, the cynic feels that such tasks take time away from a manager's real job of developing plans and setting goals for the people being supervised.

Cynics may express their doubts openly, especially if they are top managers. Or they may only pretend to be interested in microcomputers, when actually they are not interested at all.

Naïveté

Many naïve people are unfamiliar with computers. Thus, they may think computers are magic boxes capable of solving all kinds of problems that computers really can't handle. (See Figure 15-4.) In contrast, some naïve persons are actually quite familiar with computers. However, such people underestimate the difficulty of changing computer systems or of generating information.

Frustration

The frustrated person may already be quite busy and may hate having to take time to learn about microcomputers. Such a person feels it is an imposition to have to learn something new. Often she or he is too impatient to try to understand the manuals explaining what hardware and software are supposed to do. The result, therefore, is continual frustration. (See Figure 15-5.) Some people are frustrated because they try to do too much. Or they're

FIGURE 15-3
The cynic: "These gadgets are overrated."

FIGURE 15-4
The naïve: "Let the computer make the decision."

FIGURE 15-5
The frustrated: "This stuff doesn't make sense half the time."

frustrated because they find manuals difficult to understand. Oftentimes they feel stupid when actually the manuals are at fault.

Cynicism, naïveté, and frustration apply to all new technology. Do you see yourself reacting in any of these ways? They are actually commonplace responses—part of just being human. Knowing which, if any, of these reactions characterize you or those around you may be helpful.

CONCEPT CHECK

✔ Describe three ways people cope with technological changes in the workplace.

How You Can Be a Winner

Individuals need to stay current, develop specialties, and be alert to organizational changes and opportunities for innovation.

So far we have described how progressive organizations are using technology in the information age. Now let's concentrate on you as an individual. (See Making IT Work for You: Online Job Opportunities on pages 368–69.) How can you stay ahead? Here are some ideas.

Stay Current

Whatever their particular line of work, successful professionals keep up both with their own fields and with the times. We don't mean you should try to become a computer expert and read a lot of technical magazines. Rather, you should concentrate on your profession and learn how computer technology is being used within it.

Every field has trade journals, whether the field is interior design, personnel management, advertising, or whatever. Most such journals regularly present articles about the uses of computers. It's important that you also belong to a trade or industry association and go to its meetings. Many associations sponsor seminars and conferences that describe the latest information and techniques. Another way to stay current is by participating electronically with special-interest newsgroups on the Internet.

MAKING **IT** WORK FOR YOU

Locating Job Opportunities Online

Did you know that you could use the Internet to find a job? You can browse through job listings, post resumes for prospective employers, and even use special agents to continually search for that job that's just right for you.

Individual

Job Web site

Organization

Resumes

Jobs

How It Works

There are several Web sites designed to bring together prospective employers and employees. These sites maintain a database of available jobs and a database of resumes. Individuals are able to post resumes and to search through available jobs. Organizations are able to post job opportunities and to search through individual resumes.

Browsing Jobs Listings

Three well-known job search sites on the Web are hotjobs.com, yahoo!jobs, and monster.com. You can connect to these sites and browse through job opportunities.

For example, after connecting to monster.com, you can search for a job by following the steps similar to those shown below.

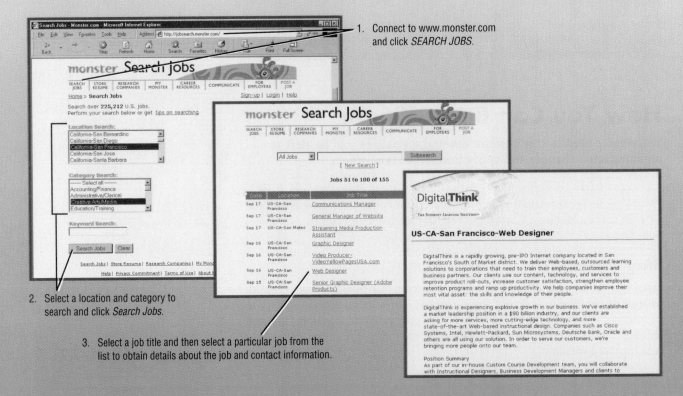

1. Connect to www.monster.com and click *SEARCH JOBS*.

2. Select a location and category to search and click *Search Jobs*.

3. Select a job title and then select a particular job from the list to obtain details about the job and contact information.

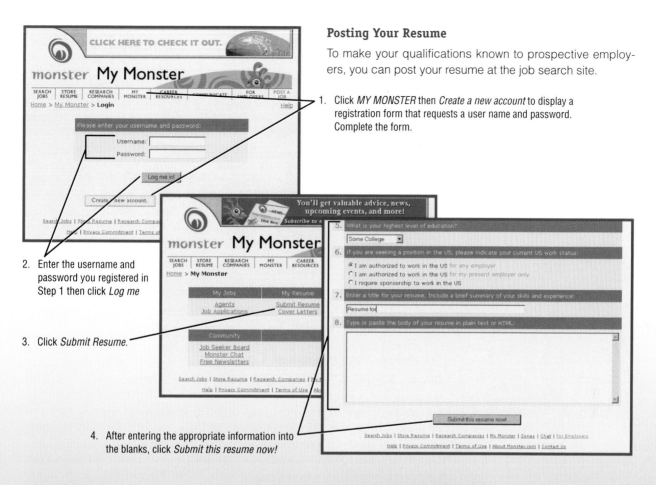

Posting Your Resume

To make your qualifications known to prospective employers, you can post your resume at the job search site.

1. Click *MY MONSTER* then *Create a new account* to display a registration form that requests a user name and password. Complete the form.

2. Enter the username and password you registered in Step 1 then click *Log me*

3. Click *Submit Resume.*

4. After entering the appropriate information into the blanks, click *Submit this resume now!*

Creating an Agent

You can automate the job search process by creating an agent. The agent monitors new job listings and notifies you by email whenever a new job fits your criteria.

1. Click *Agents.*

2. Enter keywords that describe the job you want and select the location where you want to work.

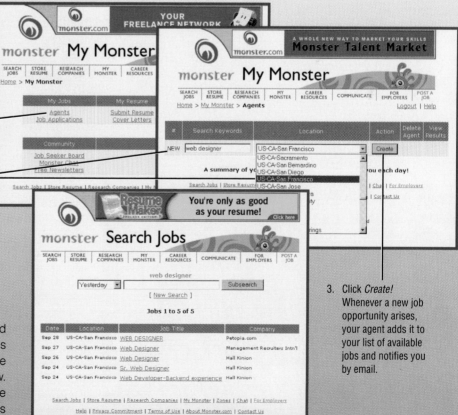

3. Click *Create!* Whenever a new job opportunity arises, your agent adds it to your list of available jobs and notifies you by email.

The Web is continually changing and some of the specifics presented in this Making IT Work for You may have changed. See our Web site at http://www.mhhe.com/it/oleary/IT.mhtml for possible changes and to learn more about this application of technology.

FIGURE 15-6
PC World on the Web

Maintain Your Computer Competence

Actually, you should try to stay *ahead* of the technology. Books, journals, and trade associations are the best sources of information about new technology that applies to your field. The general business press—*Business Week, Fortune, Inc., The Wall Street Journal,* and the business section of your local newspaper—also carries computer-related articles.

However, if you wish, you can subscribe to a magazine that covers microcomputers and information more specifically. Examples are *InfoWorld, PC World,* and *MacWorld.* You may also find it useful to look at newspapers and magazines that cover the computer industry as a whole. An example of such a periodical is *ComputerWorld.* Most of these magazines also have an online version available on the Web. (See Figure 15-6.)

Develop Professional Contacts

Besides being members of professional associations, successful people make it a point to maintain contact with others in their field. They stay in touch by telephone, e-mail, and newsgroups and go to lunch with others in their line of work. Doing this lets them learn what other people are doing in their jobs. It tells them what other firms are doing and what tasks are being automated. Developing professional contacts can keep you abreast not only of new information but also of new job possibilities. (See Figure 15-7.) It also offers social benefits. An example of a professional organization found in many areas is the local association of realtors.

Develop Specialties

Develop specific as well as general skills. You want to be well-rounded within your field, but certainly not a "jack of all trades, master of none." Master a trade or two *within* your profession. At the same time, don't become identified with a specific technological skill that might very well become obsolete.

The best advice is to specialize to some extent. However, don't make your specialty so tied to technology that you'll be in trouble if the technology shifts. For example, if your career is in marketing or graphics design, it makes sense to learn about desktop publishing and Web page design. (See Figure 15-8.) That way you can learn to make high-quality, inexpensive graphics layouts. It would not make as much sense for you to become an expert on, say, the various types of monitors used to display the graphics layouts, because such monitors are continually changing.

Expect to take classes during your working life to keep up with developments in your field. Some professions require more keeping up than others—a computer specialist, for example, compared to a human resources manager. Whatever the training required, always look for ways to adapt and improve your skills to become more productive and marketable. There may be times when you are tempted

FIGURE 15-7
Professional organizations and contacts help you keep up in your field

to start all over again and learn completely new skills. However, a better course of action may be to use emerging technology to improve your present base of skills. This way you can build on your current strong points and then branch out to other fields from a position of strength.

Be Alert for Organizational Change

Every organization has formal lines of communication—for example, supervisor to middle manager to top manager. However, there is also the *grapevine*—informal lines of communication. (See Figure 15-9.) Some service departments will serve many layers of management and be abreast of the news on all levels. For instance, the art director for advertising may be aware of several aspects of a companywide marketing campaign. Secretaries and administrative assistants know what is going on in more than one area.

Being part of the office grapevine can alert you to important changes—for instance, new job openings—that can benefit you. However, you always have to assess the validity of what you hear on the grapevine. Moreover, it's not advisable to be a contributor to office gossip. Behind-the-back criticisms of other people have a way of getting back to the person criticized.

Be especially alert for new trends within the organization—about future hiring, layoffs, automation, mergers with other companies, and the like. Notice which areas are receiving the greatest attention from top management. One tip-off is to see what kind of outside consultants are being brought in. Independent consultants are usually invited in because a company believes it needs advice in an area with which it has insufficient experience.

Look for Innovative Opportunities

You may understand your job better than anyone—even if you've only been there a few months. Look for ways to make it more efficient. How can present procedures be automated? How can new technology make your tasks easier? Discuss your ideas with your supervisor, the training director, or the head of the information systems department. Or discuss them with someone else who can see that you get the recognition you deserve. (Co-workers may or may not be receptive and may or may not try to take credit themselves.)

A good approach is to present your ideas in terms of saving money rather than "improving information." (See Figure 15-10.) Managers are generally more impressed with ideas that can save dollars than with ideas that seem like potential breakthroughs in the quality of decisions.

In general, it's best to concentrate on the business and organizational problems that need solving. Then look for a technological way of solving them. That is, avoid becoming too enthusiastic about a particular technology and then trying to make it fit the work situation.

FIGURE 15-8
Desktop publishing: a good specialty to develop for certain careers

FIGURE 15-9
Informal communication can alert you to important organizational changes

FIGURE 15-10
Present your ideas as saving money rather than "improving information"

CONCEPT CHECK

✔ Outline the strategies you can use to stay ahead and be successful in your career.

Consider a Career in Information Systems

Employment opportunities in information systems are excellent.

Being a winner does not necessarily mean having a career in information systems. There are, however, several jobs within information systems that you might like to consider. Some are *systems analyst, Webmaster, database administrator, programmer, network manager, technical writer,* and *computer trainer.*

Systems Analyst

The occupation of **systems analyst** is one of the fastest growing and is expected to almost double by the year 2005. As a systems analyst, you would work with other individuals within an organization to evaluate their information needs, design computer software and hardware to meet those needs, and then implement the information systems.

Webmaster

One of the newest and highest-demand occupations is **Webmaster.** As a Webmaster your duties would focus on designing, creating, monitoring, and evaluating corporate Web sites. As organizations are using more intranets and extranets, the importance and demand for Webmasters will continue to increase. Webmasters combine technical Internet skills with design and layout expertise.

Database Administrator

Database administrators play a critical role in large organizations. They are responsible for structuring, coordinating, linking, and maintaining internal databases. Additionally, they are involved with selecting and monitoring external databases including Internet databases. As a database administrator, you likely would use specialized database techniques such as data mining and data warehousing.

Programmer

Another high-demand profession is that of **programmer.** Programmers typically work closely with a systems analyst to either create new software or to revise existing programs. As a programmer you likely would use programming languages like C++ and Java.

Technical Writer

Technical writers explain in writing how a computer program works. As a technical writer you would likely work closely with systems analysts and users to document an information system and to create clearly written user manuals.

Network Manager

FIGURE 15-11
Network managers monitor and develop new communication systems

Nearly all information systems within an organization are connected by networks. **Network managers** ensure that existing information and communication systems are operating effectively and that new communication systems are implemented as needed. (See Figure 15-11.) The importance of this occupation within most organizations is increasing dramatically as the Internet plays a larger role in corporate communications. As a network manager, you would also be responsible for ensuring computer security and individual privacy.

Computer Trainer

One of the most important steps in creating a new information system is to prepare and train users. As a **computer trainer,** you would provide classes for users giving them an opportunity to explore a new system, to ask questions, and to try out common tasks. (See Figure 15-12.)

For a summary of careers in information systems, see Figure 15-13.

FIGURE 15-12
Computer trainers teach others about new systems and software

---

CONCEPT CHECK

✔ What are some of the employment opportunities available within the field of information systems?

Title	Description
Systems analyst	Analyzes, designs, and implements information systems
Webmaster	Designs, creates, monitors, and evaluates corporate Web sites
Database administrator	Structures, coordinates, links, and maintains databases
Programmer	Revises existing software and creates new software
Network manager	Monitors existing networks and implements new networks
Technical writer	Creates user manuals and documentation for information systems
Computer trainer	Provides classes and support for computer users

FIGURE 15-13
Careers in information systems

A LOOK TO THE FUTURE

Being computer competent means taking positive control.

The Rest of Your Life

This is not the end; it is the beginning. Being a skilled computer end user—being computer competent— is not a matter of thinking "Someday I'll . . ." ("Someday I'll have to learn all about that.") It is a matter of living in the present and keeping an eye on the future. It is also a matter of having the discipline to keep up with the prevailing technology. It is not a matter of focusing on vague "what ifs." It is a matter of concentrating on your goals and learning how the computer can help you achieve them. Being an end user, in short, is not about trying to avoid failure. Rather, it is about always moving toward success—about taking control over the exciting new tools available to you.

VISUAL SUMMARY

Your Future and Information Technology

TECHNOLOGY AND ORGANIZATIONS

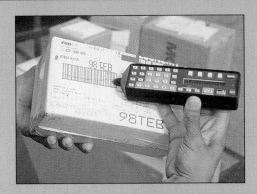

Technology can introduce new ways for businesses to compete with each other. They can compete by creating *new products,* establishing *new enterprises,* and developing *new customer and supplier relationships.*

New Products
Technology creates products that operate faster, are priced more cheaply, are often better quality, or are wholly new. New products can be individually tailored to a particular customer's needs.

New Enterprises
Technology can build entire new businesses. Two examples:
- Internet service providers—just a few years ago, only a few Internet service providers like America Online were available. Now, thousands of national and local providers are available.
- Web site development companies—thousands of small companies specializing in developing Web sites have sprung up in just the past three years.

New Customer and Supplier Relationships
Businesses that make their information systems easily available may make their customers less likely to take their business elsewhere (e.g., overnight delivery services closely track packages and bills).

TECHNOLOGY AND PEOPLE

People have different coping styles when it comes to technology. Three common reactions to new technology are *cynicism, naïveté,* and *frustration.*

Cynicism
The cynics feel that new technology is overrated and too troublesome to learn. Some cynics openly express their doubts. Others pretend to be interested.

Naïveté
Naïve people may be unfamiliar or quite familiar with computers. Unfamiliar ones tend to think of computers as magic boxes. Even those familiar with technology often underestimate the time and difficulty of using technology to generate information.

Frustration
Frustrated users are impatient and irritated about taking time to learn new technology. Often these people have too much to do, find manuals difficult to understand, and/or feel stupid.

To stay competent, you need to recognize the impact of technological change on organizations and people. You need to know how to use change to your advantage and to become a winner. Although you do not need to be a specialist in information technology, you should be aware of career opportunities in the area.

HOW YOU CAN BE A WINNER

Six ongoing activities that can help you be successful are the following.

Stay Current
Read trade journals and the general business press, join professional associations, and participate in interest groups on the Internet.

Maintain Your Computer Competency
Stay current by being alert for computer-related articles in the general business press and in trade journals in your particular profession.

Develop Professional Contacts
Stay active in your profession and meet people in your field. This provides information about other people, firms, job opportunities, and social contacts.

Develop Specialties
Develop specific as well as general skills. Expect to take classes periodically to stay current with your field and technology.

Be Alert for Organizational Change
Use formal and informal lines of communication. Be alert for new trends within the organization.

Look for Innovative Opportunities
Look for ways to increase efficiency. Present ideas in terms of saving money rather than "improving information."

CAREERS IN INFORMATION SYSTEMS

Being a winner does not necessarily mean having a career in information systems. There are, however, some excellent employment opportunities in information systems, including the following.

System Analyst
Systems analysts work with others to determine their information needs, design hardware and software systems, and implement them.

Webmaster
Webmasters focus on designing, creating, monitoring, and evaluating corporate Web sites.

Database Administrator
Database administrators are responsible for structuring, coordinating, linking, and maintaining internal and external databases.

Programmer
Programmers work with systems analysts to create new software and revise existing programs.

Network Managers
Network managers monitor existing networks and implement new ones.

Technical Writer
Technical writers create documents to explain how systems work.

Computer Trainer
Computer trainers present classes to users on new systems.

Key Terms

2001
2002

computer trainer (373)
database administrator (372)
network manager (372)
programmer (372)

systems analyst (372)
technical writer (372)
Webmaster (372)

Chapter Review

LEVEL 1

Reviewing Facts and Terms
Matching

Match each numbered item with the most closely related lettered item. Write your answers in the spaces provided.

1. You explain in writing how a computer program works. ____

2. You typically work closely with a systems analyst to either create new software or to revise existing programs. ____

3. You provide classes for users giving them an opportunity to explore a new system, ask questions, and try out common tasks. ____

4. You focus on designing, creating, monitoring, and evaluating corporate Web sites. ____

5. You are responsible for structuring, coordinating, linking, and maintaining internal databases. ____

6. You ensure that existing information and communication systems are implemented as needed. ____

7. You evaluate an organization's information needs, design computer software and hardware to meet those needs, and then implement the information systems. ____

a. Computer trainer
b. Database administrator
c. Network manager
d. Programmer
e. Systems analyst
f. Technical writer
g. Webmaster

True/False

In the spaces provided, write T or F to indicate whether the statement is true or false.

1. Most businesses are making formal plans to track and to implement technology into their competitive strategies. ____

2. Businesses never allow customers access to their information systems. ____

3. In all fields, successful professionals have to be experts in their own field as well as in computer technology. ____

4. *Infoworld*, *PC World*, and *MacWorld* are magazines that specifically cover microcomputers and information. ____

5. The office grapevine can be a good source to alert you to organizational changes. ____

Multiple Choice

Circle the letter of the correct answer.

1. The real issue with new technology is:
   a. how to make it better
   b. which printer is better
   c. how to control it
   d. how to integrate it with people
   e. managing its impact on government

2001
2002

2. By giving their customers access to their package tracking information system, Federal Express is developing new:
   a. global computer facilities
   b. customer relationships
   c. airline reservation procedures
   d. serious security problems
   e. government delivery systems

3. The type of person who underestimates the difficulty of changing computer systems or of generating information is:
   a. a cynic
   b. frustrated
   c. naïve
   d. a loser
   e. a winner

4. By staying in touch with others in your field, you are:
   a. developing professional contacts
   b. staying current
   c. developing specialties
   d. maintaining computer competence
   e. being alert to organizational changes

5. A good idea is to present your innovative ideas:
   a. in terms of improving decision making
   b. to co-workers
   c. in terms of improving information
   d. to the union chief
   e. in terms of saving money

Completion

Complete each statement in the spaces provided.

1. ATM cards, home banking, and programs to analyze cursive writing are examples of how some banks are looking to use technology in their competitive _____.

2. The person who thinks that microcomputers are overrated can be classified as a _____.

3. Reading trade journals about the use of technology is a good way to stay _____.

4. _____ is another name for the informal lines of communication within an organization.

5. Systems _____ are expected to double by the year 2005.

Reviewing Concepts

LEVEL 2

Open-Ended

On a separate sheet of paper, respond to each question and statement.

1. How do you become and stay a winner in the information age?

2. Give an example of how technology can change the nature of competition.

3. What are the three responses or attitudes that people are apt to have when confronted by new technology?

4. Describe six strategies individuals should follow in order to be successful in the information age.

5. What periodicals might you read in order to keep current on changes in microcomputer technology?

Concept Mapping

On a separate sheet of paper, draw a concept map or a flowchart showing how the following terms are related. Show all relationships. Include any additional terms you can think of.

Systems analyst   Technical writer
Webmaster   Network manager
Database administrator   Computer trainer
Programmer

Critical Thinking Questions and Projects

LEVEL 3

Read each exercise and answer the related questions on a separate sheet of paper.

1. *Volunteering your computer skills:* What would you do if you had an old but still useful microcomputer? It might not be something you want or even something you can sell. Still, someone can benefit from it. There are several groups that collect donated hardware and software for nonprofit organizations, such as conservation, veterans, arts, and child care groups.

   These groups also provide volunteers to assist nonprofits in learning to use their new systems. Perhaps this is a case where you can lend your own experience to a good cause. Contact one of the following or a similar organization, which you may learn about through local computer users' groups, to see how you can help:

   a. *Boston:* CONNECT, Technical Development Corporation, 30 Federal St., 5th floor, Boston, MA 02110 (telephone: 617-728-9151).

   b. *Chicago:* Information Technology Resource Center, 59 East Van Buren, Suite 2020, Chicago, IL 60605-1219 (telephone: 312-939-8050).

   c. *Dallas:* Technology Learning Center, Center for Nonprofit Management, 2900 Live Oak St., Dallas, TX 75024 (telephone: 214-823-8097).

   d. *New York:* Nonprofit Computer Exchange, Fund for the City of New York, 121 Sixth Ave., 6th floor, New York, NY 10013 (telephone: 212-925-5101).

   e. *San Francisco:* CompuMentor, 89 Stillman St., San Francisco, CA 94107 (telephone: 415-512-7784).

2. *Being careful about technology predictions:* Technology forecasts have a way of often being so wide of the mark that in looking back we may wonder how the experts could have erred so badly. For instance, nuclear-powered airplanes, household robots, and widespread use of electric cars have never realized the rosy promises of the forecasters.

   Editor Herb Brody of *Technology Review* suggests some guidelines for reducing erroneous predictions. Among them are the following.

   • Be wary of forecasts based on information from vested interests, such as technology developers needing financing, who may in turn exert undue influence on market-forecasting firms, the news media, and investors.

   • Expect existing technologies to keep on improving, but don't expect people to abandon what they have for something only somewhat better.

   • Expect truly revolutionary technologies to take 10 to 25 years to gain widespread use.

   Given these guidelines, describe some future uses and assess the popularity you would expect for the following: neural-network computers; pen-based computers; shirt-pocket telephones; hypermedia; computer-generated virtual realities; flat-panel display TVs to hang on living-room walls.

3. *Security, privacy and ethics:* Most people agree that one of the major challenges of the 21st century is to determine and to define the role that computers have in our lives. Specifically, how do computers affect organizational security, individual privacy, and personal ethics? Throughout this book we have discussed these issues related to a variety of different situations. Now it's your turn. Create one list that presents the top five security issues and challenges for the 21st century, a second list that presents the top five privacy issues, and a third list that presents the five most significant ethical challenges.

4. *Locating Job Opportunities Online:* More and more people today are using the Internet to locate employment opportunities and to post resumes. (See Making IT Work for You: Locating Job Opportunities Online on pages 368–69.).

   a. Have you used the Internet to search for a job? If so, describe how and why you used it and whether you found it useful.

   b. Discuss the advantages and disadvantages of using the Internet rather than traditional methods to locate job opportunities and to post resumes.

   c. If you were looking for a job and had to choose between using the Internet or a traditional approach, which would you choose? Why?

On the Web Exercises

1. Computer Innovations

It is important to stay current with computer innovations in order to maintain your computer competency. To obtain help with this task, visit our Web site at http://www.mhhe.com/it/oleary/exercise.mhtml to link to a commercial news service site. Once connected to that site, look for articles relating to technology. Select one article of interest to you; print it and write a paragraph summarizing the content.

2. Job Hunting

The Web is a good tool to learn how to look for a job. Visit the Yahoo site at http://www.yahoo.com, and explore the category of "Business and Economy Employment: Employment and Work" and using the keywords "resume," "job hunting," and "employment." Print out the first page of results from your search. Write a paragraph summarizing the most useful advice and information you found.

3. Motion Pictures

Selecting movies and movie theaters can be a real hassle. The Web can tell you which movies are being offered at your local theaters and provide times and movie reviews. Visit our Web site at http://www.mhhe.com/it/oleary/exercise.mhtml to link to one of these motion picture sites. Once you have connected to that site, find a movie playing at a theater near you that you might like to see. Print out the show time for today. Check out one or more reviews on the movie, and write a paragraph summarizing your findings.

4. JobWeb

There is little doubt that the Web will be important in your future career. It could also be important in finding your next job. Visit our Web site at http://www.mhhe.com/it/oleary/exercise.mhtml to link to one of the Web's largest employment resource pages. Once connected to that site, conduct a search for jobs in your field or interest. Print out the first page of results. Write a paragraph describing the value of this site and discussing how you might use the Web in the future when searching for a job.

The Evolution of the Computer Age

Many of you probably can't remember a world without computers, but for some of us, computers were virtually unknown when we were born and have rapidly come of age during our lifetime.

Although there are many predecessors to what we think of as the modern computer—reaching as far back as the 18th century, when Joseph Marie Jacquard created a loom programmed to weave cloth and Charles Babbage created the first fully modern computer design (which he could never get to work)—the Computer Age did not really begin until the first computer was made available to the public in 1951.

The modern age of computers spans almost 50 years (thus far), which is typically broken down into five generations. Each generation is marked by a significant advancement in technology:

- **First Generation (1951–57):** During this generation, computers were built with vacuum tubes—electronic tubes that were made of glass and were about the size of light bulbs.

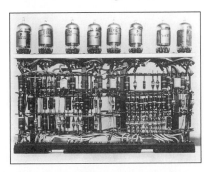

- **Second Generation (1958–63):** This generation begins with the first computers built with transistors—small devices that transfer electronic signals across a resistor. Because transistors are much smaller, use less power, and create less heat than vacuum tubes, the new computers were faster, smaller, and more reliable than the first-generation machines were.

- **Third Generation (1964–69):** In 1964, computer manufacturers began replacing transistors with integrated circuits. An integrated circuit (IC) is a complete electronic circuit on a small chip made of silicon (one of the most abundant elements in the earth's crust). These computers were more reliable and compact than computers made with transistors, and they cost less to manufacture.

- **Fourth Generation (1970–90):** There are many key advancements that were made during this generation, the most significant of which was the use of the microprocessor—a specialized chip developed for computer memory and logic. This revolutionized the computer industry by making it possible to use a single chip to create a smaller "personal" computer (as well as digital watches, pocket calculators, copy machines, and so on).

- **Fifth Generation (1991–2000 and beyond):** Our current generation has been referred to as the "Connected Generation" because of the industry's massive effort to increase the connectivity of computers. The rapidly expanding Internet, World Wide Web, and intranets have created an information superhighway that has enabled both computer professionals and home computer users to communicate with others across the globe.

This appendix provides you with a timeline that describes some of the most significant events in each generation.

First Generation
The Vacuum Tube Age

1951: Dr. John W. Mauchly and J. Presper Eckert, Jr., introduce the first commercially available electronic digital computer—the UNIVAC—built with vacuum tubes. This computer was based on their earlier ENIAC (Electronic Numerical Integrator and Computer) design completed in 1946.

1951–53: IBM adds computers to its business equipment products and sells over 1,000 IBM 650 systems.

1951	1952	1953	1954	1955	1956	1957

1957: Introduction of first high-level programming language—FORTRAN (FORmula TRANslator).

1952: Development team led by Dr. Grace Hopper, former U.S. Navy programmer, introduces the A6 Compiler—the first example of software that converts high-level language symbols into instructions that a computer can execute.

Second Generation
The Transistor Age

1958: Introduction of computers built with transistors—a 1947 Bell Laboratories invention.

1959: Introduction of the removable disk pack, providing users with fast access to stored data.

1960: Introduction of first business application programming language—COBOL (Common Business Oriented Language)—based on English-like phrases.

| 1958 | 1959 | 1960 | 1961 | 1962 | 1963 |

1963: Introduction of the first computer industry standard character set—ASCII (American Standard Code for Information Interchange) that enables computers to exchange information.

1959: General Electric Corporation introduces ERMA (Electronic Recording Machine Accounting)—the first technology that can read special characters, such as digitized information.

Third Generation
The Integrated Circuit Age

1964: Introduction of computers built with an integrated circuit (IC), which incorporates multiple transistors and electronic circuits on a single silicon chip.

1965: Digital Equipment Corporation (DEC) introduces the first minicomputer.

1969: Introduction of ARPANET and the beginning of the Internet.

| 1964 | 1965 | 1966 | 1967 | 1968 | 1969 |

1965: Introduction of the BASIC programming language.

1969: IBM announces its decision to offer unbundled software, priced and sold separately from the hardware.

1964: IBM introduces its System/360 line of compatible computers, which can all use the same programs and peripherals.

Fourth Generation
The Microprocessor Age

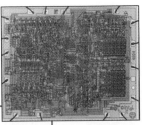

1970: Introduction of computers built with chips that used LSI (large-scale integration).

1975: First local area network (LAN)—Ethernet—developed at Xerox PARC (Palo Alto Research Center).

1977:
Apple Computer, Inc., founded by Steve Wozniak and Steve Jobs, and Apple I introduced as an easy-to-use "hobbyist" computer.

1970 1971 1972 1973 1974 1975 1976 1977 1978 1979

1971: Dr. Ted Hoff of Intel Corporation develops a microprogrammable computer chip—the Intel 4004 microprocessor.

1975: The MITS, Inc. Altair becomes the first commercially successful microcomputer, selling for less than $400 a kit.

1979: Introduction of the first public information services—Compuserve and the Source.

1980: IBM asks Microsoft founder, Bill Gates, to develop an operating system—MS DOS—for the soon-to-be released IBM personal computer.

1981: Introduction of the IBM PC, which contains an Intel microprocessor chip and Microsoft's MS-DOS operating system.

1989: Introduction of Intel 486—the first 1,000,000 transistor microprocessor.

1980 1981 1982 1983 1984 1985 1986 1987 1988 1989 1990

1984: Apple introduces the Macintosh Computer, with a unique, easy-to-use graphical user interface.

1985: Microsoft introduces their Windows graphical user interface.

1990: Microsoft releases Windows 3.0, with an enhanced graphical user interface and the ability to run multiple applications.

Fifth Generation
The Age of Connectivity

1991: Release of World Wide Web standards that describe the framework of linking documents on different computers.

1992: Apple introduces the Newton MessagePad—a personal digital assistant (PDA) that incorporates a pen interface and wireless communications.

1993: Introduction of computer systems built with Intel's Pentium microprocessor.

1995: Intel begins shipping the Pentium Pro microprocessor.

1991   1992   1993   1994   1995

1993: Introduction of the Mosaic graphical Web browser, which led to the organization of Netscape Communications Corporation.

1995: Microsoft releases Windows 95, a major upgrade to its Windows operating system.

1991: Linus Torvalds, a graduate student at the University of Helsinki, develops a version of UNIX called the Linux operating system.

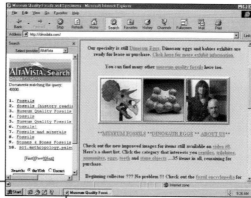

1998: Microsoft releases Office 98 integrated with Internet Explorer 4.0—antitrust legislation intensifies.

1997: Microsoft releases Office 97 with major Web enhancements integrated into its applications.

By 2002: A private Internet—Internet2— expected to be completed with higher speed, limited access, and tighter security— expected to include advanced virtual reality interfaces called nanomanipulators.

1996   1997   1998   1999   2000+

1997: The number of Internet and World Wide Web users estimated at 50 million.

1999: Microsoft releases Office 2000 featuring extensive Web integration and document collaboration—soon followed by Windows 2000.

1996: More than 50 million PCs sold and more than 250 million PCs in use worldwide.

The Buyer's Guide
How to Buy Your Own Microcomputer System

Some people make snap judgments about some of the biggest purchases in their lives: cars, college educations, houses. People have been known to buy such things based solely on an ad, a brief conversation, or a one-time look. And they may be making an impulsive decision about something costing thousands of dollars. Who is to blame, then, if they are disappointed later? They simply didn't take time to check it out.

The same concerns apply in buying a microcomputer system. You can make your choice on the basis of a friend's enthusiasm or a salesperson's promises. Or you can proceed more deliberately, as you would, say, in looking for a job.

Four Steps in Buying a Microcomputer System

The following is not intended to make buying a microcomputer an exhausting experience. Rather, it is to help you clarify your thinking about what you need and can afford.

The four steps in buying a microcomputer system are presented on the following pages. We divided each step into two parts on the assumption that your needs may change, but so may the money you have to spend on a microcomputer. For instance, later in your college career or after college graduation, you may want a far more powerful computer system than you need now. At that point you may have more money to spend. Or you may not need to spend money at all, if your employer provides you with a computer.

STEP 1
What Needs Do I Want a Computer to Serve?

The trick is to distinguish between your needs and your wants. Sure, you *want* a cutting-edge system powerful enough to hold every conceivable program you'll ever need. And you want a system fast enough to process them all at the speed of light. But do you *need* this? Your main concern is to address the two-part question:

- What do I need a computer system to do for me today?
- What will I need it to do for me in another year or two?

The questionnaire at the end of this guide will help you determine the answers to both questions.

Suggestions

The first thing to establish is whether you need a computer at all. Some colleges offer computer facilities at the library or in some dormitories. Or perhaps you can borrow a roommate's. The problem, however, is that when you are up against a term-paper deadline, many others may be also. Then the machine you want may not be available. To determine the availability of campus computers and network support, call the computer center or the dean of students' office.

Another matter on which you might want advice is what type of computer is popular on campus. Some schools favor Apple Macintoshes, others favor IBMs or IBM-compatibles. If you own a system that's incompatible with most others on campus, you may be stuck if your computer breaks down. Ask someone knowledgeable who is a year or two ahead of you if your school favors one system over another.

Finally, look ahead and determine whether your major requires a computer. Business and engineering students may find one a necessity, physical education and drama majors may not. Your major may also determine the kind of computer that's best. A journalism major may want an IBM or IBM-compatible notebook that can be set up anywhere. An architecture major may want a powerful desktop Macintosh with a laser printer that can produce elaborate drawings. Ask your academic advisor for some recommendations.

Example

Suppose you are a college student beginning your sophomore year, with no major declared. Looking at the courses you will likely take this year, you decide you will probably need a computer mainly for word processing. That is, you need a system that will help you write short (10- to 20-page) papers for a variety of courses.

By this time next year, however, you may be an accounting major. Having talked to some juniors and seniors, you find that courses in this major, such as financial accounting, will require you to use elaborate spreadsheets. Or maybe you will be a fine arts or architecture major. Then you may be required to submit projects for which drawing and painting desktop publishing software would be helpful. Or perhaps you will be out in the job market and will be writing application letters and résumés. In that case, you'll want them to have a professional appearance.

STEP 2
How Much Money Do I Have to Spend on a Computer System?

When you buy your first computer, you are not necessarily buying your last. Thus, you can think about spending just the bare-bones amount for a system that meets your needs while in college. Then you might plan to get another system later on. After all, most college students who own cars (quite often used cars) don't consider those the last cars they'll own.

You know what kind of money you have to spend. Your main concern is to answer this two-part question:

- How much am I prepared to spend on a computer system today?
- How much am I prepared to spend in another year or two?

The questionnaire at the end of this guide asks you this.

Suggestions

You can probably buy a good used computer of some sort for under $400 and a printer for under $100. On the other hand, you might spend $1,500 to $3,500 on a new state-of-the-art system. When upgraded, this computer could meet your needs for the next five years.

There is nothing wrong with getting a used system, if you have a way of checking it out. For a reasonable fee, a computer-repair shop can examine it prior to your purchase. Look at newspaper ads and notices on campus bulletin boards for good buys on used equipment. Also try the Internet. Often the sellers will include a great deal of software and other items (disks, reference materials) with the package. If you stay with recognized brands such as Apple, IBM, Compaq, or Dell, you probably won't have any difficulties. The exception may be with printers, which, since they are principally mechanical devices, may get a lot of wear and tear. This is even more reason to tell the seller you want a repair shop to examine the equipment before you buy.

If you're buying new equipment, be sure to look for student discounts. Most college bookstores, for instance, offer special prices to students. Mail-order houses also steeply discount their products. These firms run ads in such periodicals as *Computer Shopper* (sold on newsstands) and other magazines as well as the Internet. However, using mail and telephone for repairs and support can be a nuisance. Often you can use the prices advertised by a mail-order house to get local retail computer stores to lower their prices.

Example

Perhaps you have access to a microcomputer at the campus student computing center, the library, or the dormitory. Or you can borrow a friend's. However, this computer isn't always available when it's convenient for you. Moreover, you're not only going to college but also working, so both time and money are tight. Having your own computer would enable you to write papers when it's convenient for you. Spending more than $500 might cause real hardship, so a new microcomputer system may be out of the question. You'll need to shop the newspaper classified ads or the campus bulletin boards to find a used but workable computer system.

Or, maybe you can afford to spend more now—say, between $1,000 and $2,000—but probably only $500 next year. By this time next year, however, you'll know your major and how your computer needs have changed. For example, if you're going to be a finance major, you need to have a lot more computer memory (primary storage). This will hold the massive amounts of data you'll be working with in your spreadsheets. Or maybe you'll be an architecture major or graduating and looking for a job. In that case, you'll need a laser printer to produce attractive-looking designs or application letters. Thus, whatever system you buy this year, you'll want to upgrade it next year.

STEP 3
What Kind of Software Will Best Serve My Needs?

Most computer experts urge that you determine what software you need before you buy the hardware. The reasoning here is that some hardware simply won't run the software that is important to you. This is certainly true once you get into *sophisticated* software. Examples include specialized programs available for certain professions (such as certain agricultural or retail-management programs). However, if all you are interested in today are the basic tools of software—word processing, spreadsheet, and communications programs—these are available for nearly all microcomputers. The main caution is that some more recent versions of application software won't run on older hardware. Still, if someone offers you a free computer, don't say no because you feel you have to decide what software you need first. You will no doubt find it sufficient for many general purposes, especially during the early years in college.

That said, you are better served if you follow step 3 after step 2—namely, finding the answers to the two-part question:

- What kind of software will best serve my needs today?
- What kind will best serve my needs in another year or two?

The questionnaire at the end of this guide may help you determine your answers.

Suggestions

No doubt some kinds of application software are more popular on your campus—and in certain departments on your campus—than others. Are freshman and sophomore students mainly writing their term papers in Word, WordPerfect, or Word Pro? Which spreadsheet is most often used by business students: Excel, Lotus 1-2-3, or Quattro Pro? Which desktop publishing program is most favored by graphic arts majors: PageMaker, Ventura Publisher, or Freehand? Do many students use their microcomputers to access the Internet, and, if so, which communications software is the favorite? Do engineering and architecture majors use their own machines for CAD/CAM applications? Start by asking other students and your academic advisor.

If you're looking to buy state-of-the-art software, you'll find plenty of advice in various computer magazines. Several of them rate the quality of newly issued programs. Such periodicals include *InfoWorld, PC World, PC/Computing,* and *MacWorld.*

Example

Suppose you determine that all you need is software to help you write short papers. In that case, nearly any kind of word processing program would do. You could even get by with some older versions or off-brand kinds of word processing software. This might happen if such software was included in the sale of a used microcomputer that you bought at a bargain price.

But will this software be sufficient a year or two from now? Looking ahead, you guess that you'll major in theater arts, and minor in screenwriting, which you may pursue as a career. At that point a simple word processing program won't do. You learn from juniors and seniors in that department that screenplays are written using special screenwriting programs. This is software that's not available for some computers. Or, as an advertising and marketing major, you're expected to turn word-processed promotional pieces into brochures. For this, you need desktop publishing software. Or, as a physics major, you discover you will need to write reports on a word processor that can handle equations. In short, you need to look at your software needs not just for today but also for the near future. You especially want to consider what programs will be useful to you in building your career.

STEP 4
What Kind of Hardware Will Best Serve My Needs?

A bare-bones hardware system might include a three-year-old desktop or notebook computer with a 3½-inch floppy disk drive and a hard-disk drive. It should also include a monitor and a printer. With a newer system, the sky's the limit. On the one hand, as a student—unless you're involved in some very specialized activities—it's doubtful you'll really need such things as voice-input devices, touch screens, scanners, and the like. On the other hand, you will probably need speakers and a CD-ROM or DVD-ROM drive. The choices of equipment are vast.

As with the other steps, the main task is to find the answers to a two-part question:

- What kind of hardware will best serve my needs today?
- What kind will best serve my needs in another year or two?

There are several questions on the questionnaire at the end of this guide to help you determine answers to these concerns.

Suggestions

Clearly, you should let the software be your guide in determining your choice of hardware. Perhaps you've found that the most popular software in your department runs on a Macintosh rather than an IBM-compatible. If so, that would seem to determine your general brand of hardware.

Whether you buy IBM or Macintosh, a desktop or a notebook, we suggest you get a 3½-inch floppy disk drive, a hard-disk drive with at least 2 gigabytes of storage, a DVD-ROM drive, at least 128 megabytes of memory, and an ink-jet printer.

As with software, several computer magazines not only describe new hardware but also issue ratings. See *InfoWorld, PC World,* and *MacWorld,* for example.

Example

Right now, let's say, you're mainly interested in using a computer to write papers, so almost anything would do. But you need to look ahead.

Suppose you find that Word seems to be the software of choice around your campus. You find that Word 97 will run well on a Pentium machine with 8 megabytes of memory and a 1-gigabyte hard disk. Although this equipment is now outdated, you find from looking at classified ads that there are many such used machines around. Plus, they cost very little—well under $500 for a complete system.

If you're a history or philosophy major, maybe this is all the hardware and software you need. Indeed, this configuration may be just fine all the way through college. However, some majors, and the careers following them, may require more sophisticated equipment. Your choice then becomes: Should I buy an inexpensive system now that can't be upgraded, then sell it later and buy a better one? Or should I buy at least some of the components of a good system now and upgrade it over the next year or so?

As an advertising major, you see the value of learning desktop publishing. This will be a useful if not essential skill once you embark on a career. In exploring the software, you learn that Word includes some desktop publishing capabilities. However, the hardware you previously considered simply isn't sufficient. Moreover, you learn from reading about software and talking to people in your major that there are better

desktop publishing programs. Specialized desktop publishing programs like Ventura Publisher are considered more versatile than Word. Probably the best software arrangement, in fact, is to have Word as a word processing program and Ventura Publisher for a desktop publishing program.

To be sure, the campus makes computers that will run this software available to students. If you can afford it, however, you're better off having your own. Now, however, we're talking about a major expense. A computer running a Pentium III microprocessor, with 256 megabytes of memory, a 3½-inch disk drive, a DVD-ROM disk drive, and a 10-gigabyte hard disk, plus a modem, color monitor, and laser printer, could cost in excess of $1,500.

Perhaps the best idea is to buy now, knowing how you would like your system to grow in the future. That is, you will buy a microcomputer with a Pentium III microprocessor. But at this point, you will buy only an ink-jet printer and not buy a DVD-ROM drive. Next year or the year following, you might sell off the less sophisticated peripheral devices and add a DVD-ROM drive and a laser printer.

Developing a Philosophy about Computer Purchasing

It's important not to develop a case of "computer envy." Even if you bought the latest, most expensive microcomputer system, in a matter of months, something better will come along. Computer technology is still in a very dynamic state, with more powerful, versatile, and compact systems constantly hitting the marketplace. So what if your friends have the hottest new piece of software or hardware? The main question is: Do you need it to solve the tasks required of you or to keep up in your field? Or can you get along with something simpler but equally serviceable?

VISUAL SUMMARY

The Buyer's Guide:
How to Buy Your Own Microcomputer System

NEEDS

What do I need a computer system to do for me today? In another year or two?

I wish to use the computer for:

	Today	1–2 years
Word processing—writing papers, letters, memos, or reports	❏	❏
Business or financial applications—balance sheets, sales projections, expense budgets, or accounting problems	❏	❏
Record-keeping and sorting—research bibliographies, scientific data, or address files	❏	❏
Graphic presentations—of business, scientific, or social science data	❏	❏
Online information retrieval—to campus networks, service providers, or the Internet	❏	❏
Publications, design, or drawing—for printed newsletters, architectural drawing, or graphic arts	❏	❏
Multimedia—for video games, viewing, creating, presenting, or research	❏	❏
Other—(specify): _____	❏	❏

BUDGET

How much am I prepared to spend on a system today? In another year or two?

I can spend:

	Today	1–2 years
Under $500	❏	❏
Up to $1,000	❏	❏
Up to $1,500	❏	❏
Up to $2,000	❏	❏
Up to $2,500	❏	❏
Over $3,000 (specify): _____	❏	❏

Buying a Microcomputer System

Step	Questions
1	*My needs:* What do I need a computer system to do for me today? In another year or two?
2	*My budget:* How much am I prepared to spend on a system today? In another year or two?
3	*My software:* What kind of software will best serve my needs today? In another year or two?
4	*My hardware:* What kind of hardware will best serve my needs today? In another year or two?

To help clarify your thinking about buying a
microcomputer system, complete the questionnaire below
by checking the appropriate boxes.

SOFTWARE

What kinds of software will best serve my needs
today? In another year or two?

The application software I need includes:

	Today	1–2 years
Word processing—Word, WordPerfect, Word Pro, or other (specify): _____	❑	❑
Spreadsheet—Excel, Lotus 1-2-3, Quattro Pro, or other (specify): _____	❑	❑
Database—Access, Paradox, Approach, or other (specify): _____	❑	❑
Presentation graphics—PowerPoint, Freelance, CorelPresentations, or other (specify): _____	❑	❑
Browsers—Netscape Navigator, Microsoft Internet Explorer, or other (specify): _____	❑	❑
Other—integrated packages, software suites, graphics, multimedia, Web Authoring, CAD/CAM, other (specify): _____	❑	❑

The system software I need:

	Today	1–2 years
Windows 95	❑	❑
Windows 98	❑	❑
Windows 2000	❑	❑
Mac OS	❑	❑
Unix	❑	❑
Other (specify): _____	❑	❑

HARDWARE

What kinds of hardware will best serve my needs
today? In another year or two?

The hardware I need includes:

	Today	1–2 years
Microprocessor—Pentium III, Pentium II, Power PC, other (specify): _____	❑	❑
Memory—(specify amount): _____	❑	❑
Monitor—size (specify): _____	❑	❑
Floppy disk drives—3½" and/or Zip (specify): _____	❑	❑
Optical disk drive—CD-ROM, DVD-ROM (specify type, speed, and capacity): _____	❑	❑
Hard-disk drive—(specify capacity): _____	❑	❑
Portable computer—laptop, notebook, subnotebook, personal digital assistant (specify): _____	❑	❑
Printer—ink-jet, laser, color (specify): _____	❑	❑
Other—modem, speakers, fax, surge protector (specify): _____	❑	❑

The Upgrader's Guide
How to Upgrade Your Microcomputer System

If you own a microcomputer, chances are that your machine is not the latest and greatest. Microcomputers are always getting better—more powerful and faster. While that is a good thing, it can be frustrating trying to keep up.

What can you do? If you have lots of money, you can simply buy a new one. Another alternative is to upgrade or add new components to increase the power and speed of your current microcomputer. You probably can increase your system's performance at a fraction of the cost of a new one.

Three Steps in Upgrading a Microcomputer System

The following is not intended to detail specific hardware upgrades. Rather, it is intended to help you clarify your thinking about what you need and can afford.

The three steps in upgrading a microcomputer system are presented on the following pages. Each step begins by asking a key question and then provides some suggestions or factors to consider when responding to the question.

STEP 1
Is It Time to Upgrade?

Almost any upgrade you make will provide some benefit; the trick is determining if it is worth the monetary investment. It is rarely practical to rebuild an older computer system into a newer model piece by piece. The cost of a complete upgrade typically far exceeds the purchase price of a new system. But if your system is just a piece or two away from meeting your needs, an upgrade may be in order.

Clearly defining what you hope to gain with an upgrade of some of your system's hardware will enable you to make the most relevant and cost-effective selections. Before deciding what to buy, decide what goal you hope to accomplish. Do you want to speed up your computer's performance? Do you need more space to save your files? Is it a new component you'd like to add, such as a DVD-ROM drive?

Suggestions

A good place to start is the documentation on the packaging of any software you use or plan to use. Software manufacturers clearly label the minimum requirements to use their product. These requirements are usually broken down into categories that relate to specific pieces of hardware. For instance, how much RAM (random access memory) does a new program require? How much hard disk space is needed? Keep in mind these ratings are typically the bare minimum. If your system comes very close to the baseline in any particular category, you should still consider an upgrade.

Another thing to investigate is whether there is a software solution that will better serve your needs. For instance, if you are looking to enhance performance or make more room on your hard disk, there are diagnostic and disk optimization utility programs that may solve your problem. In Chapter 3, we discuss a variety of utility programs, such as Norton SystemWorks, that monitor, evaluate, and enhance system performance and storage capacity.

Typical objectives of an upgrade are to improve system performance, increase storage capacity, or add new technology.

STEP 2
What Should I Upgrade?

Once you have clearly defined your objectives, the focus shifts to identifying specific components to meet those objectives.

Suggestions

If your objective is to improve *performance*, three components to consider are RAM, the microprocessor, and expansion cards. If your objective is to increase *storage capacity*, two components to consider are hard disk drive and Zip disk drive. If you are adding *new technology*, consider the capability of your current system to support new devices.

Performance

If you want to increase the speed of your computer, consider increasing the amount of RAM. In most cases, this upgrade is relatively inexpensive and will yield the highest performance result per dollar invested. How much your system's performance will increase depends on how much RAM you start with, the size of programs you run, and how often you run large programs.

Another way to increase speed is to replace your system's microprocessor. Processor speed is measured in megahertz (MHz). This rating is not a direct measurement of how fast the processor works, but rather gives you a general idea of how it compares to other processors. (Computing magazines such as *PC World* often publish articles comparing the relative effectiveness of different processors.) The concept behind a microprocessor upgrade is simple: a faster processor will process faster. This is often an expensive upgrade and not as cost-effective as increasing RAM.

If you are looking at upgrading for a specific type of application, perhaps an expansion card is your answer. Expansion cards connect to slots on the system board, provide specialized support, and often free up resources and increase overall system performance. For example, if you run graphics-intensive programs, such as a drafting program or a video game, a video-card upgrade may be a good buy. An upgraded video card can be used to support higher resolution displays, handle all video data, and speed up overall system performance.

Storage Capacity

It's not hard to know when it's time to upgrade your storage capacity. If you frequently have to delete old files to make way for new ones, then it is probably time for more space. A larger or an additional hard drive is usually the solution. Two things to consider when comparing new hard drives are (1) size, which is usually rated in gigabytes (GB) of data the drive can hold, and (2) seek time, which is a rating of the average time it takes the drive to access any particular piece of data.

If you are storing a lot of data that you no longer use, such as old term papers, you might consider adding a Zip drive. This is usually cheaper and is a good way to archive and transport data. Access time is slower than from a hard drive, so this is an option best suited for infrequently used data or for backing up data.

New Technology

Perhaps you are not looking to modify existing hardware, but would like to add a new device. Examples include large high-resolution monitors, DVD-ROM drives, and high-speed printers.

The key consideration is whether the new device will work with your existing hardware. The requirements for these devices are typically printed on the outside packaging. If not, then refer to the product's operating manuals. Obviously, if your current system cannot support the new technology, you need to evaluate the cost of the new device plus the necessary additional hardware upgrades.

STEP 3
Who Should Do the Upgrade?

Once you've decided that the cost of the upgrade is justified and you know what you want to upgrade, the final decision is who is going to do it. Basically, there are two choices. You can either do it yourself or pay for professional installation.

Suggestions

The easiest way, and many times the best way, is to have a professional perform the upgrade. If you select this option, be sure to include the cost of installation in your analysis. If you have had some prior hardware experience or are a bit adventurous, you may want to save some money and do it yourself.

Visit a few computer stores that carry the upgrades you have selected. Most stores that provide the parts will install them as well. Talk with their technical people, describe your system (better yet, bring your system unit to the store), and determine the cost of professional installation. If you are thinking of doing it yourself, ask for their advice. Ask if they will provide assistance if you need it.

If you decide to have the components professionally installed, get the total price in writing and inquire about any guarantees that might exist. Before leaving your system be sure that it is carefully tagged with your name and address. After the service has been completed, pay by credit card and thoroughly test the upgrade. If it does not perform satisfactorily, contact the store and ask for assistance. If the store's service is not satisfactory, you may be able to have your credit card company help to mediate any disputes.

VISUAL SUMMARY

The Upgrader's Guide:
How to Upgrade Your Microcomputer System

Upgrading a Microcomputer System

Step	Questions
1	*Needs:* Is it time to upgrade? What do I need that my current system is unable to deliver?
2	*Analysis:* What should I upgrade? Will the upgrade meet my needs and will it be cost-effective?
3	*Action:* Who should do the upgrade? Should I pay a professional or do it myself?

NEEDS

Is it time to upgrade? What do I need that my current system is unable to deliver?

I am considering an upgrade to:

❏ **Improve performance because**
  ❏ My programs run too slow
  ❏ I cannot run some programs I need

❏ **Increase storage capacity because:**
  ❏ I don't have enough space to store all my files
  ❏ I don't have enough space to install new programs
  ❏ I need a secure place to back up important files
  ❏ I'd like to download large files from the Internet

❏ **Add new technology:**
  ❏ Zip Drive
  ❏ DVD-ROM
  ❏ High-performance monitor
  ❏ Printer
  ❏ TV tuner card
  ❏ Enhanced video card
  ❏ Enhanced sound card
  ❏ Other _____

ANALYSIS

What should I upgrade? Will the upgrade meet my needs and will it be cost effective?

I will improve:
❏ **Performance by**
  ❏ Adding random-access memory (RAM)
    Current RAM (MB) _____
    Upgrade to _____
    Cost $_____
    Expected improvement _____
    Other factors _____
  ❏ Replacing the current microprocessor
    Current processor _____
    Upgrade processor _____
    Cost $_____
    Expected improvement _____
    Other factors _____
  ❏ Adding an expansion card
    Type _____
    Purpose _____
    Cost $_____
    Expected improvement _____
    Other factors _____

❏ **Storage capacity by**
  ❏ Adding a hard disk drive
    Current size (GB) _____
    Upgrade size (GB) _____
    Cost $_____
    Expected improvement _____
    Other factors _____
  ❏ Adding a Zip disk drive
    Upgrade size (GB) _____
    Cost $_____
    Type _____
    Expected improvement _____
    Other factors _____

❏ **Functionality by adding**
    New technology _____
    System requirements _____
    Cost $_____
    Expected improvement _____
    Other factors _____

To help clarify your thinking about upgrading a
microcomputer system, complete the questionnaire below
by checking the appropriate boxes.

ACTION

Who should do the upgrade? Should I do it myself or should I pay a professional?

The two choices are:

❑ **Professional installation**

The easiest way, and many times the best way, is to
have a professional perform the upgrade. If you select
this option, be sure to include the cost of installation in
you analysis. Pay with a credit card, and make sure
your system is tagged with your name and address
before you part with it.

❑ **Do-it-yourself installation**

If you have had some prior hardware experience or
are a bit adventurous, you may want to save some
money and do it yourself. Avoid touching sensitive
electronic parts and be sure to ground yourself by
touching an unpainted metal surface in your
computer.

GLOSSARY

Access: Refers to the responsibility of those having data to control who is able to use that data.

Access time: The period between the time the computer requests data from a secondary storage device and the time the transfer of data is completed.

Accounts payable: The activity that shows the money a company owes to its suppliers for the materials and services it has received.

Accounts receivable: The activity that shows what money has been received or is owed by customers.

Accuracy: Relates to the responsibility of those who collect data to ensure that the data is correct.

Active desktop: The Windows dektop view in which "active content" from Web pages is displayed.

Active-matrix monitor: Monitor in which each pixel is independently activated. More colors with better clarity can be displayed.

Ada: Procedural language named after an English countess regarded as the first programmer; was originally designed for weapons systems but has commercial uses as well. Because of its structured design, modules of a large program can be written, compiled, and tested separately before the entire program is put together.

Adapter card: *See* Expansion card.

Add-on: *See* Helper applications.

Advanced Research Project Agency Network (ARPANET): A national computer network from which the Internet developed.

Agent: Program for updating search engine. Also called bot or spider.

ALU: *See* Arithmetic-logic unit.

American Standard Code for Information Interchange: *See* ASCII.

Analog signal: Signal that represents a range of frequencies, such as the human voice.

Analytical graphs: Form of graphics used to put numeric data into forms that are easier to analyze, such as bar charts, line graphs, and pie charts.

Animation: Feature involving special visual and sound effects.

Antivirus programs: Programs that guard a computer system from viruses or other damaging programs.

Applets: Java programs used on Web pages.

Application generator: Software with modules that have been preprogrammed to accomplish various tasks, such as calculation of overtime pay.

Applications: Programs such as word processing and spreadsheets.

Application software: Software that can perform useful work, such as word processing, cost estimating, or accounting tasks.

Arithmetic-logic unit (ALU): The part of the CPU that performs arithmetic and logical operations.

ARPANET: *See* Advanced Research Project Agency Network.

Artificial intelligence (AI): A field of computer science that attempts to develop computer systems that can mimic or simulate human thought processes and actions.

Artificial reality: *See* Virtual reality.

ASCII (American Standard Code for Information Interchange): Binary coding scheme widely used on all computers, including microcomputers.

Assembly language: Second generation of programming languages. These languages use abbreviations for program instructions.

Asynchronous communications port: *See* Serial port.

Asynchronous transmission: Method whereby data is sent and received one byte at a time.

Auction house sites: Web sites that operate like a traditional auction to sell merchandise to bidders.

Automated design tool: Software package that evaluates hardware and software alternatives according to requirements given by the systems analyst.

Backup programs: Programs that make copies of files to be used in case the originals are lost or damaged.

Backup tape cartridge unit: *See* Magnetic tape streamer.

Bandwidth: Bits-per-second transmission capability of a channel.

Bar code: Code consisting of vertical zebra-striped marks printed on product containers; read with a bar-code reader.

Bar-code reader: Photoelectric scanner that reads bar codes for processing.

BASIC (Beginner's All-purpose Symbolic Instruction Code): Easy-to-learn procedural programming language widely used on microcomputers.

Basic applications: *See* General-purpose applications.

Batch processing: Processing performed all at once on data that has been collected over several days.

Binary system: Numbering system in which all numbers consist of only two digits—0 and 1.

Bit (binary digit): A 0 or 1 in the binary system.

Bitmap file: Graphic file in which image is made up of thousands of dots (pixels).

Bot: *See* Agent.

Broadband: Bandwidth that includes microwave, satellite, coaxial cable, and fiber-optic channels. It is used for very high speed computers.

Browser: Special Internet software that allows users to effortlessly jump from one computer's resources to another computer's resources.

Browsing: Surfing or navigating the Web.

Bus: A data roadway along which bits travel.

Bus line: Electronic data roadway, connecting the parts of the CPU to each other and linking the CPU with other important hardware, along which bits travel. Also, the common connecting cable in a bus network.

Bus network: Network in which all communications travel along a common path. Each device in the network handles its own communications control. There is no host computer or file server.

Button: A special area you can click to make links and "navigate" through a presentation.

Byte: Unit consisting of eight bits. There are 256 possible bit combinations in a byte.

C: General-purpose procedural language originally designed for writing operating systems. Widely used and portable.

C++: One of the most widely used object-oriented programming languages.

Cable: Cords used to connect input and output devices to the system unit.

Cable modem: Allows all digital communication; speed of 27 million bps.

Cache memory: Area of random-access memory (RAM) set aside to store the most frequently accessed information. Acts as a temporary high-speed holding zone between memory and CPU.

CAD: *See* Computer-aided design.

CAM: *See* Computed-aided manufacturing.

Carder: Criminal who steals credit cards over the Internet.

Carpal tunnel syndrome: Disorder found among frequent computer users, consisting of damage to nerves and tendons in the hands. *See also* Repetitive strain injury.

CASE tool: *See* Automated design tool.

Cathode-ray tube (CRT): Desktop-type monitor built in the same way as a television set. The most common type of monitor for the office and the home. These monitors are typically placed directly on the system unit or on the top of the desk.

CD: *See* compact disc.

CD-R: Optical disk that can be written to once. After that it can be read many times without deterioration and cannot be written on or erased.

CD-ROM (compact disc–read-only memory): Optical disk that allows data to be read but not recorded.

CD-RW (compact disc–rewritable): Optical disk that is not permanently altered when data is recorded.

Center for European Nuclear Research (CERN): In Switzerland; where the Web was introduced in 1992.

Central processing unit (CPU): Part of the computer that holds data and program instructions for processing the data. The CPU consists of the control unit and the arithmetic-logic unit. In a microcomputer, the CPU is on a single electronic component, the microprocessor chip.

Channel: Topic for discussion in a chat group.

Character: A single letter, number, or special character such as a punctuation mark or $.

Checklist: List of questions that helps in guiding the systems analyst and end user through key issues for the present system.

Child node: A node one level below the node being considered in a hierarchical database or network.

Chip: A tiny circuit board etched on a small square of sandlike material called silicon.

Chlorofluorocarbons (CFCs): Toxic chemicals found in solvents and cleaning agents. Chlorofluorocarbons can travel into the atmosphere and deplete the earth's ozone layer.

CISC: *See* complex instruction set computer chip.

Client: A node that requests and uses resources available from other nodes.

Client/server network system: Network in which one powerful computer coordinates and supplies services to all other nodes on the network. Server nodes coordinate and supply specialized services, and client nodes request the services.

Closed architecture: Computer manufactured in such a way that users cannot easily add new devices.

CMOS: *See* complementary metal-oxide semiconductor.

Coaxial cable: High-frequency transmission cable that replaces the multiple wires of telephone lines with a single solid-copper core.

COBOL (Common Business-Oriented Language): Procedural language most frequently used in business.

Coding: Actual writing of a program.

Cold site: Special emergency facility in which hardware must be installed but which is available to a company in the event of a disaster to its computer system. *Compare* Hot site.

Combination key: Keys such as the *Ctrl* key that perform an action when held down in combination with another key.

Commerce server: *See* Web storefront creation packages.

Common operational database: Integrated collection of records that contain details about the operations of a company.

Common user database: Company database that contains selected information both from the common operational database and from outside (proprietary) databases.

Communications hardware: Sends and receives data and programs from one computer or secondary storage device to another.

Compact disc (CD): Widely used optical disk format.

Company database: Collection of integrated records shared throughout a company or other organization.

Compiler: Software that converts the programmer's procedural-language program (source code) into machine language (object code).

Complementary metal-oxide semiconductor (CMOS): A CMOS chip provides flexibility and expandability for a computer system; unlike RAM, it does not lose its contents if power is turned off; unlike ROM, its contents can be changed.

Complex instruction set computer (CISC) chip: The most common type of microprocessor that has thousands of programs written specifically for it.

Computer Abuse Amendments Act of 1994: Outlaws transmission of viruses and other harmful computer code.

Computer-aided design (CAD): Type of program that manipulates images on a screen.

Computer-aided manufacturing (CAM): Type of program that controls automated factory equipment, including machine tools and robots.

Computer-aided software engineering (CASE) tool: *See* Automated design tool

Computer competent: Being able to use a computer to meet one's information needs.

Computer crime: Illegal action in which a perpetrator uses special knowledge of computer technology. Criminals may be employees, outside users, hackers and crackers, and organized crime members.

Computer Fraud and Abuse Act of 1986: Law allowing prosecution of unauthorized access to computers and databases.

Computer Matching and Privacy Protection Act of 1988: Law setting procedures for computer matching of federal data for verifying eligibility for federal benefits or for recovering delinquent debts.

Computer network: Communications system connecting two or more computers and their peripheral devices.

Computer trainer: Computer professional who provides classes to instruct users.

Connectivity: Capability of the microcomputer to use information from the world beyond one's desk. Data and information can be sent over telephone or cable lines and through the air.

Continuous-speech recognition system: Voice-recognition system used to control a microcomputer's operations and to issue commands to special application programs.

Controller card: *See* Expansion card.

Control unit: Section of the CPU that tells the rest of the computer how to carry out program instructions.

Conversion: *See* Systems implementation.

Cookies: Programs that record information on Web site visitors.

CPU: *See* Central processing unit.

Cracker: One who gains unauthorized access to a computer system for malicious purposes.

Critical path: In a PERT chart, sequence of tasks that takes the longest to complete.

CRT monitor: *See* Cathode-ray tube (CRT) monitor.

Cumulative trauma disorder: *See* Repetitive strain injury.

Cybercash: *See* Electronic cash.

Cyberspace: The space of electronic movement of ideas and information.

Data: Raw, unprocessed facts that are input to a computer system.

Database: A collection of related files.

Database administrator (DBA): Person responsible for structuring, coordinating, linking, and maintaining databases.

Database file: File containing highly structured and organized data.

Database management system (DBMS): *See* Database manager.

Database manager: Software package used to set up, or structure, a database.

Data bus: *See* Bus line.

Data communications system: Electronic system that transmits data over communications lines from one location to another.

Data dictionary: Dictionary containing a description of the structure of the data in a database.

Data flow diagram: Diagram showing the data or information flow within an information system.

Data integrity: A database characteristic relating to the consistency and accuracy of data.

Data mining: Technique of searching data warehouses for related information and patterns.

Data processing system (DPS): Transaction processing system that keeps track of routine operations and records these events in a database.

Data redundancy: A common database problem in which data is duplicated and stored in different files.

Data security: Protection of software and data from unauthorized tampering or damage.

Data warehouse: Special type of database that supports data mining.

DAT drive: *See* Digital audiotape (DAT) drive.

Debugging: Programmer's word for testing and then eliminating errors in a program.

Decision model: Strategic, tactical, or operational model that gives the decision support system its analytical capabilities.

Decision support system (DSS): Flexible tool for analysis that helps managers make decisions about unstructured problems, such as the effect of events and trends outside the organization.

Decision table: Table showing the decision rules that apply when certain conditions occur and what action should take place as a result.

Demand report: The opposite of a scheduled report. A demand report is produced on request.

Demodulation: Process performed by a modem in converting analog signals to digital signals.

Desk checking: Process of checking out a computer program by studying the program listing line by line, looking for syntax and logic errors.

Desktop: Windows user interface.

Desktop computer: Computer small enough to fit on top or along the side of a desk and yet too big to carry around.

Desktop managers: Programs that provide electronic organizational tools.

Desktop publishing: Program that allows you to mix text and graphics to create publications of professional quality.

Device driver: Specialized program designed to allow particular input or output devices to communicate with the rest of the computer program.

Dial-up service: Method of accessing the Internet using a high-speed modem and standard telephone lines.

Digital audiotape (DAT) drive: Backup technology that uses 2- by 3-inch cassettes that store 1.3 gigabytes or more.

Digital camera: Similar to a traditional camera except that images are recorded digitally in the camera's memory rather than on film.

Digital cash: *See* Electronic cash.

Digital notebook: An input device that records and stores pen movements.

Digital signal: Signal that represents the presence or absence of an electronic pulse.

Digital subscriber line (DSL): Provides high-speed connection using existing telephone lines.

Digital video camera: Input device that records motion digitally.

Digital video disc: *See* DVD (digital versatile disc).

Digitizer: Device that can be used to trace or copy a drawing or photograph. The shape is converted to digital data that can be represented on a screen or printed on paper.

Digitizing tablet: Device that enables the user to create images using a special stylus.

Direct approach: Approach for systems implementation whereby the old system is simply abandoned for the new.

Direct entry: Form of input that does not require data to be keyed by someone sitting at a keyboard. Direct-entry devices create machine-readable data on paper or magnetic media or feed it directly into the CPU.

Direct file organization: File organization that makes use of key fields to go directly to the record being sought rather than reading records one after another.

Direct-imaging plotter: Plotter that creates images using heat-sensitive paper and electrically heated pins.

Disaster recovery plan: Plan used by large organizations describing ways to continue operations following a disaster until normal computer operations can be restored.

Discrete-speech recognition system: Voice-recognition system that allows users to dictate directly into a microcomputer using a microphone.

Disk: *See* Floppy disk; Hard disk; Optical disk.

Disk caching: Method of improving hard-disk performance by anticipating data needs. It requires a combination of hardware and software.

Disk cleanup: Windows utility that eliminates nonessential files.

Disk defragmenter: Windows utility that optimizes disk performance by eliminating unnecessary fragments and by rearranging files.

Diskette: *See* Floppy disk.

Distributed database: Database that can be made accessible through a variety of communications networks, which allows portions of the database to be located in different places.

Distributed processing: System in which computing power is located and shared at different locations.

Document: Any kind of text material.

Documentation: Written descriptions and procedures about a program and how to use it.

Document file: File created by a word processor to save documents such as letters, research papers, and memos.

Domain code: Last part of an Internet address, which identifies the geographical description or organizational identification.

Domain name: Part of an Internet address, separated from the domain code by a dot (.), that is a reference to a particular organization.

Domain name system (DNS): Internet addressing method that assigns names and numbers to people and computers.

DO UNTIL structure: Loop structure in programming that appears at the end of a loop. The DO UNTIL loop means that the loop statements will be executed at least once.

DO WHILE structure: Loop structure in programming that appears at the beginning of a loop. The DO WHILE loop will keep executing as long as there is information to be processed.

Downloading: Process of transferring information from a remote computer to the computer one is using.

Draw program: Program used to help create artwork for publications. *See also* Illustration program.

DSS: *See* Decision support system.

Dual-scan monitor: *See* Passive-matrix monitor.

Dumb terminal: Terminal that can be used to input and receive data but cannot process data independently.

DVD (digital versatile, or video, disc): Similar to CD-ROMs except that more data can be packed into the same amount of space.

DVD-R: DVD-recordable.

DVD-RAM: DVD–random-access memory.

DVD-ROM: Digital versatile disc–read-only memory.

DVD-RW: DVD-rewritable

EBCDIC (Extended Binary Coded Decimal Interchange Code): Binary coding scheme that is a standard for minicomputers and mainframe computers.

Economic feasibility: Condition in which costs of designing a new sytem will be justified by the benefits it will provide.

EIS: *See* Executive information system.

Electronic cash (e-cash): Currency for Internet purchases.

Electronic commerce (e-commerce): Buying and selling of goods over the Internet.

Electronic Communications Privacy Act of 1986: Law protecting the privacy of users on public electronic-mail systems.

Electronic mail (e-mail): Similar to an electronic bulletin board, but provides confidentiality and may use special communications rather than telephone lines.

Electronic spreadsheet: *See* Spreadsheet.

Electrostatic plotter: Plotter that uses electrostatic charges to create images made up of tiny dots on specially treated paper.

E-mail: *See* Electronic mail.

Encrypting: Coding information so that only the user can read or otherwise use it.

End user: Person who uses microcomputers or has access to larger computers.

Energy Star: Program created by the Environmental Protection Agency to discourage waste in the microcomputer industry.

Enter key: Key used to enter a command after it has been typed into the computer.

Enterprise computing: Integrating an organization's networks.

Erasable optical disk: Optical disk on which the disk drive can write information and also erase and rewrite information.

Ergonomics: Study of human factors related to things people use.

ESS: *See* Executive support system.

Ethics: Standards of moral conduct.

Exception report: Report that calls attention to unusual events.

Executive information system (EIS): Sophisticated software that can draw together data from an organization's databases in meaningful patterns.

Executive support system (ESS): *See* Executive information system.

Expansion card: Optional device that plugs into a slot inside the system unit. Ports on the board allow cables to be connected from the expansion board to devices outside the system unit.

Expert system: Computer program that provides advice to decision makers who would otherwise rely on human experts.

Extended Binary Coded Decimal Interchange Code: *See* EBCDIC.

External modem: Modem that stands apart from the computer and is connected by a cable to the computer's serial port. Another cable connects the modem to the telephone wall jack.

Extranet: Private network that connects more than one organization.

Facsimile transmission machine: *See* Fax machine.

Fair Credit Reporting Act of 1970: Law prohibiting credit agencies from sharing credit information with anyone but authorized customers and giving consumers the right to review and correct their credit records.

Fax machine: Device that scans an image and sends it electronically over telephone lines to a receiving fax machine, which converts the electronic signals back to an image and recreates it on paper.

Fax/modem board: Expansion board that provides the independent capabilities of a fax and a modem.

Fiber-optic cable: Special transmission cable made of glass tubes that are immune to electronic interference. Data is transmitted through fiber-optic cables in the form of pulses of light.

Field: Each column of information within a record is called a field. A field contains a set of related characters.

File: A collection of related records.

File compression: Process of reducing the storage requirements for a file.

File compression programs: Programs that reduce the size of files.

File decompression: Process of expanding a compressed file.

File transfer protocol (FTP): Internet service for transferring files.

Filter: Program to block selected Web sites, set time limits, monitor use, and generate reports on use.

Find: In word processing, a command that allows the user to locate any character, word, or phrase in a document.

Firewall: Security hardware and software.

FireWire port: Used to connect high-speed printers and even videocameras to system unit.

Firmware: *See* ROM.

Fixed disk: *See* Internal hard disk.

Flash memory: *See* Flash RAM.

Flash RAM: RAM chips that retain data even when power is disrupted.

Flatbed scanner: An input device similar to a copying machine.

Flat-panel monitor: Monitor that lies flat instead of standing upright.

Flexible disk: *See* Floppy disk.

Floppy: *See* Floppy disk.

Floppy disk: Flat, circular piece of magnetically treated mylar plastic that rotates within a jacket.

Floppy-disk cartridge: Include Zip disks, SuperDisks, and HiFD disks, all competing to become the next higher capacity floppy disk standard.

Formatting toolbar: A collection of buttons used as shortcuts to commands that enhance the appearance of a document.

Formula: Instructions for calculations in a spreadsheet.

FORTRAN (FORmula TRANslation): Most widely used scientific and mathematical procedural language.

Freedom of Information Act of 1970: Law giving citizens the right to examine data about them in federal government files, except for that restricted for national security reasons.

FTP: *See* File transfer protocol.

Full-duplex communication: Mode of communication in which data is transmitted back and forth at the same time.

Function: In a spreadsheet, a built-in formula that performs calculations automatically.

Fuzzy logic: Used by expert systems to allow users to respond by using qualitative terms such as *great* and *OK*.

Gantt chart: Chart using bars and lines to indicate the time scale of a series of tasks.

General ledger: Activity that produces income statements and balance sheets based on all transactions of a company.

General-purpose applications: Applications designed to be used by most people doing the most common tasks, such as browsers and word processors.

Generations of programming languages: The five generations are machine languages, assembly languages, procedural languages, problem-oriented languages, and natural languages.

Gopher: Program that helps individuals on the Internet to access other computers on the Internet.

Gopher site: Internet computer that provides menus describing its available resources and direct links to the resources.

Grammar checker: In word processing, a tool that identifies poorly worded sentences and incorrect grammar.

Graphical map: Diagram of a Web site's overall design.

Graphical user interface (GUI): Special screen that allows software commands to be issued through the use of graphic symbols (icons) or pull-down menus.

Graphics suite: Group of graphics programs offered at lower cost than if purchased separately.

Green PC: Microcomputer industry concept of an environmentally friendly, low-power-consuming machine.

Grid chart: Chart that shows the relationship between input and output documents.

Group decision support system (GDSS): System used to support the collective work of a team addressing large problems.

GUI: *See* Graphical user interface.

Hacker: Person who gains unauthorized access to a computer system for the fun and challenge of it.

Half-duplex communication: Mode of communication in which data flows in both directions, but not simultaneously.

Handheld computer: *See* Personal digital assistant (PDA).

Hard copy: Images output on paper by a printer or plotter.

Hard disk: Enclosed disk drive that contains one or more metallic disks. A hard disk has many times the capacity of a floppy disk.

Hard-disk cartridge: Hard disk that is easily removed.

Hard-disk pack: Several platters aligned one above the other, thereby offering much greater storage capacity.

Hardware: Equipment that includes a keyboard, monitor, printer, the computer itself, and other devices.

Hashing: Program that uses mathematical operations to convert the key field's numeric value to a particular storage address.

Head crash: Occurs when the surface of the read-write head or particles on its surface contact the magnetic disk surface.

Header: The first element of an e-mail message.

Help: A feature in most application software providing options that typically include an index, a glossary, and a search feature to locate reference information about specific commands.

Helper applications: Independent programs that can be executed by a browser. Also called add-ons.

Hierarchical database: Database in which fields or records are structured in nodes.

Hierarchical network: Network consisting of several computers linked to a central host computer. The computers linked to the host are themselves hosts to other computers or devices.

HiFD disk: High-capacity floppy disk manufactured by Sony.

High-definition television (HDTV): All-digital television that delivers a much clearer and more detailed wide-screen picture.

History file: Created by browser to store information on Web sites visited.

Home page: Top-level or opening page of Web site.

Home PC: Specialized large-screen system with high-quality audio.

Horizontal portal: Web portal designed to appeal to mass audiences.

Host computer: A large centralized computer. A common way of accessing the Internet. The host computer is connected to the Internet and provides a path or connection for individuals to access the Internet.

Hot site: Special emergency facility consisting of a fully equipped computer center available to a company in the event of disaster to its computer system. *Compare* Cold site.

HTML: *See* Hypertext Markup Language (HTML).

HTML editor: *See* Web authoring program.

Hybrid network: *See* Hierarchical network.

Hyperlink: A hypertext link.

Hypermedia: *See* Multimedia.

Hypertext link: These create connections between information references within a document or between documents.

Hypertext Markup Language (HTML): Programming language for the document files that are used to display Web pages.

Icon: Graphic object on the desktop used to represent commonly used features.

IF-THEN-ELSE structure: Logical selection structure whereby one of two paths is followed according to IF, THEN, and ELSE statements in a program.

Illustration program: Used to modify vector images and thus create line art, 3-D models, and virtual reality. Also called draw program.

Image editor: Used to create and modify bitmap files. Also called paint program.

Image scanner: Device that identifies images on a page and automatically converts them to electronic signals that can be stored in a computer.

Index sequential file organization: Compromise between sequential and direct file organization. Records are stored sequentially, but an index is used to access a group of records directly.

Individual database: Collection of integrated records useful mainly by just one person.

Industry Standard Architecture (ISA): Bus-line standard developed for the IBM Personal Computer. It first consisted of an 8-bit-wide data path, then a 16-bit-wide data path.

Information: Data that has been processed by a computer system.

Information pusher: Program that automatically gathers information on topics of your choice and saves it on your hard disk. Also called push product and Web Broadcaster utility.

Information system: Collection of hardware, software, people, data, and procedures that work together to provide information essential to running an organization.

Information worker: Employee that creates, distributes, and communicates information.

Ink-jet plotter: Plotter that forms images by spraying droplets of ink onto paper.

Ink-jet printer: Printer that sprays small droplets of ink at high speed onto the surface of the paper.

Input device: Piece of equipment that puts data into a form a computer can process.

Instant messaging: Communication and collaboration tool for direct, "live," connections over the Internet.

Integrated circuit: *See* Silicon chip.

Integrated package: Collection of computer programs that work together and share information.

Intelligent terminal: Terminal that includes a processing unit, memory, secondary storage, communications software, and a telephone hookup or other communications link.

Interactivity: User participation in a multimedia presentation.

Interface card: *See* Expansion card.

Internal hard disk: Storage device consisting of one or more metallic platters sealed inside a container. Internal hard disks are installed inside the system cabinet of a microcomputer.

Internet: A huge computer network available to nearly everyone with a microcomputer and a means to connect to it. It is a resource for information about an infinite number of topics.

Internet relay chat (IRC): Leading type of chat group service.

Internet service provider (ISP): Provides access to the Internet.

Internet terminal: Provides access to the Internet and displays Web pages on a standard television set. Also called Web terminal.

Interpreter: Software that converts a procedural language one statement at a time into machine language just before the statement is executed.

Intranet: Like the Internet but privately owned by an organization.

Inventory: Material or products that a company has in stock.

ISA: *See* Industry Standard Architecture.

ISDN: *See* Integrated services digital network.

ISP: *See* Internet service provider.

Java: A portable programming language.

Joystick: Popular input device for computer games.

Keyboard: Input device that looks like a typewriter keyboard but has additional keys.

Key field: The common field by which tables in a database are related to each other.

Knowledge base: *See* Expert system.

Knowledge work system (KWS): Specialized information system used to create information in a specific area of expertise.

Label: Column or row heading in a spreadsheet.

Lands and pits: Flat and bumpy areas, respectively, that represent 1s and 0s on the optical disk surface to be read by a laser.

LAN: *See* Local area network.

Language translator: Converts programming instructions into a machine language that can be processed by a computer.

Laser printer: Printer that creates dotlike images on a drum, using a laser beam light source.

Layout files: Sample presentation files.

LCD: *See* Liquid crystal display.

Levels of programming languages: Levels used by computer professionals to describe programming languages. A lower-level language is one closer to the language the computer uses itself. A higher-level language is closer to the language humans use.

Light pen: Light-sensitive penlike device used with a special monitor to enter commands by touching the monitor with the pen.

Link: A connection to related information.

Linux: Type of Unix operating system initially developed by Linus Torvalds.

Liquid crystal display (LCD): Display consisting of liquid crystal molecules whose optical properties can be altered by an applied electric field.

List address: Internet mailing list address. Members of a mailing list communicate by sending messages to the list address.

Local area network (LAN): Network consisting of computers and other devices that are physically near each other, such as within the same building.

Logic error: Error that occurs when a programmer has used an incorrect calculation or left out a programming procedure.

Logic structure: Structure that controls the logical sequence in which computer program instructions are executed. The three structures are sequence, selection, and loop.

Loop structure: Logic structure in which a process may be repeated as long as a certain condition remains true.

Lurking: Observing or reading communications from others on an Internet discussion group without participating.

Machine language: Language in which data is represented in 1s and 0s.

Mac OS: Operating system designed for Macintosh computers.

Magnetic-ink character recognition (MICR): Direct-entry scanning device used in banks. This technology is used to automatically read the futuristic-looking numbers on the bottom of checks.

Magnetic tape drive: Device used to read data from and store data on magnetic tape.

Magnetic tape streamer: Device that allows duplication (backup) of the data stored on a microcomputer hard disk.

Magnetic tape unit: *See* Magnetic tape drive.

Mailing list: Type of discussion group available on the Internet.

Main board: *See* System board.

Mainframe: Computer that can process several million program instructions per second. Large organizations rely on these room-size systems to handle large programs with lots of data.

MAN: *See* Metropolitan area network.

Management information system (MIS): Computer-based information system that produces standardized reports in summarized, structured form. It is used to support middle managers.

Mark sensing: *See* Optical-character recognition.

Master file: Complete file containing all records current up to the last update.

Medium band: Bandwidth of special leased lines, used mainly with minicomputers and mainframe computers.

Megahertz (MHz): Unit representing 1 million beats (cycles) per second.

Memory: Part of the microcomputer that holds data for processing, instructions for processing the data, and information (processed data) waiting to be output or sent to secondary storage.

Menu: List of commands.

Message: The content portion of e-mail correspondence.

Metasearch utility: Program that automatically submits your search request to several indices and search engines, then creates an index from received information. Also called metasearch program.

Metropolitan area network (MAN): Network linking office buildings in a city.

MHz: *See* Megahertz.

MICR: *See* Magnetic-ink character recognition.

Microcomputer: Small, low-cost computer designed for individual users.

Microcomputer database: *See* Individual database.

Microprocessor: The central processing unit of a microcomputer. The microprocessor is contained on a single integrated circuit chip.

Microsecond: One-millionth of a second.

Microwave communication: Communication using high-frequency radio waves that travel in straight lines through the air.

Middle-level manager: Manager dealing with control and planning. The manager implements the long-term goals of the organization.

Midrange computer: *See* Minicomputer.

Minicomputer: Desk-sized machine falling in between microcomputers and mainframes in processing speed and data-storing capacity.

MIS: *See* Management information system.

Modem: Communications device that translates the electronic signals from a computer into electronic signals that can travel over a telephone line.

Modulation: Process of converting digital signals to analog signals.

Module: *See* Program module.

Monitor: Output device like a television screen that displays data processed by the computer.

Morphing: Special effect in which one image seems to melt into another.

Motherboard: *See* System board.

Mouse: Device that typically rolls on the desktop and directs the cursor on the display screen.

Multimedia: Technology that can link all sorts of media into one form of presentation.

Multimedia authoring program: Program to create a multimedia presentation by bringing together text, graphics, audio, and other elements into an interactive framework.

Multitasking: Operating system that allows a single user to run several application programs at the same time.

Nanomanipulator: Advanced virtual reality interface.

National Information Infrastructure Protection Act of 1996: Provides penalties for trespassing on computer systems, threats made against computer networks, and theft of information.

Natural language: Language designed to give people a more human connection with computers.

Navigating: Moving from one Web site to another.

Net personal computer (Net PC): Low-cost type of intelligent terminal that typically has only one type of secondary storage and no expansion slots.

Network adapter card: Connects the sytem unit to a cable that connects to other devices on the network.

Network architecture: Describes how a network is arranged and how the resources are shared.

Network bridge: Connects networks of the same configuration.

Network computer: *See* Network terminal.

Network database: Database with a hierarchical arrangement of nodes, except that each child node may have more than one parent node.

Network gateway: Connection by which a local area network may be linked to other local area networks or to larger networks.

Network manager: Computer professional who ensures that existing information and communication systems are operating effectively and that new ones are implemented as needed. Also responsible for meeting security and privacy requirements.

Network operating system (NOS): Software that interacts with applications and computers, and also coordinate activities between computers on a network.

Network terminal: Low-cost alternative to intelligent terminal; relies on host computer or server for software. *Also* called network computer.

Newsgroup: Most popular type of Internet discussion group. Newsgroups use a special network of computers called the UseNet. Each of these computers maintains the newsgroup listing. The newsgroups are organized into major topic areas that are further subdivided into hierarchies.

Node: Any device connected to a network. Also, points in a hierarchical database connected like the branches of an upside-down tree.

No Electronic Theft (NET) Act of 1997: Provides penalties for unathorized distribution of copyrighted materials (software) on the Internet.

Nonvolatile storage: Permanent storage used to preserve data and programs.

Norton AntiVirus: Collection of antivirus programs from Symantec.

Norton CleanSweep: Collection of programs to safely remove files and programs from Symantec.

Norton CrashGuard: Collection of programs that protect against system crashes from Symantec.

Norton Utilities: Collection of 17 troubleshooting utilities from Symantec.

Norton Web Services: Service from Symantec that monitors a computer system for out-of-date software.

Notebook: Portable computer weighing between 5 and 10 pounds.

Object linking and embedding (OLE): Powerful feature of many application programs that allows sharing of information.

Object-oriented database: Keep track of objects, which are entities that contain both data and the action that can be taken on data.

Object-oriented programming (OOP): Methodology in which a program is organized into objects, each containing both the data and processing operations necessary to perform a task.

Object-oriented software development: Software development approach that focuses less on the tasks and more on defining the relationships between previously defined procedures or objects.

OCR: *See* Optical-character recognition.

Off-line browser: Program that automatically connects to selected Web sites, downloads HTML documents, and saves them to your hard disk. Also called pull product and Web-downloading utility.

Office automation system (OAS): System designed primarily to support data workers. It focuses on managing documents, communicating, and scheduling.

OMR: *See* Optical-mark recognition.

1.44 MB 3½-inch disk: The most widely used floppy disk.

Online service provider: Provides access to Internet plus other services.

OOP: *See* Object-oriented programming.

Open architecture: Microcomputer architecture allowing users to expand their systems by inserting optional devices known as expansion cards.

Operating system: Software that interacts between application software and the computer. The operating system handles such details as running programs, storing data and programs, and processing data.

Operational feasibility: Condition in which the design of a new system will be able to function within the existing framework of an organization.

Optical-character recognition (OCR): Scanning device that uses special preprinted characters, such as those printed on utility bills, that can be read by a light source and changed into machine-readable code.

Optical disk: Storage device that can hold 650 megabytes of data. Lasers are used to record and read data on the disk.

Optical-mark recognition (OMR): Device that senses the presence or absence of a mark, such as a pencil mark.

Organization chart: Chart showing the levels of management and formal lines of authority in an organization.

Output device: Equipment that translates processed information from the central processing unit into a form that can be understood.

Packet: Before a message is sent on the Internet, it is broken down into small parts called packets. Each packet is then sent separately over the Internet. At the receiving end, the packets are reassembled into the correct order.

Paint program: *See* Image editor.

Palmtop computer: *See* Personal digital assistant (PDA).

Parallel approach: Systems implementation in which old and new systems are operated side by side until the new one has shown it is reliable.

Parallel data transmission: Data transmission in which bits flow through separate lines simultaneously.

Parallel port: Used to connect external devices that send or receive a lot of data over a short distance. Mostly used to connect printers to system unit.

Parent node: Node one level above the node being considered in a hierarchical database or network.

Pascal: Procedural programming language widely used on microcomputers.

Passive-matrix monitor: Monitor that creates images by scanning the entire screen.

Password: Special sequence of numbers or letters that limits access to information, such as electronic mail.

Payroll: Activity concerned with calculating employee paychecks.

PC card: *See* PCMCIA card.

PC/TV: The merger of microcomputers and television.

Peer-to-peer network system: Network in which nodes can act as both servers and clients. For example, one microcomputer can obtain files located on another microcomputer and can also provide files to other microcomputers.

Pen plotter: Plotter that creates plots by moving a pen or pencil over drafting paper.

Pen scanner: A highly portable device for capturing text.

Periodic report: Report generated at regular intervals.

Peripheral Component Interconnect (PCI): Bus architecture that combines the capabilities of MCA and EISA with the ability to send video instructions at speeds to match the microprocessor.

Personal Computer Memory Card International Association (PCMCIA) card: Credit card–sized expansion cards developed for portable computers.

Personal digital assistant (PDA): A device that typically combines pen input, writing recognition, personal organizational tools, and communication capabilities in a very small package. Also called Handheld PC and palmtop computer.

Personal information managers (PIMs): Programs that provide electronic organizational tools.

Personal laser printer: Inexpensive laser printer widely used by single users to produce black-and-white documents.

Phased approach: Systems implementation in which the new system is implemented gradually over a period of time.

Physical security: Activity concerned with protecting hardware from possible human and natural disasters.

Pilot approach: Systems implementation in which the new system is tried out in only one part of the organization. Later it is implemented throughout the rest of the organization.

Pits: *See* Lands and pits.

Pixel: Smallest unit on the screen that can be turned on and off or made different shades.

Platform scanner: Handheld direct-entry device used to read special characters on price tags.

Plotter: Special-purpose output device for producing bar charts, maps, architectural drawings, and three-dimensional illustrations.

Plug and Play: Set of hardware and software standards developed to create operating systems, processing units, and expansion cards, as well as other devices, that are able to configure themselves.

Plug-in: Program that is automatically loaded and operates as part of a browser.

Plug-in board: *See* Expansion card.

Pointer: For a monitor, a pointer is typically displayed as an arrow and controlled by a mouse. For a database, a pointer is a connection between parent node and child node in a hierarchical database.

Pointing stick: Device used to control the pointer by directing the stick with your finger.

Point-of-sale (POS) terminal: Terminal that consists of a keyboard, screen, and printer. It is used like a cash register.

Point-to-point protocol (PPP): Software that allows your computer to become part of a client/server network.

The provider or host computer is the server providing access to the Internet. Your computer is the client. Using special client software, your computer is able to communicate with server software running on the provider's computer and on other Internet computers.

Polling: Process whereby a host computer or file server asks each connecting device whether it has a message to send and then allows the message to be sent.

Port: Connecting socket on the outside of the system unit. Used to connect input and output devices to the system unit.

Portable language: Language that can be run on more than one kind of computer.

Portable scanner: Handheld input device for scanning images and text.

POS terminal: *See* Point-of-sale (POS) terminal.

Preliminary investigation: First phase of systems analysis and design. It involves defining the problem, suggesting alternative systems, and preparing a short report.

Presentation file: A file created by presentation graphics programs to save presentation materials.

Presentation graphics: Graphics used to communicate a message or to persuade other people.

Printer: Device that produces printed paper output.

Privacy: Computer ethics issue concerning the collection and use of data about individuals.

Privacy Act of 1974: Law designed to restrict the way federal agencies share information about American citizens. It prohibits federal information collected for one purpose from being used for a different purpose.

Problem-oriented language: Programming language designed to solve specific problems.

Procedural language: Programming language designed to express the logic that can solve problems.

Procedures: Rules or guidelines to follow when using hardware, software, and data.

Processing rights: Refers to which people have access to what kind of data.

Processor: *See* Central processing unit.

Program: List of instructions for the computer to follow to process data. *See also* Software.

Program analysis: *See* Program specification.

Program definition: *See* Program specification.

Program design: Creation of a solution using programming techniques such as top-down program design, pseudocode, flowcharts, logic structures, object-oriented programming, and CASE tools.

Program Evaluation Review Technique (PERT) chart: Chart showing the timing of a project and the relationships among its tasks. The chart identifies which tasks must be completed before others can begin.

Program flowchart: Flowchart graphically presenting the detailed sequence of steps needed to solve a programming problem.

Program maintenance: Activity of updating software to correct errors, improve usability, standardize, and adjust to organizational changes.

Programmer: Computer professional who creates new software or revises existing software.

Programming: Six-step procedure for creating a program.

Program module: Logically related program statements.

Program specification: Programming step in which objectives, outputs, inputs, and processing requirements are determined.

Project: One-time operation composed of several tasks that must be completed during a stated period of time.

Project manager: Software that enables users to plan, schedule, and control the people, resources, and costs needed to complete a project on time.

Property: Computer ethics issue relating to who owns data and rights to software.

Proprietary database: Enormous database an organization develops to cover certain particular objects. Access to this type of database is usually offered for a fee.

Protocol: Rules for exchanging data between computers.

Prototyping: Building of a model (prototype) that can be modified before the actual system is installed. It allows users to find out right away how a change in the system can help their work.

Provider: *See* Host computer.

Proxy server: Computer that acts as a gateway or checkpoint in an organization's firewall.

Pseudocode: Narrative form of the logic of a computer program.

Pull product: *See* Off-line browser.

Purchasing: Buying of raw materials and services.

Push product: *See* Information pusher.

Query language: Easy-to-use language understandable to most users. It is used to search and generate reports from a database.

RAD: *See* Rapid applications development.

RAM (random-access memory): Volatile storage that holds the program and data the CPU is presently processing.

RAM cache: *See* Cache memory.

Random-access memory (RAM): *See* RAM (random-access memory).

Rapid applications development (RAD): Involves the use of powerful development software and specialized teams as an alternative to the systems development life cycle approach.

Reader/sorter: A special purpose input device that reads characters made of ink containing magnetized particles.

Read-only memory (ROM): *See* ROM (read-only memory).

Real-time processing: Processing that occurs when data is processed at the same time the transaction occurs.

Record: Each line of information in a database is a record. A record is a collection of related fields.

Reduced instruction set computer (RISC) chip: Powerful microprocessor chip found in workstations.

Redundant arrays of inexpensive disks (RAIDs): Groups of inexpensive hard-disk drives related or grouped together using networks and special software. They improve performance by expanding external storage.

Relation: Table in a relational database in which data elements are stored.

Relational database: A widely used database structure, in which data is organized into related tables.

Repetitive motion injury: *See* Repetitive strain injury.

Repetitive strain injury (RSI): Category of injuries resulting from fast, repetitive work that causes neck, wrist, hand, and arm pain.

Replace: In word processing, command that enables the user to search for a word and replace it with another.

Resolution: A measurement in pixels of a monitor's clarity.

Right to Financial Privacy Act of 1979: Law setting strict procedures that federal agencies must follow when seeking to examine customer records in banks.

Ring network: Network in which each device is connected to two other devices, forming a ring. There is no host computer, and messages are passed around the ring until they reach the correct destination.

Robot: Machine used in factories and elsewhere that can be reprogrammed to do more than one task.

Robotics: Field of study concerned with developing and using robots.

Roller ball: *See* Trackball.

ROM (read-only memory): Refers to chips that have programs built into them at the factory. The contents of such chips cannot be changed by the user.

RS-232C connector: Serial port, a port set up for serial data transmission.

RSI: *See* Repetitive strain injury.

Sales order processing: Activity that records the demands of customers for a company's products or services.

Satellite/air connection services: Connection services that use satellites and the air to download or send data to users at a rate seven times faster than dial-up connections.

Satellite communication: Communication using satellites as microwave relay stations.

Scanner: *See* Image scanner.

Search engine: Search tool that lets you search by entering key words or phrases.

Search tools: Help you locate information on the Web; there are two basic kinds: indexes and search engines.

Secondary storage: Permanent storage used to preserve programs and data, including floppy disks, hard disks, and magnetic tape.

Sector: Section shaped like a pie wedge that divides the tracks on a disk.

Security: The protection of information, hardware, and software.

Selection structure: Logic structure that determines which of two paths will be followed when a decision must be made by a program.

Semiconductor: Silicon chip through which electricity flows with some resistance.

Sequence structure: Logic structure in which one program statement follows another.

Sequential file organization: File organization in which records are stored physically one after another in predetermined order.

Serial data transmission: Method of transmission in which bits flow in a series, one after another.

Serial line Internet protocol (SLIP): Internet protocol that enables your computer to become part of a client/server network. The provider or host computer is the server providing access to the Internet. Your computer is the client.

Serial port: Used to connect external devices that send or receive data one bit at a time over a long distance. Used for mouse, keyboard, modem, and many other devices.

Server: A connection to the Internet that stores document files used to display pages.

Service program: *See* Utility.

Shared laser printer: More expensive laser printer used by a group of users to produce black-and-white documents.

Sherlock: A search feature in Mac OS 8.5 operating system.

Signature line: Provides additional information about a sender of an e-mail message, such as name, address, and telephone number.

Silicon chip: Tiny circuit board etched on a small square of sandlike material called silicon. Chips are mounted on carrier packages, which then plug into sockets on the system board.

Simplex communication: Mode of communication in which data travels in one direction only.

Slot: Area on a system board that accepts cards to expand a computer system's capabilities.

Small computer interface (SCSI) card: This card uses only one slot and can connect as many as seven devices to the system unit.

Smart card: Card about the size of a credit card containing a tiny built-in microprocessor. It can be used to hold such information as frequent flier miles.

Snoopware: Programs that record virtually every activity on a computer system.

Soft copy: Images or characters output on a monitor screen.

Software: Computer program.

Software Copyright Act of 1980: Law allowing owners of programs to make copies for backup purposes, and to modify them to make them useful, provided they are not resold or given away.

Software development: *See* Programming.

Software piracy: Unauthorized copying of programs for personal gain.

Software suite: Individual application programs that are sold together as a group.

Sorting: Arranging objects numerically or alphabetically.

Source document: Original version of a document before any processing has been performed on it.

Spelling checker: Program used with a word processor to check the spelling of typed text against an electronic dictionary.

Spider: *See* Agent.

Spike: *See* Voltage surge.

Spreadsheet: Computer-produced spreadsheet based on the traditional accounting "worksheet" that has rows and columns that can be used to present and analyze data.

Standard toolbar: A collection of buttons used as shortcuts to the most frequently used menu commands.

Star network: Network of computers or peripheral devices linked to a central computer through which all communications pass. Control is maintained by polling.

Start menu: A Windows menu listing commands used to gain access to information, change hardware settings, find information, get online help, run programs, log off a network, and shut down the computer system.

Story boards: Design tool for planning and structuring a multimedia presentaiton.

Structured programming techniques: Techniques that consist of top-down program design, pseudocode, flowcharts, and logic structures.

Structured query language (SQL): Program control language used to create sophisticated database applications.

Stylus: A special input device that traces images.

Subnotebook: Handheld or pocket-size portable computer.

Subscription address: Mailing list address. To participate in a mailing list, you must first subscribe by sending an e-mail request to the mailing list subscription address.

Supercomputer: Fastest calculating device ever invented, processing billions of program instructions per second.

SuperDisk: High-capacity floppy disk manufactured by Imation.

SVGA: Refers to a resolution standard that displays 800 by 600 pixels.

Supervisor: Manager responsible for managing and monitoring workers. Supervisors have responsibility for operational matters.

Surfing: Moving from one Web site to another.

Surge protector: Device separating the computer from the power source of the wall outlet. When a voltage surge occurs, a circuit breaker is activated, protecting the computer system.

SXGA: Refers to a resolutuion standard of 1,280 by 1,024 pixels.

Synchronous transmission: Method whereby data is transmitted several bytes or a block at a time.

Syntax error: Violation of the rules of whatever language a computer program is written in.

System: Collection of activities and elements designed to accomplish a goal.

System board: Flat board that usually contains the CPU and some memory chips.

System clock: Clock that controls how fast all the operations within a computer take place.

System flowchart: Flowchart that shows the kinds of equipment used to handle the data or information flow.

System software: "Background" software that enables the application software to interact with the computer. It includes programs that help the computer manage its own internal resources.

System unit: Part of a microcomputer that contains the CPU.

Systems analysis: Determining the requirements for a new system. Data is collected about the present system, analyzed, and new requirements are determined.

Systems analysis and design: Six-phase problem-solving procedure for examining an information system and improving it.

Systems analyst: Computer professional who studies an organization's systems to determine what actions to take and how to use computer technology to assist them.

Systems audit: A systems audit compares the performance of a new system to the original design specifications to determine if the new procedures are actually improving productivity.

Systems design: Phase consisting of designing alternative systems, selecting the best system, and writing a systems design report.

Systems development: Phase consisting of developing software, acquiring hardware, and testing the new system.

Systems implementation: Process of changing from the old system to the new and training people to use it.

Systems life cycle: Phases of systems analysis and design.

Systems maintenance: Consists of a systems audit and periodic evaluation.

Table (in database): The list of records in a database.

Technical feasibility: Condition in which hardware, software, and training will be available to facilitate the design of a new system.

Technical writer: Computer professional who explains in writing how a computer program works.

Technostress: Tension that arises when humans must unnaturally adapt to computers.

Television board: Contains a TV tuner and video converter that changes the TV signal into one that can be displayed on your monitor.

Telnet: Internet service that helps to connect computers on the Internet and to run programs on remote computers.

Templates: Model presentations.

Terminal: Form of input (and output) device that consists of a keyboard, a monitor, and a communications link.

Terminal connection: Method of accessing the Internet using a high-speed modem and standard telephone lines.

Terminal network system: Network system in which processing power is centralized in one large computer, usually a mainframe. The nodes connected to this host computer are terminals with little or no processing capabilities.

Thermal printer: Printer that uses heat elements to produce images on heat-sensitive paper.

Thin client: *See* Network terminal.

Thin film transistor (TFT): Type of flat-panel monitor that activates each pixel independently.

Time-sharing system: System that allows several users to share resources in the host computer.

Toggle key: On-off key.

T1, T2, T3, T4 lines: High-speed lines that support all digital communications, provide very high capacity, and are very expensive.

Toolbar: Bar located typically below the menu bar. It contains icons or graphical representations for commonly used commands.

Top-down analysis method: Method used to identify the top-level component of a system and break this component down into smaller components for analysis.

Top-down program design: Process of identifying the top element (module) for a program and then breaking the top element down into smaller pieces in a hierarchical fashion.

Top-level manager: Manager concerned with long-range (strategic) planning. Top-level managers supervise middle-level managers.

Topology: The configuration of a network.

Touch screen: Monitor screen that allows actions or commands to be entered by the touch of a finger.

Track: Closed, concentric ring on a disk on which data is recorded.

Trackball: Device used to control the pointer by rotating a ball with your thumb.

Transaction: Event recorded in a database.

Transaction file: File containing recent changes to records that will be used to update the master file.

Transaction processing system (TPS): System that records day-to-day transactions.

Transmission control protocol/Internet protocol (TCP/IP): The two standard protocols for all communications on the Internet.

Troubleshooting programs: Programs that recognize and correct computer related problems.

Twisted pair: Copper-wire telephone line.

2HD: Two-sided, double-density disk.

Unicode: A 16-bit code designed to support international languages like Chinese and Japanese.

Uniform resource locator (URL): Addresses of resources on the Web.

Unix: An operating system originally developed for minicomputers. It is now important because it can run on many of the more powerful microcomputers.

Uploading: Process of transferring information from the computer the user is operating to a remote computer.

Uninstall program: Programs that safely and completely remove unwanted programs and related files.

Universal serial bus (USB) port: Expected to replace serial and parallel ports.

UseNet: Special network of computers that support newsgroups.

User: *See* End user.

User interface: Means by which users interact with programs and hardware.

User name: The part of an Internet address that identifies a unique person or computer at the listed domain.

Utility: Performs specific tasks related to managing computer resources or files.

Utility program: Program that performs common repetitious tasks, such as keeping files orderly, merging, and sorting.

UXGA: Refers to a resolution standard of 1,600 by 1,200 pixels.

Value: Number contained in a cell in a spreadsheet.

Vector image: Graphics file made up of a collection of objects such as lines, rectangles, and ovals.

Vertical portal: Web portal designed to appeal to special-interest groups.

VGA (video graphics array): Circuit board that may be inserted into a microcomputer and offers up to 256 colors.

Videoconferencing systems: Computer systems that allow people located at various geographic locations to have in-person meetings.

Video display screen: *See* Monitor.

Video Privacy Protection Act of 1988: Law preventing retailers from selling or disclosing video-rental records without the customer's consent or a court order.

Virtual environment: *See* Virtual reality.

Virtual memory: Feature of an operating system that increases the amount of memory available to run programs.

Virtual reality: Interactive sensory equipment (headgear and gloves) that allows users to experience alternative realities to the physical world.

Virtual reality modeling language (VRML): Used to create real-time animated 3-D scenes.

Virus: Hidden instructions that migrate through networks and operating systems and become embedded in different programs. They may be designed to destroy data or simply to display messages.

Visual BASIC: A widely used programming language created by Microsoft.

Voiceband: Bandwidth of a standard telephone line.

Voice-input device: Direct-entry device that converts speech into a numeric code that can be processed by a computer.

Voice-messaging system: Computer system linked to telephones that converts human voice into digital bits.

Voice-output device: Device that makes sounds resembling human speech that are actually prerecorded vocalized sounds.

Voice recognition system: *See* Voice-input device.

Volatile storage: Temporary storage that destroys the current data when power is lost or new data is read.

Voltage surge (spike): Excess of electricity, which may destroy chips or other electronic computer components.

WAN: *See* Wide area network.

Wand reader: Special-purpose handheld device used to read OCR characters.

Web: A service that provides a multimedia interface to resources on the Internet. *See also* World Wide Web.

Web appliance: *See* Internet terminal.

Web auction: Similar to traditional auctions except that buyers and sellers meet only on the Web.

Web authoring: Creating a Web site.

Web authoring program: Word processing program for generating Web pages. Also called HTMLeditor; Web page editor.

Web broadcaster: *See* Information pusher.

Web cam: Specialized digital video camera for capturing images and broadcasting to the Internet.

Web crawler: *See* Search engine.

Web directory: *See* Index.

Web-downloading utility: *See* Off-line browser.

Webmaster: Person who designs, creates, monitors, and evaluates corporate Web sites.

Web page: Browsers interpret HTML documents to display Web pages.

Web page editor: *See* Web authoring program.

Web portal: Site that offers a variety of services such as e-mail, news, and sports updates.

Web site: A location on a server.

Web spider: *See* Search engine.

Web storefront creation packages: Programs for creating Web sites for virtual stores.

Web storefronts: Virtual stores where shoppers can go to inspect goods and make purchases.

Web terminal: *See* Internet terminal.

What-if analysis: Spreadsheet feature in which changing one or more numbers results in the automatic recalculation of all related formulas.

Wide area network (WAN): Countrywide network that uses microwave relays and satellites to reach users over long distances.

Window: A rectangular area containing a document or message.

Windows: An operating environment that extends the capability of DOS.

Wireless modem: Modem that connects to the serial port but does not connect to telephone lines. It receives through the air.

Word: Unit that describes the number of bits in a common unit of information.

Word processing: Use of a computer to create, edit, save, and print documents composed of text.

Word wrap: Feature of word processing that automatically moves the cursor from the end of one line to the beginning of the next.

Worksheet file: File created by an electronic spreadsheet.

World Wide Web (WWW, the Web): Internet service that uses hypertext to jump from document to document and from computer to computer. The Web is accessed by browsers.

Worm: Variant on computer virus, a destructive program that fills a computer system with self-replicating information, clogging the system so that its operations are slowed or stopped.

WORM (write once, read many): Form of optical disk that allows data to be written only once but read many times without deterioration.

Write-protect notch: Notch on a floppy disk used to prevent the computer from destroying data or information on the disk.

WWW: *See* World Wide Web.

XGA (extended graphics array): Circuit board that can be inserted into a microcomputer and offers up to 256 colors under normal circumstances and more than 65,000 colors with special equipment.

Zip disk: High capacity floppy disk manufactured by Sony.

CREDITS